Chemical Mobility and Reactivity in Soil Systems

SSSA Special Publication Number 11

Proceedings of a symposium
sponsored by Divisions S-1, S-2, and A-5
of the American Society of Agronomy
and the Soil Science Society of America
in Atlanta, Georgia, 29 Nov.–3 Dec. 1981.

Editorial Committee
D. W. Nelson, Chmn.
D. E. Elrick
K. K. Tanji

Organizing Committee
George Blake
Robert Blanchar

Managing Editor
David M. Kral

Assistant Editor
Sherri L. Hawkins

1983
Published by the
SOIL SCIENCE SOCIETY OF AMERICA
AMERICAN SOCIETY OF AGRONOMY
677 South Segoe Road
Madison, Wisconsin 53711

Copyright © 1983 by the American Society of Agronomy
Soil Science Society of America, Inc.

ALL RIGHTS RESERVED UNDER THE U.S. COPYRIGHT
ACT OF 1976 (P.L. 94-553)

Any and all uses beyond the limitations of the "fair use" provision of the law require written permission from the publisher(s) and/or the author(s); not applicable to contributions prepared by officers or employees of the U.S. Government as part of their official duties.

Second Printing 1986
Third Printing 1992

American Society of Agronomy, Inc.
Soil Science Society of America
677 South Segoe Road, Madison, WI 53711, USA

Library of Congress Catalog Card Number: 83-071451.
Standard Book Number: 0-89118-771-5

Printed in the United States of America

DEDICATION

This Special Publication is dedicated to Dr. Roscoe Ellis, Jr., Professor of Agronomy at Kansas State University and Editor-in-Chief of the *Soil Science Society of America Journal* at the time of his death on 9 Sept. 1982.

Dr. Ellis taught courses in soil and plant analysis applications, chemical properties of soils, and soil physical chemistry. He directed research programs for graduate students and among many soil chemistry contributions, he and his graduate students provided a better understanding of the effect of gas on soils, the grass tetany phenomena, and the rate of phosphorus fertilizer-reaction products and their availability in soils.

The achievements of Dr. Ellis were many and his zeal for scientific competence and integrity was unsurpassed.

Roscoe Ellis, Jr.
1920–1982

Contents

	Page
Foreword	vii
Preface	ix

PART I
PRINCIPLES OF CHEMICAL MOBILITY AND REACTIVITY IN SOIL

1. Ion Activity Products in Soil Solutions
   H. L. Bohn 1
2. Effects of Aeration on Reactivity and Mobility of Soil Constituents
   K. R. Reddy and W. H. Patrick, Jr. 11
3. Physical Processes Influencing Water and Solute Transport in Soils
   J. M. Davidson, P. S. C. Rao and P. Nkedi-Kizza 35
4. Chemical Transport Modeling: Current Approaches and Unresolved Problems
   W. A. Jury 49
5. Spatial Soil Variability and Mass Transfers from Agricultural Soils
   D. R. Nielsen, P. J. Wierenga, and J. W. Biggar 65

PART II
NUTRIENT AND SALT MOVEMENT IN SOIL

6. Assessing Nitrogen Movement in the Field
   J. L. Starr 79
7. The Movement of Phosphorus in Soil
   C. G. Enfield and Roscoe Ellis, Jr. 93
8. The Movement of Micronutrients in Soils
   B. G. Ellis, B. D. Knezek, and L. W. Jacobs 109
9. Principles of Salt Movement in Soils
   R. J. Wagenet 123
10. Irrigation Management and Crop Production as Related to Nitrate Mobility
    R. J. Hanks, D. W. James, and D. W. Watts 141

PART III
BIOLOGICAL ACTIVITY AND CHEMICAL MOBILITY IN SOIL

11. Principles of Microbial Processes of Chemical Degradation, Assimilation, and Accumulation
    D. R. Keeney.. 153

12. Incorporation of the Rhizosphere Into Plant Root Models
    J. H. Cushman 165

13. Sorption and Movement of Pesticides and Other Toxic Organic Substances in Soils
    P. S. C. Rao and R. E. Jessup 183

PART IV
ENVIRONMENTAL IMPACT OF TOXIC CHEMICAL TRANSPORT IN SOIL SYSTEMS

14. Mobility of Radionuclides in Soil Systems
    G. W. Gee, Dhanpat Rai, and R. J. Serne 203

15. Movement of Heavy Metals in Soils
    R. H. Dowdy and V. V. Volk......................... 229

16. Movement of Phosphorus and Nitrogen in Soil Following Application of Municipal Wastewater
    J. E. Hook 241

17. Management of Soil Systems for Industrial Wastes
    J. C. Corey 257

Foreword

Soil is the most fundamental of all natural resources, forming as it does the medium for root growth, supplying most of the needed nutrients used by plants, and slowing the flow of rainwaters to the sea, thus making them available over the annual cycle for use by plants and animals. Additionally, the composition of the above- and below-ground air is modified by processes which take place in soil. And, surface and sub-surface waters carry dissolved and suspended natural mineral and organic materials, as well as man-made substances, some of which are toxic to plants and animals. Availability of plant nutrients in soil water and the composition of waters used by mankind depend almost entirely upon physical and chemical processes, including biochemical interactions, involving soil.

An understanding of the mobility and reactivity of substances in the soil system is essential to the practical management of soil as a medium for sustained crop production and as it becomes involved in transmitting or retaining substances which affect man's environment. This book, *Chemical Mobility and Reactivity of Soil Systems,* deals in a most effective manner with the state-of-the-art of the many processes involved given from the point of view of a few of the many outstanding scientists working in several sub-fields of soil science.

Charles F. Eno, President
American Society of Agronomy

Walter H. Gardner, President
Soil Science Society of America

Preface

Although soil chemists and soil physicists frequently conduct research on related subjects, there has been relatively little interchange of ideas or interdisciplinary studies conducted by these two groups of the soil scientists. To encourage better communication between soil scientists the Program Chairmen for the Soil Physics (S-1) and Soil Chemistry (S-2) Divisions organized an expanded symposium dealing with "Chemical Mobility and Reactivity in Soil Systems" that was presented over 4 days at the 1981 Annual Meetings of the American Society of Agronomy (ASA) and Soil Science Society of America (SSSA) held in Atlanta, Georgia. The symposium was co-sponsored by Div. S-1, S-2, and A-5. Nineteen eminent soil scientists were invited to give 20 minute presentations on their research dealing with mobility and reactivity of specific constituents in soils. Attendance at each of the four sessions comprising the symposium was large indicating that a variety of soil scientists had significant interest in the material that was presented.

Chemical Mobility and Reactivity in Soil Systems is the written version of the papers that were presented at the symposium. The special publication is divided into four parts: (1) Principles of chemical mobility and reactivity in soil, (2) Nutrient and salt movement in soil, (3) Biological activity and chemical mobility in soil, and (4) Environmental impact of toxic chemical transport in soil systems. The four parts correspond to the four portions of the symposium. All presentations made at the 1981 Annual Meetings are included except those by Dr. Willard Lindsay and Dr. Donald Kaufmann entitled "Speciation and Chemical Mobility in Soils" and "Breakdown of Organic Chemicals in Soil Profiles", respectively. Much of the material covered by Dr. Lindsay is included in his recent book entitled *Chemical Equilibria in Soils* (John Wiley and Sons). The authors were asked to prepare rather short papers that gave the state-of-the-art of their subject rather than a thorough literature review. We also requested that the manuscripts be written in a manner readily understandable by most soil scientists and that they attempt to integrate soil chemistry and physics concepts as they apply to the subject under discussion.

We are grateful for all of the effort put forth by the authors. Their papers will be important milestones in interdisciplinary efforts within soil science and may stimulate increased cooperation and interchange between soil chemists and physicists. We also wish to thank the large number of scientists who contributed suggestions for the symposium and reviewed manuscripts. We also appreciate the excellent work of the ASA Headquarters staff in the publication process.

Organizing Committee
George Blake
Robert Blanchar

Editorial Committee
Darrell W. Nelson, Chairman
David E. Elrick, Member
Kenneth K. Tanji, Member

Chapter 1

Ion Activity Products in Soil Solutions[1]

HINRICH L. BOHN[2]

Ion activity products (IAP) have been measured and calculated in soils to identify the components that control the composition of the soil solution. The IAP is usually less than the solubility product (Ksp) of pure substances (Hendrickson and Corey, 1981) and varies from soil to soil. Those substances in soil solids are in a more stable state than as pure solids. Predictions of the composition of soil solutions and other natural waters are unreliable due largely to the uncertain solubility of soil and sediment components (Nordstrom et al., 1979).

The difference between an IAP in soil solution and the Ksp of the pure substance has been attributed to adsorption and nonequilibrium in the soil. A third reason could be the effects of impurities, solid solutions, and mixing on coprecipitation and solid solubility. Soil minerals are usually impure and the amorphous fraction is presumably an intimate mixture of many substances. Mixing and impurities affect the chemical activity and solubility of solids in soils but their effects have not been understood. Vaslow and Boyd (1952), Wright and Peech (1960), Crockett and Winchester (1966), Murrman and Peech (1968), and Thorstenson and Plummer (1977) considered the coprecipitation of two solid components—$Ag(Cl,Br)$, $(Al,Fe)PO_4$, $(Ca,Zn)CO_3$, $Ca_5(OH,F)(PO_4)_3$, and $(Ca,Mg)CO_3$, respectively—on their chemical activities and solubilities. Stumm and Morgan (1981) and Sposito (1981) also considered the activity of a two-component solid phase. Orville (1972) and Saxena (1973) investigated two-component solid mixtures at high temperatures. This paper is an attempt to extend their work to a multicomponent mixture such as soil.

---

[1] Joint contribution from the Univ. of Arizona and the U.S. Geological Survey.
[2] Professor, Dep. of Soils, Water, and Eng., Univ. of Arizona, Tucson, AZ 85721.

Copyright © 1983 ASA, SSSA, 677 South Segoe Road, Madison, WI 53711. *Chemical Mobility and Reactivity in Soil Systems.*

THEORY

Soil IAP measurements presume equilibrium which in the strictest sense is inaccurate. Nonetheless equilibrium seems to describe the behavior of many ions in soil solutions and natural waters (Nordstrom et al., 1979). Under some conditions the background rate of change of silicates, organic carbon, and other compounds can be assumed negligible compared to the reaction rate in question. Helgeson (1969) described such a slowly changing matrix in which some components are at, or very nearly at, equilibrium as a partial equilibrium. Although not always tested thoroughly, soil reactions in the laboratory can attain what is generally accepted to be such an equilibrium.

The Ksp, IAP, and ion exchange and adsorption coefficients are various expressions for ions distributing or partitioning between solids and aqueous solutions (McIntire, 1963) at equilibrium. One statement of equilibrium partitioning between solid and solution is that the chemical potentials of a component are the same in both phases

$$\mu_{a(solid)} = \mu_{a(aq)}. \qquad [1]$$

Wright and Peech (1960) pointed out that IAP and Ksp are expressions of $\mu_{(aq)}$. If the IAP of a component in the soil solution equals Ksp, then

$$\mu_{a(solid)} = \mu^{\circ}_{a(solid)} = 1, \qquad [2]$$

where the standard state is the pure solid. Equation [2] should be generally invalid in solid mixtures. The chemical potential or activity of a solid varies with its concentration in the solid phase. For one component mixed or coprecipitated in a two-component solid, the change of activity with composition is (Vaslow and Boyd, 1952)

$$\mu_{a(solid)} = \mu^{\circ}_{a(solid)} + RT \ln f_a X_{a(solid)}, \qquad [3]$$

where f is the activity coefficient of the solid phase, X is the mole fraction and

$$\lim_{X \to 1} f = 1 \qquad [4]$$

Wright and Peech (1960) modified Eq. [3] to

$$IAP_a = f_{a(solid)} X_{a(solid)} Ksp_a. \qquad [5]$$

At equilibrium, the IAP in solution of a component in a solid mixture is proportional to its mole fraction in the solid phase.

Equations [3] and [5] have been examined only for mixtures whose mole fractions are clearly defined. The mole fractions of substances in soils are not so obvious. The mole fraction in soil could be defined by a complete knowledge of the soil's composition, which as yet is difficult if not impossible to obtain, or be defined arbitrarily. My calculation of the mole fractions of soil components is arbitrary and based on the moles of soil oxides. The justifications are that oxygen is virtually the only negatively-charged ion in soils and that no other basis presented itself. Table 1 shows the composition of three widely-different soils. The total number of metal oxides in these and nine other soils and horizons averaged 30.7 ± 1.8 mol of oxides/kg of soil. Soils were assumed to contain 30 moles of oxides/kg because the accuracy of the data used later justifies only one significant figure.

The numerator of an oxide mole fraction in soils is also arbitrary. The amounts of soil components are only crudely measureable and were not published in the IAP data used. For simplicity, the number of moles of soil components were calculated as if all of each cation existed as a hydroxyoxide which was miscible and homogeneous with the rest of the soil. The data to calculate these mole numbers came from Table 1 or data of typical total soil composition in Bohn et al. (1979). For example, the chemical potential or IAP of $(Al)(OH)^3$ in solution was considered to be in equilibrium with an Al hydroxyoxide whose actual chemical state altered the potential of the hydroxyoxide. For a second example, the mole numbers of $AlPO_4$ and $FePO_4$ were calculated from typical soil phosphate contents assigning half of the phosphate to Al and half to Fe. A typical soil contains 0.1% P or 0.03 mol P/kg of soil. This yielded 0.0015 mols $AlPO_4$/kg or an "$AlPO_4$ mole fraction" of 0.0005 in the soil. The justification for this crude calculation is the absence of any other way to adapt soil data to Eqs. [3] and [5].

This treatment assumes that the soil components have come to internal equilibrium i.e., that the soil is homogeneous. This is contrary to prevailing opinion among soil chemists and appears to disregard the dedicated efforts of clay mineralogists to identify specific minerals in soils. The disagreement may be reconcilable, however, by relaxing the defini-

Table 1. Composition of soil, as mols of oxides per kg of soil.

	Barnes loam South Dakota		Cecil sandy clay loam, N. Carolina		Columbiana clay, Costa Rica	
	Mass percent†	mols of oxides/ kg of soil	Mass percent†	mols of oxides/ kg of soil	Mass percent†	mols of oxides/ kg of soil
SiO_2	77	25.6	80	27	26	8.7
Al_2O_3	13	3.8	13	3.8	49	14.4
Fe_2O_3	4	0.75	5	0.94	20	3.8
TiO_2	0.6	0.15	1	0.25	3	0.75
MnO	0.2	0.028	0.3	0.04	0.4	0.56
CaO	2	0.36	0.2	0.04	0.3	0.54
MgO	1	0.25	<0.1	--	0.7	1.7
K_2O	2	0.21	0.6	0.06	0.1	0.01
Na_2O	1	0.16	0.2	0.03	0.3	0.05
P_2O_5	0.2	0.07	0.2	0.07	0.4	0.14
SO_3	0.1	0.04	--	--	0.3	0.14
Total		31.4		32.0		30.8

† From Bohn et al., 1979.

tions of equilibrium, homogeneity, and phase and by not requiring that these states be totally satisfied. Partial satisfaction may be sufficient. Such modifications are not unknown in adapting the thermodynamics of pure systems to nature.

A typical pure definition is, "A phase is a region of uniformity...uniform chemical composition and uniform physical properties" (Castellan, 1971). A single pure crystal is therefore a homogeneous phase, as are a crystal with regular isomorphous solution and a completely random ion array, because their properties are uniform with distance. On an atomic-size scale of distance, however, this definition of homogeneous and phase is imprecise. A pure crystal (a regular ion array) and a random ion array are uniform only if averaged over many ion diameters. Both solids are quite nonuniform and nonhomogeneous over distances of several ion diameters. Cheseworth and Dejou (1980) and Babcock and Doner (1981) also discussed the problems of dimension in defining a mineral or colloidal system in terms of equilibrium and homogeneity.

Soil solids are neither pure crystals nor completely random ion arrays but may approach a random array closely enough to apply Eq. [5] as a first approximation. The first degrees of mixing decrease the chemical potential most because concentration and mixing affect the chemical potential logarithmically, Eq. [3]. Further increments of mixing toward complete homogeneity achieve progressively less change of chemical potential. The soil is in a sense a mixture of Si, Al, Fe, Mg, and other oxide complexes. Although many of these complexes are arranged into identifiable structures, these structures are mixtures of oxides. So the concept of soil as a relatively random ion array may not be totally far-fetched.

The soil is also not a complete equilibrium, but, as with homogeneity, the real question is whether the soil is close enough to equilibrium to accept it as a working hypothesis. The consensus supports a partial equilibrium in soils because the deviation from partial equilibrium is often rather small in many soil-water and sediment-water reactions. The compositions of soil and natural solutions are limited to small ranges, compared to laboratory conditions, and the composition range of a particular soil is even narrower. The solids formed in a soil were at equilibrium with the solution from which they formed, a solution probably not much different from the present soil solution. Since the change of chemical potential of the soil solution is therefore small, there is reason to assume at least a near-equilibrium with respect to soil solids and solution. The same reasoning of narrow range of soil solution composition also supports the concept that the soil is relatively homogeneous.

Flexibility is necessary to apply thermodynamics to soils, but modifying and loosening definitions is controversial. The alternative of accepting only strict definitions leads to little progress. The above approach was used to try to (i) rationalize the differences between IAP in soil solutions and Ksp of pure compounds and (ii) calculate the chemical activities and activity coefficients of those substances in soil.

RESULTS

Soil IAP measurements from the literature were applied to Eq. [5] with the calculated oxide mole fractions.

The IAP/Ksp ratio in several reports was the average distance between plotted IAP data points and the line representing Ksp. Those data were plotted logarithmically so this distance was the average logarithm rather than the average arithmetic value. This error is probably no greater than the errors from the other assumptions and was neglected. Table 2 shows the IAP/Ksp ratio, oxide mole fraction, and solid activity coefficient for some hydroxyoxides, phosphates, and arsenate in soil. The activity coefficients ranged from 0.1 to 50 while the IAP/Ksp ratios ranged from 1 to 10^{-6}. The narrower range of the activity coefficient suggests that Eqs. [3] and [5] normalize the IAP in soil solutions to a considerable degree. The variation of the activity coefficient in Table 2 was deemed small because (i) the Ksp values of pure substances vary by as much as three orders of magnitude (Nordstrom et al., 1979), (ii) the actual mole fractions in the soil are unknown, (iii) equilibrium was unproven, (iv) soil IAPs are pH dependent (Bohn, 1970; Blanchar and Scrivner, 1972; Turner and Singh, 1971) (v) investigators understandably choose Ksp values from the literature which are close to their measured

Table 2. Ion activity products and activity coefficients f of hydroxyoxides, phosphates, and arsenate in soils.

Soil component	IAP/Ksp	Approximate mole fraction	Solid activity coefficient, f	Source
$Al(OH)_3$	0.7	0.1	7	Frink and Peech (1962)
$Al(OH)_3$	0.3	0.1	3	Lindsay and Norvell (1969)
$Al(OH)_3$	0.25 (minimum)	0.1	2	Misra et al. (1974)
$Al(OH)_3$	0.26	0.1	2	Lindsay (1979)
$Fe(OH)_3$	1	0.02	50	Bohn (1967)
$Fe(OH)_3$	0.3	0.02	20	Lindsay and Norvell (1969)
$Fe(OH)_3$	0.3	0.02	20	Blanchar and Scrivner (1972)
$MnO_{1.5-2}$	10^{-4} to 10^{-2}	10^{-3}	0.1–10	Bohn (1967)
$Zn(OH)_2$	1.5×10^{-5}	10^{-5} (Zn EDTA added)	1	Lindsay and Norvell (1969)
$Zn(OH)_2$	0.01	0.005 (Zn added)	20	Udo et al. (1970)
$Cu(OH)_2$	0.001	10^{-4} (Cu added)	10	Cavallaro and McBride (1980)
$AlPO_4$	0.067	0.0005 (as ½ of PO_4)	30	Wright and Peech (1960)
$FePO_4$	0.07	0.0005 (as ½ of PO_4)	30	Wright and Peech (1960)
$AlAsO_4$	2×10^{-3}	5×10^{-3} (as AsO_4)	0.4	Livesey and Huang (1981)
	1×10^{-6}	1×10^{-6} (as AsO_4)	1	

IAPs, and (vi) the ions in solution were assumed to be in equilibrium with the entire soil matrix rather than just the soil surface. A value of $f = 10$ may roughly represent the activity coefficients of the soil components listed in Table 2.

The low f values for Mn hydroxyoxide may represent nonequilibrium. The Mn(III-IV) oxidation state in the solid must reach redox equilibrium with Mn^{2+} in solution. Coprecipitation reactions requiring a simultaneous change of oxidation state could be more prone to nonequilibrium because both electron transfer and solid-solution reactions must reach equilibrium. Many redox couples have large kinetic hindrances to electron transfer.

Table 3 shows similar data for carbonate, sulfate, and calcium compounds in soils. The range and absolute values of f are much greater than in Table 2. The IAP/Ksp ratios of $CaCO_3$ and $Ca_5(OH)(PO_4)_3$ in soils exceed one, i.e., they are more soluble in the soil than in the pure state. The variability of hydroxyapatite IAPs measured in soils is high but Norvell's (1974) values consistently exceed Ksp, as do two of Marion and Babcock's (1977) three IAPs. Calcium, carbonate, and sulfate compounds may exhibit this high solubility because those ions are appreciably different structurally from other common soil ions of the same valences (McBride, 1980), as described below.

Tables 2 and 3 include all such data found in many literature search. Many reports undoubtedly were overlooked and I would be much obliged if they were brought to my attention.

DISCUSSION

Equation [3], although familiar in the liquid and gas phases, has found little application to soils. Since the IAP in soil solution has been routinely compared to Ksp, illustrating a more familiar example of the

Table 3. Ion activity products and solid activity coefficients of calcium, carbonate, and sulfate compounds in soils.

Component	IAP/Ksp	Approximate mole fraction	Solid activity coefficient, f	Source
$ZnCO_3$	approx. 1	0.005 (Zn added)	200	Udo et al. (1970)
$CdCO_3$	0.2	7×10^{-6} to 1.5×10^{-2} (Cd + Pb added)	10 (highest addition) to 3×10^4 (lowest addition)	Santillan-Medrano and Jurinak (1976)
$PbCO_3$	0.2	7×10^{-6} to 1.5×10^{-2} (Cd + Pb added)		Santillan-Medrano and Jurinak (1976)
$CaCO_3$	4	0.007	600	Marion and Babcock (1977)
$CaSO_4$	1.1	7×10^{-4} (gypsum added)	2×10^3	Bennett and Adams (1972)
$Ca_5(OH)(PO_4)_3$	3000	0.001 (PO_4 added)	3×10^8	Norvell (1974); Marion and Babcock (1977) also indicate f is high for hydroxyapatite.

implications of mixing and Eq. [3] on the properties of solids and liquids may be worthwhile. Figure 1 shows the effect of mixing two liquids A and B on their vapor pressures. Babcock and Doner (1981) discussed the close analogy of vapor pressure to the chemical potentials of ions in solutions. For the purposes of the present paper, vapor pressure is the partition coefficient between gas and liquid; solubility is the partition coefficient between solid and liquid. Ions dissolving from a solid are analogous to molecules evaporating from a liquid. The liquids are immiscible in the concentration range C to D. The dashed line is the ideal (Raoult's law) change of vapor pressure with concentration. The vapor pressures of both A and B decrease as their concentrations in the mixture decrease. The effect of mixing on the partition coefficient is most apparent when the two liquids are miscible.

The lower vapor pressures in the mixture is due to the higher entropy of the mixture. The molecules in the mixture are dispersed among molecules of the other liquid. Increased dispersion or randomness increases entropy. The subsequent entropy gain by evaporation is therefore less than by evaporating from the pure liquid so the vapor pressure of each liquid in the mixture decreases.

Similarly, solid mixtures, solid solutions, and isomorphously-substituted ions are at a higher entropy state than pure or single minerals and therefore their solubility and chemical potential will tend to be lower than pure minerals. Figure 1 suggests that the decrease of solid solubility should be more pronounced if the mixing is "miscible", as in solid solutions and isomorphous substitution. Miscible mixing, however, also changes the bonds between molecules in liquids and between ions in solids. The new bonds formed may be less stable and would then destabilize the mixture and raise the chemical potentials of the components of the mixture. This instability or stress induced by mixing counteracts the decrease of chemical potential due to the entropy of mixing. Because the electrostatic bonding between ions in solids is more energetic than the intermolecular bonding between molecules of liquids, the stress of solid

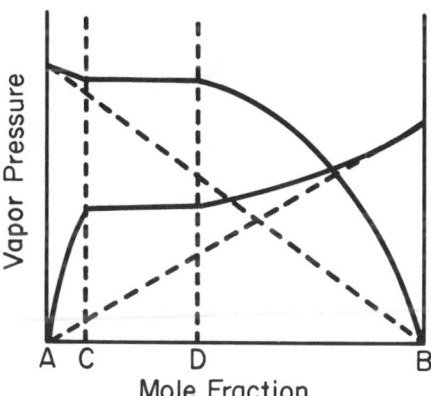

Fig. 1. Vapor pressures of a mixture of liquids A and B, immiscible over the composition range C to D.

mixing could result in high activity coefficients, so high that the components in the mixture could be more soluble than in their pure states.

The literature gives little indication of the trends of solid activity coefficients with composition. The relatively small activity coefficients in Table 2 indicate that for Al, transition metal, and phosphate ions the entropy of mixing is not overshadowed by the stress of mixing. The large solid activity coefficients in Table 3 indicate that calcium, carbonate, and sulfate compounds have a high chemical potential in soils. McBride (1979, 1980) suggested that this could be due to a poor fit of these ions in the tetrahedral and octahedral oxygen coordination of soil silicates.

The effects of mixing dissimilar cations has been more clearly measured by the adsorption and coprecipitation of ions with calcite. Assuming homogeneity, McBride's data for Cd^{2+} adsorbed by calcite yields $f_{CdCO_3} = 10^3$ at an apparent $X_{CdCO_3} = 10^{-4}$. For Mn^{2+} adsorbed by calcite, his data yielded $f_{MnCO_3} = 10^4$ over the range of apparent mole fractions $X_{MnCO_3} = 10^{-6}$ to 10^{-3}. If the adsorption had been homogeneous coprecipitation, his data probably would have yielded lower activity coefficients because the Cd^{2+} and Mn^{2+} concentrations on the calcite surfaces were higher than if that same amount had been coprecipitated. Crockett and Winchester (1966) coprecipitated trace amounts of Zn^{2+} in calcite and found that $f_{ZnCO_3} = 10$. For higher concentrations of Mg^{2+} coprecipitated in calcite, Thorstenson and Plummer (1977) reported that f_{MgCO_3} was as large as 10^9 over the range of $X_{MgCO_3} = 0.01$ to 0.25.

Using adsorption data for comparison is questionable because the surface ion concentrations are almost surely in disequilibrium with the solid phase as a whole. The ion mixing on the surfaces, however, is a first step toward the asymptotic increase of the entropy of mixing and toward a totally random ion array. Santillan-Medrano and Jurinak's (1975) data indicate that the activity coefficient decreases with increasing soil content of the component, as does Hendrickson and Corey's (1981) similar selectivity coefficient for adsorption data, and in accord with Eq. [4].

The familiar activity coefficients of solutes in aqueous solutions generally are limited to values between 0.2 and one. The activity coefficient in other mixtures may have a much wider range. For "regular solutions" of liquids and solids, the change of activity coefficient f with mole fraction X is (Garrels and Christ, 1965)

$$\ln f_a = \frac{B_a}{RT} (1 - X_a)^2 = B'_a (1 - X_a)^2, \qquad [6]$$

where B_a is an empirical and little-understood constant. Garrels and Christ suggest that B' varies within ± 5, i.e., that f could vary from 0.01 to 100 in such solutions. A regular solution is defined as not changing entropy when a small amount of one of its components is transferred to it from an ideal solution of the same composition, the total volume remaining unchanged (Hildebrand et al., 1970). The limits of activity coefficients for solid irregular solutions are uncertain.

The treatment of the data in Tables 2 and 3 suggest that the difference between the IAP of soil solutions and the Ksp of pure solids may be due to solid mixing or coprecipitation rather than adsorption. Much ion-soil interaction ascribed to adsorption should perhaps be re-evaluated in terms of Eq. [5]. The ability of a soil component to create an IAP in soil solutions close to Ksp appears to be related to: a) a high concentration of the component in the soil; b) immiscibility with other soil solids, i.e., the ions do not fit well in soil aluminosilicate and hydroxyoxide structures; and c) favorable kinetics, including redox reactions. To attain a solid-phase activity coefficient of about f = 10, the ion or component apparently should be: a) present in trace quantities, b) similar to its native concentration, and c) similar in change and ion size to major soil components.

The approach used here to understand the IAPs in soil solutions is only a crude and tentative first step. The approach is lacking for results (e.g., Santillan-Medrano and Jurinak, 1975; Emmerick et al., 1982) in which the trace element concentration in soil solutions is independent of pH, and therefore independent of the Ksp of hydroxyoxides and carbonates. The approach shows some promise, however, that further study of coprecipitation may improve our understanding of the soil's ion retention and our predictions of the composition of the soil solution.

ACKNOWLEDGMENT

I thank Donald Thorstenson and Eric Sundquist of the U.S. Geological Survey and anomymous reviewers for valuable criticism. The disagreement centers around definitions of a homogeneous phase on macro- and atomic-size scales which I have herein tried to reconcile.

LITERATURE CITED

Babcock, K. L., and H. E. Doner. 1981. A macroscopic basis for soil chemistry. Soil Sci. 131: 276-283.

Bennett, A. C., and F. Adams. 1972. Solubility and solubility product of gypsum in soil solution and other aqueous solutions. Soil Sci. Soc. Am. Proc. 36:288-291.

Blanchar, R. W., and C. L. Scrivner. 1972. Aluminum and iron ion products in acid extracts of samples from various depths in a Menfro soil. Soil Sci. Soc. Am. Proc. 36.897-901.

Bohn, H. L. 1967. The $(Fe)(OH)^3$ ion product in suspensions of acid soils. Soil Sci. Soc. Am. Proc. 31:641-644.

----. 1970. Comparisons of measured and theoretical Mn^{2+} concentrations in aqueous solutions and soil extracts. Soil Sci. Soc. Am. Proc. 34:195-197.

----, B. L. McNeal, and G. A. O'Connor. 1979. Soil chemistry. Wiley-Interscience, N.Y.

Castellan, G. W. 1971. Physical chemistry. 2nd ed. Addison-Wesley, Reading, Mass.

Cavallaro, N., and M. B. McBride. 1980. Activities of Cu^{2+} and Cd^{2+} in soil solutions as affected by pH. Soil Sci. Soc. Am. J. 44:729-732.

Chesworth, W., and J. Dejou. 1980. Are considerations of mineralogical equilibrium relevant to pedology? Evidence from a weathered granite in central France. Soil Sci. 130: 290-292.

Crockett, J. H., and J. W. Winchester. 1966. Coprecipitation of zinc with calcium carbonate. Geochim. Cosmochim. Acta 30:1093-1109.

Emmerick, W. E., L. J. Lund, A. L. Page, and A. C. Chang. 1982. Predicted solution phase forms of heavy metals in sewage-treated soils. J. Environ. Qual. 11:182-186.

Frink, C. R., and M. Peech. 1962. The solubility of gibbsite in aqueous solutions and soil extracts. Soil Sci. Soc. Am. Proc. 26:346–347.

Garrels, R. M., and C. L. Christ. 1965. Solutions, minerals and equilibria. Harper and Row, N.Y.

Helgeson, H. C. 1968. Evaluation of irreversible reactions in geochemical processes involving minerals and aqueous solutions—I: Thermodynamic relations. Geochim. Cosmochim. Acta. 32:853–877.

Hendrickson, L. L., and R. B. Corey. 1981. Effect of equilibrium metal concentrations on apparent selectivity coefficients of soil complexes. Soil Sci. 131:163–171.

Hildebrand, J. H., J. M. Prausnitz, and R. L. Scott. 1970. Regular and related solutions: The solubility of gases, liquids, and solids. Van Nostrand Reinhold, N.Y.

Lindsay, W. L. 1979. Chemical equilibria in soils. Wiley-Interscience, N.Y.

----, and W. A. Norvell. 1969. Equilibrium relationships of Zn^{2+}, Fe^{3+}, Ca^{2+}, and H^+ with EDTA and DTPA in soils. Soil Sci. Soc. Am. Proc. 33:62–68.

Livesey, N. T., and P. H. Huang. 1981. Adsorption and arsenate by soils and its relation to selected chemical properties and anions. Soil Sci. 131:88–94.

Marion, G. M., and K. L. Babcock. 1977. The solubilities of carbonates and phosphates in calcareous soil suspensions. Soil Sci. Soc. Am. J. 41:724–728.

McBride, M. B. 1979. Chemisorption and precipitation of Mn^{2+} at $CaCO_3$ surfaces. Soil Sci. Soc. Am. J. 43:693–698.

----. 1980. Chemisorption of Cd^{2+} on calcite surfaces. Soil Sci. Soc. Am. J. 44:26–29.

McIntire, W. L. 1963. Trace element partition coefficients—a review of theory and applications to geology. Geochim. Cosmochim. Acta. 27:407–417.

Misra, V. K., R. W. Blanchar, and W. J. Upchurch. 1974. Aluminum content of soil extracts as a function of pH and ionic strength. Soil Sci. Soc. Am. J. 38:897–902.

Murrman, R. P., and M. Peech. 1968. Reaction products of applied phosphate in limed soils. Soil Sci. Soc. Am. Proc. 32:493–496.

Nordstrom, D. K., et al. 1979. A comparison of computerized chemical models for equilibrium calculations in aqueous systems. In E. A. Jenne (ed.) Chemical modeling in aqueous systems. Am. Chem. Soc. Symp. Series 93:857–897.

Norvell, W.A. 1974. Insolubilization of inorganic phosphate by anoxic lake sediment. Soil Sci. Soc. Am. J. 38:441–445.

Orville, P. M. 1972. Plagioclase cation exchange equilibria with aqueous chloride solution: results at 200°C and 2000 bars in the presence of quartz. Am. J. Sci. 272:234–272.

Santillan-Medrano, J., and J. J. Jurinak. 1975. The chemistry of lead and cadmium in soil: solid phase formation. Soil Sci. Soc. Am. Proc. 39:851–856.

Saxena, S. K. 1973. Thermodynamics of rock-forming crystalline solutions. Springer-Verlag, Berlin.

Sposito, G. 1981. The thermodynamics of soil solutions. Oxford Press, N.Y.

Stumm, W., and J. J. Morgan. 1981. Aquatic chemistry. 2nd ed. Wiley, N.Y.

Thorstenson, D. C., and L. m. Plummer. 1977. Equilibrium criteria for two-component solids reacting with fixed composition in an aqueous phase—example: the magnesian calcites. Am. J. Sci. 277:1203–1223.

Turner, R. C., and J. E. Brydon. 1965. Factors affecting the solubility of $Al(OH)_3$ precipitated in the presence of montmorillonite. Soil Sci. 100:176–181.

----, and S. S. Singh. 1971. The role of sparingly soluble solids and cation exchange reactions in controlling conditions in solutions. Soil Sci. Soc. Am. Proc. 35:445–449.

Udo, E. J., H. L. Bohn, and T. C. Tucker. 1970. Zinc adsorption by calcareous soils. Soil Sci. Soc. Am. Proc. 34:405–407.

Vaslow, F., and G. E. Boyd. 1952. Thermodynamics of coprecipitation: Dilute solid solutions of AgBr in AgCl. J. Am. Chem. Soc. 74:4691–4695.

Wright, B. C., and M. Peech. 1960. Characterization of phosphate reaction products in acid soils by the application of solubility criteria. Soil Sci. 90:32–43.

Chapter 2

Effects of Aeration on Reactivity and Mobility of Soil Constituents[1]

K. R. REDDY AND W. H. PATRICK, JR.[2]

Aeration status of a soil primarily refers to the O_2 concentration in the soil atmosphere. The principal gases of the soil atmosphere and some typical values for an aerated soil are N_2 (79.2%), O_2 (20.6%), and CO_2 (0.25%). The percentages for atmospheric air are N_2 (79.0%), O_2 (20.97%), and CO_2 (0.03%). The composition of the soil atmosphere is determined by the physical, chemical, and biological conditions of the soil and the plant-root density. For example, incorporation of organic residues into soil can deplete soil O_2 and increase CO_2 concentration. Flooding a soil can displace soil air, and increase the concentrations of CO_2 and CH_4 as a result of anaerobic organic matter decomposition.

Adequate soil O_2 is essential for normal root growth for most plants, for growth of most microorganisms, and to activate several essential biochemical reactions in the soil. The optimum soil O_2 requirements for plant-root development greatly varies with species and age of the plant (Kramer, 1965, 1969). The optimum soil O_2 requirements of microorganisms vary with the type of biochemical reactions functioning in the soil system. A few reviews have appeared in the literature on soil aeration and its effects on plant growth (Grable, 1966; Stolzy, 1974; Meek and Stolzy, 1978). These reviews, however, were limited to plant growth and development and provide very little or no information on the effects of soil aeration on the reactivity of the soil processes associated with the move-

---

[1] Contribution from the Univ. of Florida, Institute of Food and Agricultural Sciences, and Louisiana State Univ. Florida Agric. Exp. Stns. J. Series No. 4567.

[2] Associate professor, Univ. of Florida, IFAS, Agric. Res. and Education Center, P.O. Box 909, Sanford, FL 32771; and Boyd professor, Center for Wetland Resources, Louisiana State Univ., Baton Rouge, LA 70803, respectively.

Copyright © 1983 ASA, SSSA, 677 South Segoe Road, Madison, WI 53711. *Chemical Mobility and Reactivity in Soil Systems.*

ment of nutrients. The purpose of this paper is to examine the effects of soil aeration (primarily O_2) on the reactivity and mobility of nutrients in well-aerated, moderately aerated, poorly aerated, and flooded soils.

SOIL OXYGEN

Oxygen status of a soil is controlled by the gaseous interchange between the atmosphere above the soil surface and the soil atmosphere. The atmosphere above the soil has an abundant supply of O_2, and this O_2 is continuously supplied to the soil through diffusion, although changes in soil temperature, volume of pore space, barometric pressure, and wind speed can also result in minor contributions through mass flow (Russell, 1952).

Diffusion is the random thermal motion of molecules of a gas or of a liquid. Molecular sizes and weights are different for various gases, which will enable them to diffuse at different rates. Net movement of gases by diffusion is also a function of the concentration or partial pressure gradient, the cross sectional area available for the diffusion, the properties of the material through which the gases are diffusing, and the time. Soil O_2 content is generally lower than the atmospheric O_2, and the net diffusion of O_2 will be from atmosphere to the soil. Soil CO_2 concentration is generally higher than atmospheric CO_2, hence the net diffusion of CO_2 will be from soil to the atmosphere.

Rate of O_2 diffusion in an unflooded soil is dependent on the effective air-filled porosity (Kohnke, 1968). For every soil there appears to be an optimum effective air-filled porosity. Flocker et al. (1959) observed an air-filled porosity of 10% to be critical and below this value O_2 diffusion into soil was negligible. Their study also showed that optimum growth of plants occurred with air-filled porosities between 20 and 35%. Grable and Siemer (1968) found that O_2 diffusion decreased to zero at or near air-filled porosities of 10 to 20% in the soil they studied.

Air-filled porosity of a soil changes with volumetric water content and soil-water suction. In a field study, Patrick et al. (1973) have shown the volume distribution of soil solids, air space, and water in several soils of Louisiana. Bruin (coarse-silty, mixed, thermic Fluvaquentic Eutrochrept), Dundee (fine-silty, mixed, thermic Aeric Ochraqualf), and Tensas (fine, montmorillonitic, thermic Vertic Ochraqualf) soils had air-filled porosities generally >10%. Mhoon (fine-silty, mixed, nonacid, thermic Typic Fluvaquent) and Tunica (clayey over loamy, montmorillonitic, nonacid, thermic Vertic Haplaquept) soils had air-filled porosities of <10%, a value at which normal exchange of gases do not occur (Fig. 1). The effective pore space for air movement within a given soil is inversely related to soil water content. As the soil-water content approaches saturated condition, air-filled porosity of the soil will approach zero, resulting in interruption of the normal processes of gaseous exchange between soil and air. At this point, O_2 diffusion into a saturated or a flooded soil will depend on consumption at the soil surface and depth of the overlying water. Oxygen diffusion in these systems was found to be about 10 000 times slower than diffusion in gas-filled pores (Greenwood, 1961).

Incorporation of organic wastes (crop residues, animal and municipal wastes) can increase soil O_2 demand, thus resulting in decreased O_2 concentration (Meek and Grass, 1975). In most well aerated soils, O_2 consumption rates are usually lower than the potential O_2 diffusion rate from the atmosphere. As a result, the soil is maintained under aerobic (oxidized) conditions. In poorly aerated soils, O_2 diffusion cannot keep up with demand, resulting in slower renewal of O_2 as compared to consumption. This results in the formation of anoxic microsites in the soil. In flooded soils, O_2 diffusion is extremely slow, and the slow O_2 supply rate as compared to demand in the soil, results in depletion of soil O_2 in less than 36 h (Turner and Patrick, 1968), thus creating anaerobic conditions.

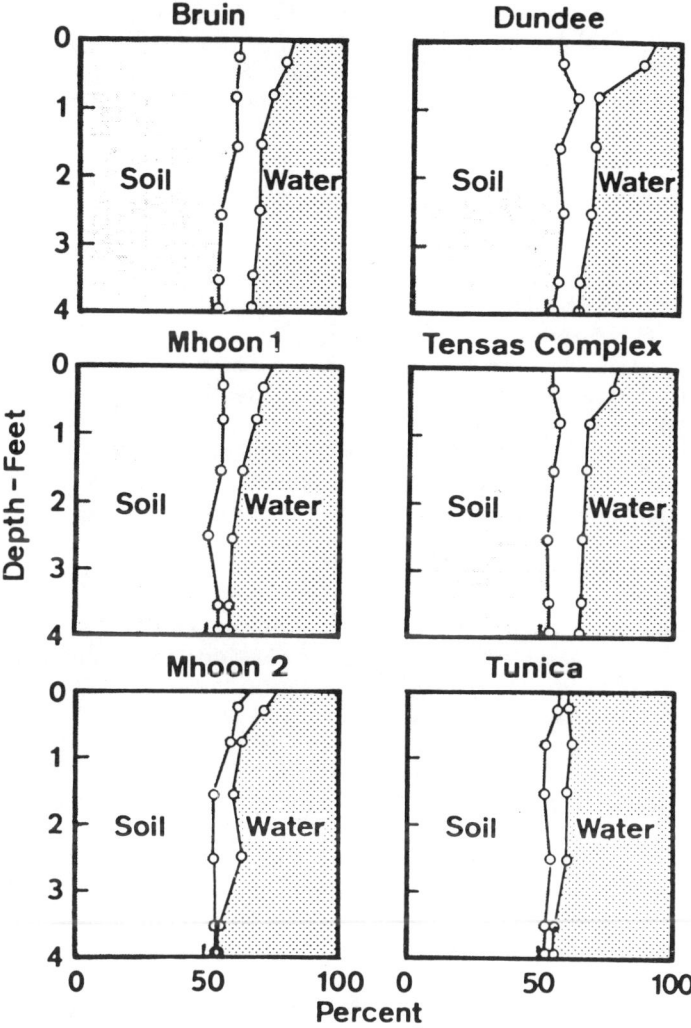

Fig. 1. Volume distribution of soil solids, air space, and water at beginning of 1970 growing season for Bruin, Mhoon 1, and Mhoon 2 soils at the Northeast Louisiana Exp. Stn. and Dundee, Tensas Complex and Tunica soils at Highland Plantation (Patrick et al., 1973).

The greater potential O_2 consumption by the flooded soil, compared to the slow supply rate through the floodwater results in the development of two distinctly different soil layers: 1) an oxidized or aerobic surface soil layer where O_2 is present; and 2) an underlying anaerobic or reduced soil layer, where no free O_2 is present (Pearsall, 1938; Mortimer, 1941; Alberda, 1953).

The thickness of the oxidized soil layer is controlled by the O_2 concentration in the overlying water and O_2 consumption by the underlying sediment. Howeler and Bouldin (1971) using swamp sediment (6.2% organic matter), measured an oxidized layer thickness of 0.48 mm at an O_2 concentration of 4%, compared to 2.4 mm at an O_2 concentration of 21% (Fig. 2a). By varying the sediment organic matter content (adding varying amounts of C source), Engler and Patrick (1974) showed an inverse relationship between the oxidized layer and sediment organic matter content (Fig. 2b). The role of oxidized and reduced layers on reactivity and mobility of nutrients in flooded soils is discussed in the latter part of this paper.

Oxygen consumption in well-aerated, poorly aerated, and flooded soils depends on source of O_2-demanding components (energy source), temperature, and soil texture (Meek and Stolzy, 1978). Incorporation of organic wastes (crop residues, animal manures, municipal, and industrial wastes) into soil increases microbial activity, thus exerting greater demand on soil O_2. During aerobic microbial respiration, O_2 is used as an electron acceptor and organic C is used as an energy source. Oxygen is also consumed during the nitrification process where nitrifying organisms use O_2 and derive energy from the oxidation of NH_4^+ and NO_2^-. These reactions, therefore, can potentially use large quantities of O_2 and nitrification can at times deplete O_2 in soils receiving large amounts of nitrogenous wastes, although other microbial processes compete favorably for O_2 under such conditions. In flooded soils, O_2 reaching the soil surface is consumed by (i) heterotrophic microbial respiration in the aerobic layer; (ii) biological autotrophic oxidation of NH_4^+ in the aerobic layer; and (iii) chemical oxidation of reduced Fe^{+2}, and Mn^{+2}, and sulfides. Several mathematical models for describing O_2 diffusion and consumption in flooded soils were presented by Bouldin (1968). Models which include the consumption of O_2 during microbial respiration in the aerobic zone and oxidation of mobile and nonmobile reductants such as Fe^{+2} (Howeler and Bouldin, 1971) were found to be best suited for describing O_2 consumption in flooded soils. In a recent study (Reddy et al., 1980b), O_2 consumption by 37 anaerobic soils was also found to be a two phase process. The rate of O_2 consumption during Phase I was approximately the same for all soils ($k_I = 0.15$ h^{-1}), where O_2 consumption during Phase II ranged from 0.003 to 0.054 h^{-1}. The first phase represented chemical oxidation of Fe^{+2}, while the second phase represented chemical and biological oxidation. Soil temperature affects O_2 status directly by its effect on movement and indirectly by its effect on plants and microbial activity. Several researchers (Drobnik, 1962; Moureux, 1967; Pal et al., 1975) calculated Q_{10} values in the range of 1.5 to 2.0 for a temperature range of 5 to 37°C during aerobic decomposition of organic matter. Luxmoore and Stolzy (1972) calculated Q_{10} values of 1.5 to 1.8 for a 5 to 25°C range during respiration rate of root tips.

CHARACTERIZATION OF SOIL AERATION

Aeration status of a soil can be characterized by measuring the (1) composition of soil atmosphere, (2) gaseous diffusion (O_2 diffusion rates),

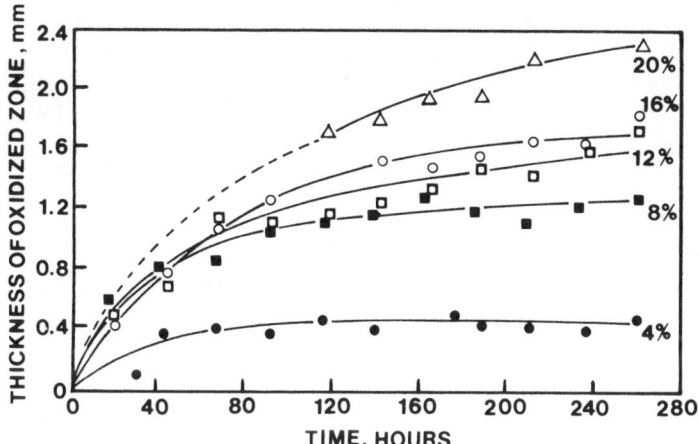

Fig. 2a. Thickness of oxidized layer in flooded soil, as influenced by varying levels of oxygen in the atmosphere above floodwater (Howeler and Bouldin, 1971).

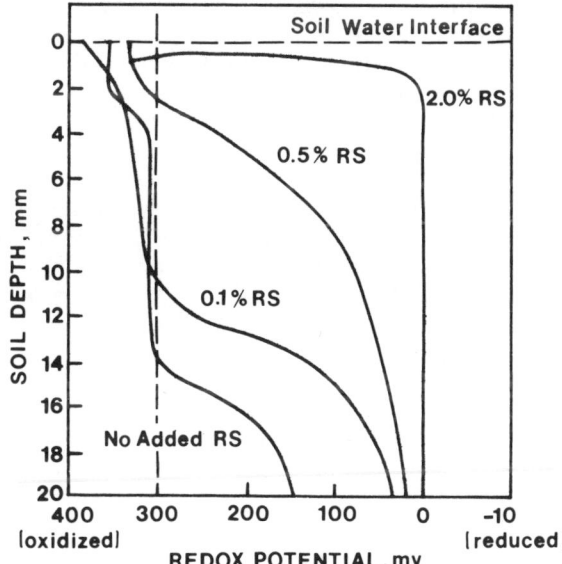

Fig. 2b. Thickness of oxidized layer in flooded soil, as influenced by energy source (organic matter) (Engler and Patrick, 1974).

and (3) oxidation-reduction potential (Eh). Measuring O_2 uptake and respiratory quotient (Bridge and Rixon, 1976) and measuring reduced chemical forms in soil solution (Meek and Grass, 1975) have been reported as possible alternate methods for assessing soil aeration status.

Composition of Soil Atmosphere

Aeration status of a soil can be established by measuring the composition of the soil atmosphere. With good aeration, O_2 content of the soil may approach that of atmospheric air. Decreased soil O_2 concentration can result in partial or complete anaerobic conditions. These conditions can measurably increase N_2O content of the soil air as a result of denitrification. An inverse relationship was observed between soil O_2 and N_2O (Dowdell and Smith, 1974). Under highly reduced conditions, gases such as H_2S (sulfate reduction end product) and CH_4 (CO_2 reduction end product) can be detected in significant quantities in soils with high C content. In addition to CH_4, Smith and Robertson (1971) and Sheard and Leyshon (1976) have found C_2H_4 concentrations in the range of 17 to 20 ppm in soils incubated under anaerobic conditions for a period of 13 to 19 days.

Gaseous Diffusion

The technique of measuring soil O_2 diffusion rate (ODR) with a stationary platinum electrode (Lemon and Erickson, 1952) has been found to be satisfactory in the higher range of soil aeration and was found to be less sensitive in the lower range of soil aeration (Poorly drained and flooded soils). This method measures the O_2 reduction rate at the platinum electrode surface when constant electrical potential is applied (Sojka et al., 1975; West and Black, 1978). Once the O_2 present at the electrode surface is reduced, further reduction depends on the diffusion of O_2 to the electrode surface. To obtain precise measurements, the potential applied must be one at which O_2 is the only component reduced (Van Doren and Erickson, 1966). McIntyre (1970) presented a detailed review of the potential advantages and problems involved in using platinum electrodes for O_2 flux measurements. Shalhevet et al. (1969) observed ODR values in the range of 0.04 to 0.09 μg cm^{-2} min^{-1} for soil with Eh values of -270 to 90 mv, while ODR sharply increased (0.09 to 0.36 μg cm^{-2} min^{-1}) in the soil at Eh values of 90 to 200 mv.

Oxidation-Reduction Potential (Eh)

Oxidation-reduction potential or redox potential is a measure of electron availability potential in a chemical or biological system. This method was used by several researchers (Patrick and Mahapatra, 1968; Ponnamperuma, 1972, for example) to characterize the intensity of reduction and identify different forms of redox couples (e.g. Fe^{+2} and Fe^{+3}) in flooded soils and sediments. An inert electrode, usually platinum, is used to measure the potential of all redox couples. Under well-drained conditions (oxidized systems), low concentrations of redox couples reduce the stability and reproducibility of Eh measurements (Bohn, 1971), while in flooded soils and in soils with low O_2 levels, the higher concentrations of redox couples increases the sensitivity of Eh measurements, thus making this technique more applicable to anaerobic systems. Well-drained soils have characteristic Eh value of > 400 mv, and as the soil O_2 decreases, soil Eh values decrease. In some flooded soils with high C content, Eh values as low as -350 mv can be observed. Changes in the concentrations of various redox couples as a function of Eh is shown in Fig. 3. Oxygen disappeared at Eh values of about 300 mv, while NO_3^- is removed at Eh values between 200 to 300 mv followed by the reduction of Mn^{+4}, Fe^{+3}, and SO_4^{-2}. Detailed discussions of the stability of redox couples in soil systems have been presented by Patrick and Mahapatra (1968) and Ponnamperuma (1972).

REACTIVITY AND MOBILITY

In this paper, reactivity is defined as a combination of transformations (physical, chemical, and biological), simultaneously functioning in well-drained, moderately to poorly drained, and flooded soil systems, and converting non-mobile chemical species into mobile forms and mobile species to non-mobile forms. The role of these conversion processes or transformations on the mobility of several important soil constituents at varying levels of soil aeration (redox changes) will be discussed.

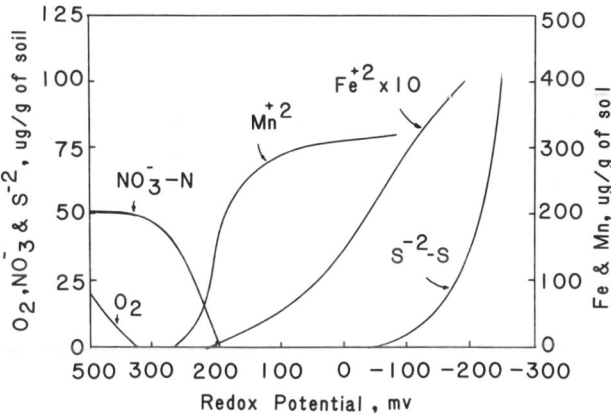

Fig. 3. Stability of various redox couples shown as a function of redox potential (data obtained from Patrick, 1964; Connell and Patrick, 1968; Turner and Patrick, 1968, Gotoh and Patrick, 1972).

Nitrogen

Most of the N in soils and sediments is in organic form with only a very small amount present in the inorganic form. A series of biochemical and physico-chemical processes are involved in transforming one source of N to another source. The most important forms of inorganic N compounds include NH_4^+, NO_3^-, NO_2^-, N_2, and N_2O. These compounds are the end products of specific biological reactions. The important microbial conversion processes functioning in a soil system are organic N $\rightarrow$ NH_4^+ (ammonification), $NH_4^+ \rightarrow NO_2^- \rightarrow NO_3^-$ (nitrification), $NO_3^- \rightarrow N_2O \rightarrow N_2$ (denitrification), $NO_3^- \rightarrow NH_4^+$ (assimilatory NO_3^- reduction), $N_2 \rightarrow$ organic N (biological fixation), and $NH_4^+ \rightarrow NH_3$ (volatilization).

Organic N conversion to NH_4^+ occurs at all levels of soil aeration, but at varying rates. Under well-aerated conditions, very little or no NH_4^+ accumulates in a soil system, because of rapid oxidation of NH_4^+ to NO_3^-. Ammonium N accumulates in O_2-deficient soils since O_2 is required for the oxidation of NH_4^+ to NO_3^-. Data shown in Fig. 4 with varying soil moisture levels present indirect evidence for the effects of varying levels of soil O_2. Ammonification was shown to occur at a maximum rate in a soil with 0.3 atm moisture tension. Although soil O_2 levels are probably high at moisture tension > 0.3 atm, the rates of these processes were lower than at low tension, probably due to available water limiting the process.

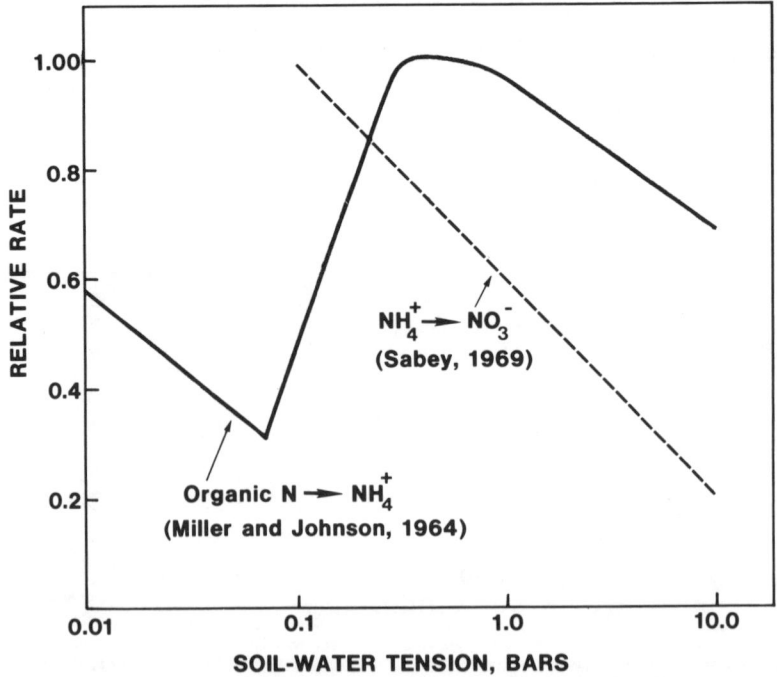

Fig. 4. Effect of soil-water tension on relative rate of ammonification and nitrification in drained soils (Miller and Johnson, 1964; Sabey, 1969).

On the other hand, at moisture tensions < 0.3 atm, soil O_2 was probably limiting because of larger fraction of soil pores filled with water. Reactions of NH_4^+ that are concentration dependent are increased in O_2-deficient soils because of high NH_4^+ concentration. Since NH_4^+ undergoes fixation reactions within the clay lattices, the increased concentration should increase fixation.

In addition to the normal reactions of NO_3^- assimilation by plants and microbes, restricted aeration causes the dissimilatory reduction of NO_3^- by facultative anaerobes. The rate of this process can be much greater than assimilation. With the exception of O_2, NO_3^- is the first redox constituent to disappear from the soil following restricted O_2 supply. In a poorly drained soil, the rate of this process is controlled by the fraction of soil pores free of O_2 and the available energy source. In soils that have both aerobic and anaerobic zones, rate of denitrification is dependent on the NO_3^- diffusion from aerobic soil pores to anaerobic microsites. Although this mechanism was not demonstrated experimentally, Reddy et al (1978) have shown the effect of NO_3^- diffusion on the denitrification process.

The above described N processes control the concentration of NH_4^+ and NO_3^- in drained, poorly drained, and flooded soils. The mobility of these ions is also related to the soil aeration. Apparent NH_4^+ diffusion was shown to increase with the soil-water content (Fig. 5). As the soil-water

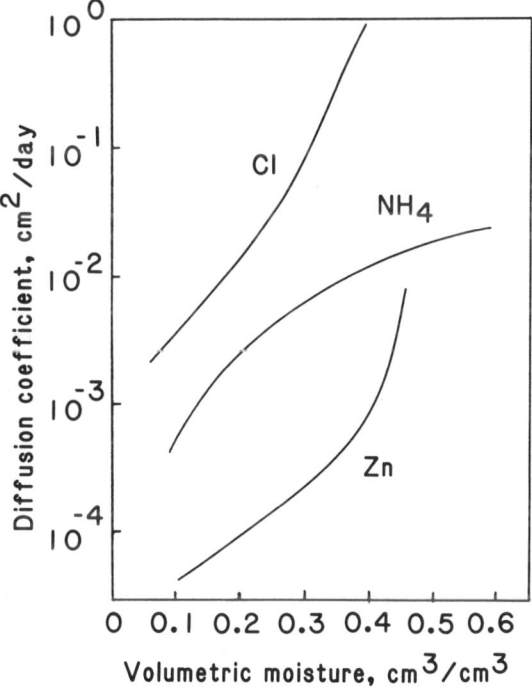

Fig. 5. Diffusion coefficient of Cl, NH_4, and Zn, as influenced by volumetric soil-water content (Cl, average of six soils, Warncke and Barber, 1972; NH_4^+, Crowley silt loam soil, Reddy et al., 1981b; Zn, Fincastle soil (fine-silty, mixed, mesic Aeric Ochraqualf), Warncke and Barber, 1972).

content increases, the liquid phase becomes more continuous and diffusion path less tortuous (Porter et al., 1960). The mobility of NH_4^+ is increased in O_2-deficient soils because of the higher concentration and also because a larger fraction of NH_4^+ is in the pore water instead of being adsorbed onto the exchange complex. This latter effect is due to displacement from the exchange complex by other cations that are mobilized as a result of reduction processes. For example, in poorly drained and flooded soils, concentration of Mn^{+2} and Fe^{+2} can increase as a result of reduction and compete with NH_4^+ and other cations for exchange sites, and may result in displacement of NH_4^+ from the exchange complex. This will decrease NH_4^+ adsorption and increases its mobility. In flooded soils, a significant amount of NH_4^+ is lost as a result of diffusion into the overlying oxygenated floodwater and surface aerobic layer. Ammonium diffused into surface soil layer or floodwater can be rapidly oxidized to NO_3^- or volatilized to NH_3. Nitrate thus formed diffused back into underlying anaerobic soil and is lost through denitrification (Reddy et al., 1976). In flooded soils and sediments, high pH conditions can exist in the overlying floodwater as a result of imbalance between photosynthesis and respiration of algae, thus creating favorable conditions for N loss through NH_3 volatilization (Mikkelsen et al., 1978).

Nitrate diffusion is also influenced by soil-water content, soil O_2, and denitrification potential of a soil. Nitrate diffusion rate increases with soil-water content. Movement of NO_3^- was found to be more rapid in a sandy soil than a loam soil (Clarke and Barley, 1968). Low soil O_2 and greater demand for electron acceptors in anaerobic sites can increase NO_3^- diffusion from aerobic sites to anaerobic microsites where it is lost through denitrification. Although the role of denitrification in anaerobic microsites has been demonstrated (Rolston and Marino, 1976); the rate of NO_3^- diffusion from aerobic sites to anaerobic microsites has not been documented.

In flooded soils and sediments, NO_3^- in floodwater and in the surface oxidized zone is derived from (1) nitrification in the oxidized soil zone or floodwater, and (2) input from external sources (drainage effluents, wastewaters). In this system, NO_3^- from floodwater diffuses into underlying anoxic sediments where it undergoes denitrification (Reddy et al., 1978). It has been shown that very little or no denitrification occurs in floodwater low in available C (Engler and Patrick, 1974), and under these conditions, floodwater NO_3^- removal is dependent on the diffusion of NO_3^- into anaerobic portions of the sediment. The flux of NO_3^- from the floodwater is controlled by (i) concentration gradient across sediment-water interface, (ii) floodwater depth, (iii) temperature, and (iv) mixing and aeration.

Phosphorus

Phosphorus in soils occurs both in organic and inorganic forms. In mineral soils, inorganic P probably is more important than organic P, primarily due to slower availability of the organic P. However, in organic soils and mineral soil high in organic matter content, organic P mineralization can play a significant role in releasing soluble inorganic P. Inorganic P is the most mobile form of P; however, under certain conditions soluble organic P can be mobile. Inorganic P in the soil can be divided into four main groups: calcium phosphate, aluminum phospahte, iron phosphate, and reductant-soluble P extracted after the removal of the first three forms. The latter two fractions are not very important in the fertility of well-aerated soils, but have been found to be of significance in poorly drained and flooded soils.

Although P itself is not normally biologically reduced in redox reactions in soils, it does undergo reactions that have a pronounced effect on its reactivity (Fig. 6). Most of this change in reactivity of P as a result of anaerobic conditions is associated with the Fe chemistry of the soil. Reduction of Fe^{3+} oxyhydroxide and Fe^{3+} phosphate compounds increase the solubility of PO_4^{-3} (HPO_4^{-2}, $H_2PO_4^-$) and make it more available to plants. The reduction-solubilization of Fe^{+3} compounds releases PO_4^{-3} in two ways: (i) by converting insoluble Fe^{+3} phosphates to more soluble Fe^{+2} phosphates, and (ii) by solubilizing the Fe^{+3} oxyhydroxide material in the soil that has occluded in its matrix forms of phosphate that are more reac-

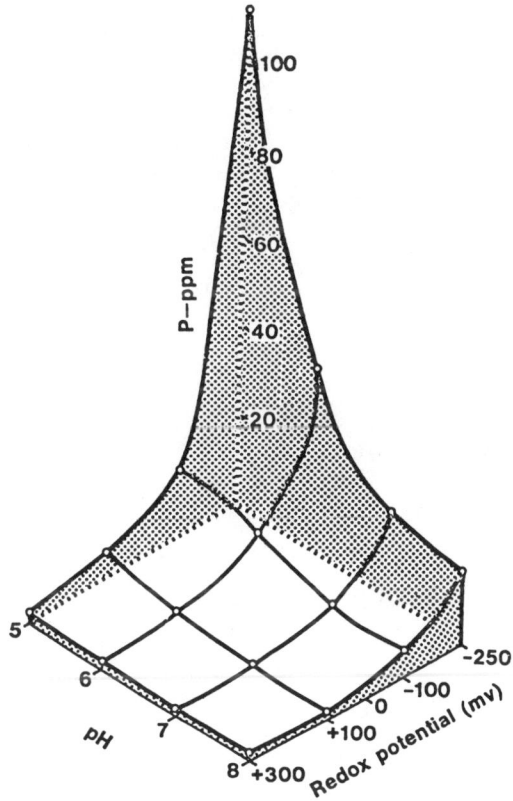

Fig. 6. Solubility of P as influenced by pH and redox potential.

tive than the occluding Fe^{+3} oxyhydroxide material (Patrick, 1964; Mahapatra and Patrick, 1969; Patrick and Khalid, 1974). Phosphorus release was also found to be significantly higher in flooded organic soils than soils maintained under drained conditions (Reddy, 1983) (Table 1). Although organic soils are low in Fe and Al, anaerobic conditions increased the solubilization of organic matter, thus increasing the availability of soluble P.

In alkaline soils, Turner and Gilliam (1976a) observed that increase in available P (expressed as anion resin-adsorbable P) under flooded conditions was not due to soil reduction of ferric phosphate but due to increased soil-water content. Reasons for not detecting an increase in available P as the alkaline soils were reduced were due to (1) lack of reductant soluble P, and (2) soluble P being controlled by the calcium system. Increased available P under water saturated conditions (oxidized or reduced) was attributed to the greater rate of diffusion to the resin particles.

The movement of soil P to plant roots occurs by mass flow and diffusion. Movement of P by mass flow is determined by the concentration in soil solution and the soil-water content as well as by the rate of water uptake. Under well-drained conditions, soil solution P is generally low, indicating that P movement to plant roots in these soils probably occurs primarily due to diffusion (Barber et al., 1963; Olsen, 1971). As discussed earlier, in most of the poorly drained and flooded soils, soluble P concentrations are usually high, which results in an increased rate of movement by diffusion and mass flow. No evidence is available on the effect of soil O_2 on P diffusion; however, several researchers (Olsen et al., 1965; Mahtab et al., 1971; Turner and Gilliam, 1976b) have observed increased rate of P diffusion with increase in soil-water content (Fig. 7). For example, Olsen et al. (1965) showed that P diffusion coefficients increased from 0.4×10^{-7} to 15.5×10^{-7} cm^2/s as the volumetric moisture increased from 0.22 to 0.55 cm^3/cm^3. In these soils, it was not clear whether increased rate diffusion was due to soil reduction (caused by high soil-water

Table 1. Phosphorus release rates in organic soils in Florida maintained under flooded and drained conditions (Reddy, 1983).

Soil type	Soluble P		Total P	
	Non-flooded	Flooded	Non-flooded	Flooded
	kg P ha^{-1} d^{-1}			
Cultivated soils				
Oklawaha muck	0.12	1.12	0.14	1.38
Monteverde muck	0.07	0.45	0.07	0.80
Lauderhill muck	0.16	0.62	0.20	0.70
Brighton muck	0.14	0.27	0.16	0.36
Pahokee muck	0.01	0.05	0.01	0.10
Virgin soils				
Monteverde muck	0.19	1.28	0.21	1.64
Pahokee muck	0.02	0.09	0.02	0.20

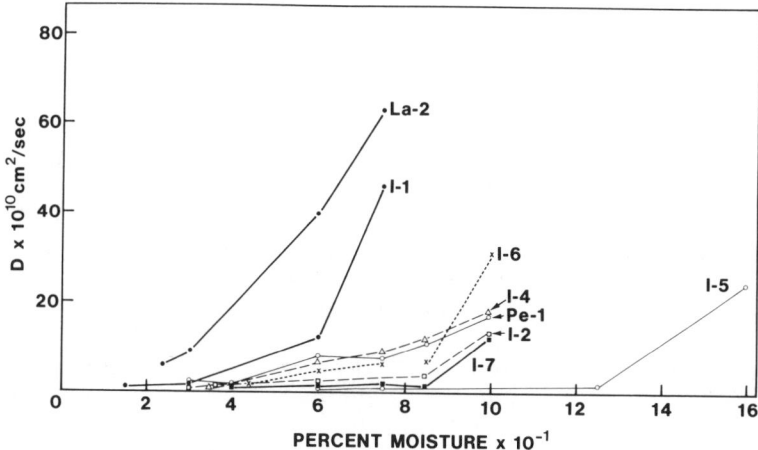

Fig. 7. Effect of soil moisture and oxidized/reduced conditions on P diffusion (reproduced from Turner and Gilliam, 1976a; Plant Soil 45:353–363).

content) or water content or both. Turner and Gilliam (1976a) have shown that for an alkaline soil increased rate of diffusion was due to saturated soil-water content (oxidized or reduced) rather than reduction as compared to moist conditions.

Rate of P diffusion is controlled by the adsorption-desorption process which controls the soil solution concentrations. Observed P adsorption behavior was shown to be influenced by oxidized and reduced soil conditions (Patrick and Khalid, 1974). In a Crowley silt loam soil, more P was adsorbed under oxidized conditions than under reduced conditions when the P concentration in the soil solution was low. This pattern was reversed at higher solution P concentration. The difference in behavior of phosphate under oxidized and reduced conditions was attributed to the change brought about in ferric oxyhydroxide by soil reduction. The greater surface area of the gel-like reduced ferros compounds in a reduced soil probably resulted in more P being solubilized where solution P was low and more solution P was adsorbed when solution P was high. These data suggested that at high solution P concentrations, P movement can occur at a more rapid rate under oxidized conditions than reduced conditions. This study was conducted at saturated soil-water content under both oxidzied and reduced conditions, and it is not clear whether similar effects will be seen if P adsorption is measured at water contents encountered under drained field conditions. Assuming a linear adsorption isotherm, Krom and Berner (1980) calculated adsorption coefficients (K_D) for the data reported in the literature on P adsorption in oxic and anoxic soils and sediments. For anoxic sediments, K_D values were in the range of 1 to 80 ml/g, while for oxic soils and sediments K_D values were in the range of 10 to 5000 ml/g. These data suggest a stronger adsorption capacity of oxic soils as compared to anoxic soils.

Potassium

Potassium is not involved in oxidation-reduction reactions and is therefore less affected by poor soil aeration and flooded conditions than are P and N. Although K occurs in several forms in soil, adsorbed and solution phases are primarily involved in mobility. The release of large amounts of Fe^{+2}, Mn^{+2}, and NH_4^+ in soils with low O_2 results in displacement of K^+ ions from the exchange complex to the soil solution. No data are available on the availability and mobility of K as influenced by soil O_2 alone. However, Pasricha and Ponnamperuma (1976) have shown increased K^+ concentrations in soil solution as a result of flooding. Potassium diffusion rates were found to be dependent on soil moisture and soil solution concentrations and a significant linear relationship was observed between K diffusion, uptake by plants, and soil moisture (Mengel and Von Braunschweig, 1972).

Calcium and Magnesium

Effects of soil aeration on reactivity and mobility of Ca^{+2} and Mg^{+2} will be of a similar nature as K^+. These nutrients are also not involved in soil oxidation-reduction reactions. However, reduction of insoluble Fe^{+3} and Mn^{+4} to soluble forms displaces other cations from the exchange complex to the soil solution. This process occurs both in poorly drained soils with low O_2 levels and in flooded soils. Under flooded conditions, increased water content can solubilize some compounds, which otherwise are relatively insoluble, and increase the concentration in soil solution. In poorly drained and flooded alkaline soils, low O_2 and high CO_2 levels can result in dissolution of $CaCO_3$ and $MgCO_3$ and increase the Ca^{+2} and Mg^{+2} concentration of the soil solution (Ponnamperuma, 1972). Low O_2 levels associated with increased soil-water content can also increase the mobility of these cations in soils. Increases in the concentration of other cations on the exchange complex can decrease the adsorption coefficient and result in rapid diffusion. Van Schaik et al. (1966) reported that the diffusion of ions is dependent upon the fraction of the adsorbed cations present in the diffuse layer and bulk solution. Their results show that approximately 70% of the exchangeable Na^+ ions were mobile, whereas only 25% of the Ca^{+2} ions were mobile.

Sulfur

Sulfur is subject to oxidation-reduction reactions and exists in soils and sediments at different valance states. The most common valances are +6 (sulfate), 0 (elemental S), and −2 (sulfide). The valance state of S in soil is governed to a large extent by the environmental conditions (O_2 and water content) that affect the composition and activity of the microflora. Detailed reviews on biogeochemical transformations of S in soils and sediments have been presented by Trudinger (1979); Krouse and McCready (1979); and Ralph (1979).

In aerated soils, the principal S transformations include (1) the oxidation of elemental S, thiosulfates, and sulfides; (2) mineralization of organic S; and (3) incorporation of S and sulfates into microbial tissue, and uptake by plants. Under conditions of O_2 free environment, sulfate reducing obligate anaerobic bacteria use SO_4^{-2} as an electron acceptor, resulting in the formation of sulfides (Alexander, 1977). These organisms are very sensitive to O_2 and cannot function in the presence of soil O_2. Sulfate reduction was shown to occur in soils with Eh < -75 mv (Harter and McLean, 1965), while Connell and Patrick (1968) reported sulfide formation in an anaerobic soil with Eh < -150 mv. The optimum pH range for maximum sulfide formation was found to be between 6.5 to 8.5 (Connell and Patrick, 1968). Besides soil O_2, sulfide formation can be retarded by the presence of NO_3^- (Fig. 8). Sulfur oxidizing bacteria, such as *Thiobacillus denitrificans* can reduce NO_3^- to N_2O and N_2, while oxidizing S to SO_4^{-2}.

In anoxic systems, S^{-2} concentrations of the soil can be decreased as a result of precipitation with metallic cations, such as Fe, Mn, Cu, Pb, Hg, Zn, Cd (Lindberg and Harriss, 1974; Krauskopf, 1967). Engler and Patrick (1975) observed that the metal sulfides were relatively unstable in an aerobic soil due to microbial oxidation. However, these sulfides were found to be relatively stable under anaerobic conditions with little or no dissolution of the ^{35}S from the metal sulfides.

Although some of the S in reduced form (S^{-2}) may accumulate in a soil system as metal sulfides and elemental S, most is eventually oxidized to SO_4^{-2}, a process in which microbial activities play a major role. The sulfate thus formed rapidly moves in the soil water. As the air-filled pore space is decreased as a result of an increase in soil water, the rate of SO_4^{-2} movement increases, primarily due to a decrease in the diffusion path

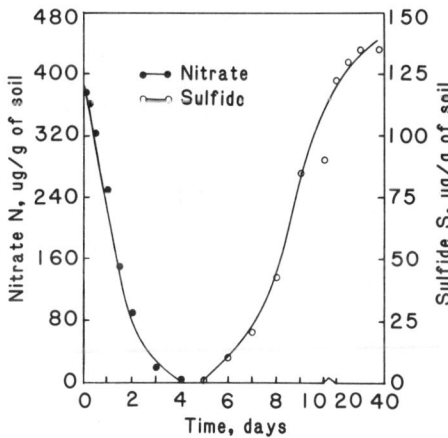

Fig. 8. Relationship between NO_3^- reduction and S^{-2} formation in a flooded soil (reproduced from Connell and Patrick, 1969. Soil Sci. Soc. Am. Proc. 33:711–715).

length. Sulfate diffusion coefficient in sediments saturated with water was reported to be 0.345 cm^2/day (Goldhaber et al., 1977). Sulfate ions can diffuse into deeper soil layers devoid of O_2, where it can be subsequently reduced to sulfides. Several researchers (Lambert et al., 1971; Donnelly et al., 1972) have reported significant quantities of sulfide formation in groundwaters. Goldhaber et al. (1977) observed a decrease in SO_4^{-2} concentration with increase in sediment depth. High SO_4^{-2} concentrations were observed at the sediment water interface and in the surface 1 to 2 cm of the sediment, while H_2S concentrations were found to increase with depth. In anoxic sediments, some of the sulfide formed in the anaerobic zones can diffuse into the surface oxidized zone where it can be oxidized to elemental S and subsequently to SO_4^{-2}. However, precipitation of sulfides with Fe^{+2} and other metals will slow this process. Sulfate thus formed can also diffuse back into the anaerobic zone where it will be reduced to sulfide.

Trace Metal Nutrients

Reactivity and mobility of trace metallic nutrients are also influenced by soil aeration. Some of the micronutrients that are influenced by the depletion of soil O_2 include Fe, Mn, Zn, Cu, Co, and Mo. In this paper, primary emphasis will be placed on the stability of Fe and Mn, while the effect on other nutrients will be discussed to limited extent.

In aerated soils, Fe and Mn exist as sparingly soluble Fe^{+3} and Mn^{+4} compounds. Decrease in soil O_2 levels as a result of temporary waterlogging, or as a result of rapid O_2 consumption due to the incorporation of high O_2 demand wastes (crop residues, animal wastes), can result in reduction of Mn^{+4} to Mn^{+2} and Fe^{+3} to Fe^{+2}. Reduction of these compounds is inhibited by the presence of soil O_2 and to some extent by NO_3^- (Fig. 2). The abundance of reduced Fe^{+2} in solution and on the exchange complex is favored by increase in soil acidity and soil reduction. The reduction of Fe and Mn has an indirect impact on the chemistry of the soil, such as (1) increases in water soluble Fe^{+2} and Mn^{+2}, (2) pH increases, (3) displacement of other cations from the exchange complex into soil solution, (4) increasing the solubility of P and Si, and (5) formation of new minerals (Ponnamperuma, 1972).

Gotoh and Patrick (1972, 1974) have reported that both water soluble and exchangeable Fe^{+2} and Mn^{+2} increased with decrease in Eh and pH. At pH 5, the effect of Eh was minimal on Mn^{+2} solubility, while between pH 6 and 8 the conversion of insoluble Mn^{+4} to Mn^{+2} was dependent on both Eh and pH. Similarly, appreciable amounts of Fe^{+2} were detectable at Eh of +300 mv and pH 5. However, reducible Fe became unstable when Eh was between +300 and +100 mv at pH 6 and 7, and −100 mv at pH 8 (Fig. 9). Application of an energy source followed by an irrigation decreased Eh values approximately 100 mv and resulted in high levels of Mn and Fe in the soil solutions obtained at 80 cm depth (Meek and Grass, 1975).

The mobility of Fe and Mn is increased greatly by reduction to the more soluble Fe^{+2} and Mn^{+2} forms. As the intensity of anaerobiosis increases, the effectiveness of the soil exchange complex in adsorbing cations will be reduced due to the larger total concentration of cations, thus increasing the mobility of Fe^{+2} and Mn^{+2}. Under reduced conditions, Mn^{+2} is subject to greater movement through both mass flow and diffusion. Because of its greater relative solubility than reduced Fe in anoxic soils, Mn^{+2} is usually depleted before Fe with a corresponding increase in the Fe/Mn ratios in the soil.

Reduction of insoluble Fe^{3+} oxyhydroxide compounds to the more soluble Fe^{2+} form greatly increases mobility of Fe^{2+}. In general, Fe^{2+} does not appear to be as mobile as Mn^{2+}, probably as a result of the easier reduction of oxidized Mn compounds to the soluble reduced forms. This condition results in the appearance of Mn^{2+} ions in the soil solution of a flooded soil before Fe^{2+} shows up. It also means that at a given reducing redox potential a greater fraction of the Mn is going to be in the reduced soluble form and subject to mobility. Another factor that decreases the mobility of Fe^{2+} relating to Mn is the difference in partitioning of the Fe^{2+} and the Mn^{2+} between the solution phase and the exchange complex at any given pH-redox conditions. A larger fraction of the reduced Fe^{2+} will be on the exchange complex as compared to the Mn^{2+} where more will remain in solution. Unless large amounts of S^{2-} are present, Fe^{2+} is quite

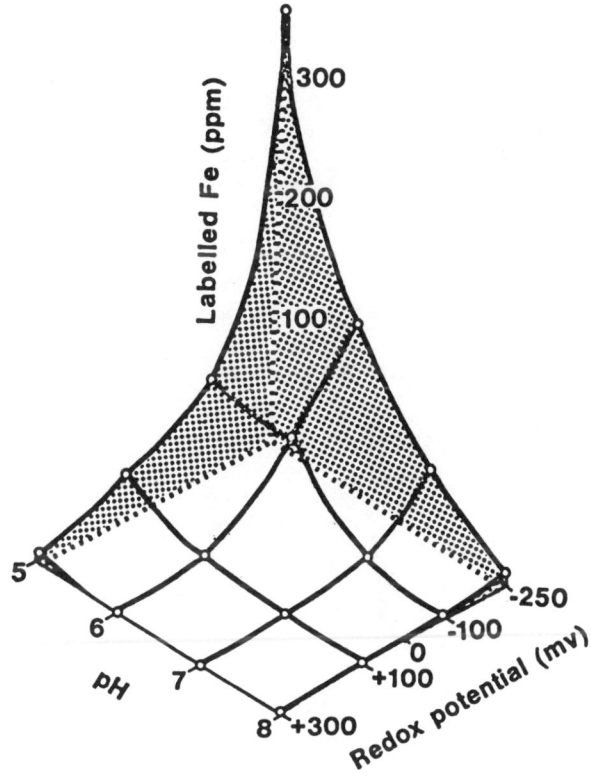

Fig. 9. Solubility of Fe as influenced by pH and redox potential.

mobile, however, and its movement in reduced soils in response to concentration gradients is evident. Ellis et al. (1970a) observed an increased diffusion coefficient of Fe^{+2} in montmorillonite clay with increased concentration of Fe^{+2} (up to 0.2 meq/g of clay). Similar results were also observed for other micronutrients (Mn, Zn, and Cu) (Ellis et al., 1970b). In the same study, diffusion coefficient of Fe^{+2} was found to be about five times larger than the diffusion coefficient of Fe^{+3}. Three examples of iron movement are (i) diffusion to aerobic surface layer where it is precipitated as Fe^{3+} oxyhydroxide (Fig. 10), (ii) diffusion and mass flow to plant roots where it is oxidized and precipitates as Fe^{3+} oxyhydroxides encrustations around the roots, and (iii) mass flow with downward percolating water to oxygenated subsoil layer where it precipitates along with Mn. This last process occurs only in soils where subsoil is oxygenated as compared to anaerobic root zone. As with Mn, reduced Fe affects mobility of other cations by disturbing the equilibrium between the soluble and exchangeable forms, resulting in an increased mobility. In tropical regions this process can result in "ferrolysis" in which Fe has displaced most of the other cations (Brinkman, 1970).

Micronutrients such as Zn, Cu, Mo, Co, and B are not as readily involved in soil oxidation-reduction reactions, but their solubility and mobility are affected by poor aeration. However, reduction can act to solubilize trace metals specifically adsorbed onto Fe^{+3} and Mn^{+4} oxides and hydrous oxides. In studies using controlled pH and Eh levels, Jugsujinda and Patrick (1977) found uptake of native added ^{65}Zn by rice (*Oryza sativa* L.) seedlings was greater from aerobic than anaerobic soil suspensions. In a recent study, Reddy and Patrick (1977) observed that chelated Zn and Cu in soil solution decreased with decreasing Eh from +500 mv (oxidized) to −200 mv (reduced). The instability of Zn and Cu chelates was attributed to chemical fixation (possibly with sulfide) of added Zn

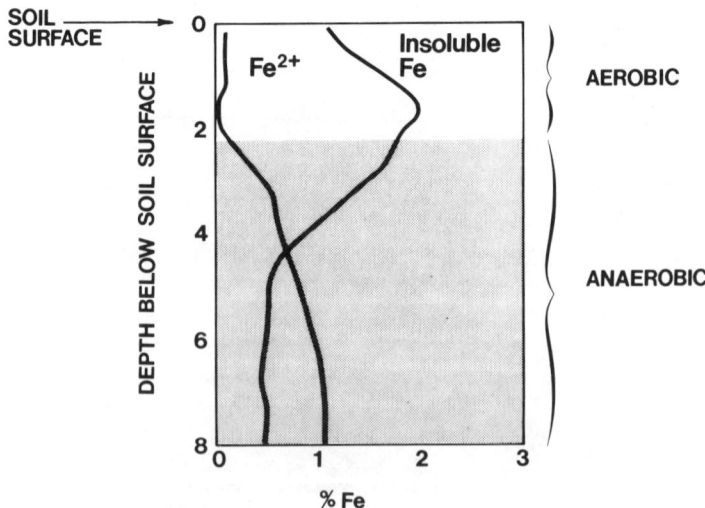

Fig. 10. Distribution of Fe^{+2} and Fe^{+3} in a flooded soil.

and Cu, and not due to chemical or microbial breakdown of the metal chelate complex. Some of these metallic cations may be less mobile under reduced conditions because of possible precipitation with sulfide (Engler and Patrick, 1975).

Mobility of metallic cations is also influenced by soil-water content. For example, rate of Zn diffusion (Fig. 5) was shown to increase with an increase in volumetric water content (Warncke and Barber, 1972). However, no information is available on the effect of varying levels of soil O_2 on diffusion of micronutrients.

CONCLUSIONS

Aeration status of a soil is controlled by the O_2 concentration in the soil atmosphere. Under well-aerated conditions, soil O_2 approaches that of atmospheric air and in poorly drained soils soil O_2 decreases. Decrease in soil O_2 is dependent on the fraction of pore volume filled with water, presence of O_2 demanding soil constituents, and soil physical properties. Under certain conditions a soil with small fraction of pore space filled with water can have O_2 concentrations near zero (example: soils receiving heavy loading of carbonaceous wastes). On the other hand, in soils with low O_2 demand, excessive soil-water content may not result in completely anoxic conditions. Although, these two cases are extreme, differentiation should be made in evaluating the aeration status of the soil.

Depletion of soil O_2 retards the root growth of several plants. The optimum soil O_2 requirement for plant-root development varies with species and age of the plant. The chemistry and microbiology of the soils are also altered as a result of changes in soil O_2. Restricted supply of O_2 to a soil results in reduction of NO_3^- to N_2O and N_2, Mn^{+4} to Mn^{+2}, Fe^{+3} to Fe^{+2}, SO_4^{-2} to S^{-2}, and CO_2 to CH_4. Reduction of these species depends on the intensity of soil anaerobiosis.

The most common method of characterizing the soil aeration is to measure (1) composition of soil atmosphere, (2) O_2 diffusion rates, and Eh. Oxygen diffusion measurements were found to be useful and more reliable in aerated soils while Eh measurements were found to be more effective in poorly drained and flooded soils.

Both reactivity and mobility of plant nutrients are influenced by the O_2 status of the soil. Biochemical processes involving aerobic bacteria were found to be slower in soils with low O_2 levels. For example, in poorly drained soils, ammonification and nitrification rates were found to be slower while denitrification rates were found to be faster. Oxygen depletion and saturated soil-water conditions increases the ionic strength of the soil solution, primarily due to increased levels of Fe^{+2}, Mn^{+2}, and NH_4^+ in solution. The release of large amounts of reduced cations results in displacement of other cations such as K^+, Ca^{+2}, and Mg^{+2} from the exchange complex, thus increasing their mobility. Mobility of some metallic cations (e.g., Zn, Cu) can be decreased as a result of reduced conditions because of possible precipitation with sulfides. However, this process occurs under a completely anoxic environment.

As discussed earlier, poor soil aeration can be the result of low O_2 levels, high water content, or the effects of both. Most of the research conducted in the past is oriented to evaluate the effect of varying levels of soil moisture on reactivity and mobility of nutrients. Data are scarce on the effects of varying levels of soil O_2 on the reactivity and mobility of nutrients. Future research should be oriented to study the effects of varying O_2 levels on nutrient transformations, adsorption-desorption processes, precipitation reactions, and diffusion mechanisms. The results from these studies will be useful in nutrient management of agricultural soils, thus, improving the utilization of nutrients by plants and reducing the pollution of surface and groundwater.

LITERATURE CITED

Alberda, T. 1953. Growth and root development of lowland rice and its relation to oxygen supply. Plant Soil 5:1–28.

Alexander, M. 1977. Introduction to soil microbiology. 2nd ed. John Wiley and Sons, Inc., N.Y.

Barber, S. A., J. M. Walker, and E. H. Vasey. 1963. Mechanisms for the movement of plant nutrients from the soil and fertilizer to the plant root. J. Agric. Food Chem. 11:204–207.

Bohn, H. L. 1971. Redox potentials. Soil Sci. 112:39.

Bouldin, D. R. 1968. Models for describing the diffusion of oxygen and other mobile constituents across the mud-water interface. J. Ecol. 56:77–87.

Bridge, B. J., and A. J. Rixon. 1976. Oxygen uptake and respiratory quotient of field soil cores in relation to their air filled pore space. J. Soil Sci. 27:279–286.

Brinkman, R. 1970. Geoderma 3:199–206.

Clarke, A. L., and K. P. Barley. 1968. The uptake of nitrogen from soils in relation to solute diffusion. Aust. J. Soil Res. 6:75–92.

Connell, W. E., and W. H. Patrick, Jr. 1968. Sulfate reduction in soil: Effects of redox potential and pH. Science 159:86–87.

----, and ----. 1969. Reduction of sulfate to sulfide in waterlogged soil. Soil Sci. Soc. Am. Proc. 33:711–715.

Donnelly, T. H., I. B. Lambert, and D. H. Dale. 1972. Sulphur isotope studies of the Mount Gunson copper deposits, Pernatty Lagoon, South Australia. Miner. Deposita 7:314–322.

Dowdell, R. J., and K. A. Smith. 1974. Field studies of the soil atmosphere, II. Occurrence of nitrous oxide. J. Soil Sci. 25:231–238.

Drobnik, J. 1962. The effect of temperature on soil respiration. Folia Microbiol. 7:132–140.

Ellis, J. H., R. E. Phillips, and R. I. Barnhisel. 1970a. Diffusion of iron in montmorillonite as determined by x-ray emission. Soil Sci. Soc. Am. Proc. 34(4):591–595.

----, R. I. Barnhisel, and R. E. Phillips. 1970b. The diffusion of copper, manganese, and zinc as affected by concentration, clay mineralogy, and associated anions. Soil Sci. Soc. Am. Proc. 34(5):866–870.

Engler, R. M., and W. H. Patrick, Jr. 1974. Nitrate removal from floodwater overlying flooded soils and sediments. J. Environ. Qual. 3:409–413.

----, and ----. 1975. Stability of sulfides of manganese, iron, zinc, copper, and mercury in flooded and nonflooded soil. Soil Sci. 119:217–221.

Flocker, W. J., J. A. Vomocil, and F. D. Howard. 1959. Some growth responses of tomatoes to soil compaction. Soil Sci. Soc. Am. Proc. 23:188–191.

Goldhaber, M. V., R. C. Aller, J. K. Cochran, J. K. Rosenfeld, C. S. Martens, and R. A. Berner. 1977. Sulfate reduction, diffusion and bioturbation in Long Island Sound sediments: Report of the FOAM group. Am. J. Sci. 277:193–237.

Gotoh, S., and W. H. Patrick, Jr. 1972. Transformations of manganese in a waterlogged soil as affected by redox potential and pH. Soil Sci. Soc. Am. Proc. 36:738.

————, and ————. 1974. Transformation of iron in a waterlogged soil as influenced by redox potential and pH. Soil Sci. Soc. Am. Proc. 38:66–71.

Grable, A. R. 1966. Soil aeration and plant growth. Adv. Agron. 18:57–106.

————, and E. G. Siemer. 1968. Effects of bulk density, aggregate size, and soil-water suction on oxygen diffusion, redox potentials and elongation of corn roots. Soil Sci. Soc. Am. Proc. 32:180–186.

Greenwood, D. J. 1961. The effect of oxygen concentration on the decomposition of organic materials in soil. Plant Soil 14:360.

Harter, R. D., and E. O. McLean. 1965. The effect of moisture level and incubation time on the chemical equilibria of a Toledo clay loam soil. Agron. J. 57:583.

Howeler, R. H., and D. R. Bouldin. 1971. The diffusion and consumption of oxygen in submerged soils. Soil Sci. Soc. Am. Proc. 35:202–208.

Jugsujinda, A., and W. H. Patrick, Jr. 1977. Growth and nutrient uptake by rice in a flooded soil under controlled aerobic-anaerobic and pH conditions. Agron. J. 69:705–710.

Kohnke, W. 1968. Soil physics. McGraw-Hill Book Co., N.Y. 347 p.

Kramer, P. J. 1965. Effects of deficient aeration on the roots of plants. p. 13–14. In Intersociety Conference on Irrigation and Drainage.

————. 1969. Plant and soil water relationship. p. 104–149. McGraw Hill Book Co., N.Y.

Krauskopf, K. B. 1967. Introduction of geochemistry. McGraw-Hill Book Co., N.Y. 721 p.

Krom, M. D., and R. A. Berner. 1980. Adsorption of phosphate in anoxic marine sediments. Limnol. Oceanogr. 25:797–806.

Krouse, H. R., and R. G. L. McCready. 1979. Biogeochemical cycling of sulfur. p. 401–430. In P. A. Trudinger and D. J. Swaine (ed.) Biogeochemical cycling of mineral forming elements. Elsevier Science Publishing Co., N.Y.

Lambert, I. B., J. McAndrew, and H. E. Jones. 1971. Geochemical and bacteriological studies of the cupriferous environment at Per Lagoon, South Australia Proc. Aust. Inst. Mining Met. 240:15–23.

Lemon, E. R., and A. E. Erickson. 1952. The measurement of oxygen diffusion in the soil with a platinum microelectrode. Soil Sci. Soc. Am. Proc. 16:160–163.

Lindberg, S. E., R. C. Harriss. 1974. Mercury-organic matter associations in estuarine sediments and interstitial water. Environ. Sci. Technol. 8:459–462.

Luxmoore, R. J., and L. H. Stolzy. 1972. Oxygen diffusion in the soil-plant system. V. Oxygen concentration and temperature effects on oxygen relations predicted for maize roots. Agron. J. 64:720–725.

Mahapatra, I. C., and W. H. Patrick, Jr. 1969. Inorganic phosphate transformations in waterlogged soils. Soil Sci. 107:281–288.

Mahtab, S. K., C. L. Godfrey, A. R. Swoboda, and G. W. Thomas. 1971. Phosphorus diffusion in soils: I. The effect of applied P, clay content, and water content. Soil Sci. Soc. Am. Proc. 35:393–397.

McIntyre, D. S. 1970. The platinum microelectrode method for soil aeration measurement. Adv. Agron. 22:235–283.

Meek, B. D., and L. B. Grass. 1975. Redox potential in irrigated desert soils as an indicator of aeration status. Soil Sci. Soc. Am. Proc. 39:870–875.

————, and L. H. Stolzy. 1978. Short-term flooding. p. 351–373. In Plant life in anaerobic environments. D. D. Hook and R. M. Crawford (ed.) Ann Arbor Sci. Pub., Inc., Ann Arbor, Mich.

Mengel, K., and L. C. Von Braunschweig. 1972. The effect of soil moisture upon the availability of potassium and its influence on the growth of young maize plants (*Zea mays* L.). Soil Sci. 114:142–148.

Mikkelsen, D. S., S. K. DeDatta, and W. N. Obcemea. 1978. Ammonia volatilization losses from flooded rice soils. Soil Sci. Soc. Am. J. 42:725–730.

Miller, R. H., and D. D. Johnson. 1964. The effect of soil moisture tension on carbon dioxide evolution, nitrification, and nitrogen mineralization. Soil Sci. Soc. Am. Proc. 28:644–646.

Mortimer, C. H. 1941. The exchange of dissolved substances between mud and water in lakes. J. Ecol. 29:280–329.

Moureaux, C. 1967. Influence de la temperature et de l'humidite sur les actvites biologiques de quelques sols ouest. Africains. Cashiers ORSTROM Pedologie, Paris. 5:393–420.

Olsen, S. R. 1971. Diffusion of nutrients. p. 497. *In* McGraw Hill, Encyclopedia of Science and Technology. Vol. 12.

----, W. D. Kemper, and J. C. VanSchaik. 1965. Self diffusion coefficient of phosphorus in soil measured by transient and steady state methods. Soil Sci. Soc. Am. Proc. 29:154–158.

Pal, D., F. E. Broadbent, and D. S. Mikkelsen. 1975. Influence of temperature on rice straw decomposition in soils. Soil Sci. 442–449.

Pasricha, N. S., and F. N. Ponnamperuma. 1976. Ionic equilibria in flooded saline, alkali soils: The $K^+ - (Ca^{+2} + Mg^{+2})$. Exchange equilibria. Soil Sci. 122:315–320.

Patrick, W. H., Jr. 1964. Extractable iron and phosphorus in a submerged soil at controlled redox potential. Trans. 8th Int. Congr. Soil Sci. 3:605–609.

----, R. D. Delaune, and R. M. Engler. 1973. Soil oxygen content and root development of cotton in Mississippi River alluvial soils. Bull. No. 673, Louisiana Agric. Exp. Stn. p. 27.

----, and R. A. Khalid. 1974. Phosphate release and sorption by soils and sediments: Effect of aerobic and anaerobic conditions. Science 186:53–55.

----, and I. C. Mahapatra. 1968. Transformation and availability of nitrogen and phosphorus in waterlogged soils. Adv. Agron. 20:323–359.

Pearsall, W. H. 1938. The soil complex in relation to plant committees. J. Ecol. 26:180–315.

Ponnamperuma, F. N. 1972. The chemistry of submerged soils. Adv. Agron. 24:29–96.

Porter, L. K., W. D. Kemper, R. D. Jackson, and B. A. Stewart. 1960. Chloride diffusion in soils as influenced by moisture content. Soil Sci. Soc. Am. Proc. 24:460–463.

Ralph, B. J. 1979. Oxidative reactions in the sulfur cycle. p. 369–400. *In* P. A. Tradinger and D. J. Swaine (ed.) Biogeochemical cycling of mineral forming elements. Elsevier Science Publishing Co., N.Y.

Reddy, C. N., and W. H. Patrick, Jr. 1977. Effect of redox potential on the stability of zinc and copper chelates in flooded soil. Soil Sci. Soc. Am. J. 41:729–732.

Reddy, K. R. 1983. Soluble phosphorus release from organic soil. Agric. Ecosystems Environ. 9:(In press).

----, W. H. Patrick, Jr., and R. E. Phillips. 1976. Ammonium diffusion as a factor in nitrogen loss from flooded soils. Soil Sci. Soc. Am. J. 40:528–533.

----, ----, and ----. 1978. The role of nitrate diffusion in determining the order and rate of denitrification in flooded soil: I. Experimental results. Soil Sci. Soc. Am. J. 42:268–272.

----, ----, and ----. 1980a. Evaluation of selected processes controlling nitrogen loss in flooded soil. Soil Sci. Soc. Am. J. 44:1241–1246.

----, P. S. C. Rao, and W. H. Patrick, Jr. 1980b. Factors influencing the oxygen consumption in flooded soils. Soil Sci. Soc. Am. J. 44:741–744.

Rolston, D. E., and A. M. Marino. 1976. Simultaneous transport of nitrate and gaseous denitrification products in soil. Soil Sci. Soc. Am. J. 40:860–865.

Russell, M. B. 1952. Soil aeration and plant growth. Adv. Agron. 2:253–301.

Sabey, B. R. 1969. Influence of soil moisture tension on nitrate accumulation in soil. Soil Sci. Soc. Am. Proc. 33:263–266.

Shalhevet, J., H. Enoch, and S. Dasberg. 1969. Response of sugar beet to soil drainage and aeration. Israel J. Agric. Res. 19:161–170.

Sheard, R. W., and A. J. Leyshon. 1976. Short-term flooding of soil: Its effect on the composition of gas and water phases of soil and on phosphorus uptake of corn. Can. J. Soil Sci. 56:9–20.

Smith, K. A., and P. D. Robertson. 1971. Effect of ethylene on the root extension of cereals. Nature 234:148–149.

Sojka, R. E., L. H. Stolzy, and M. R. Kaufmann. 1975. Wheat growth related to rhizosphere temperature and oxygen levels. Agron. J. 67:591–596.

Stolzy, L. H. 1974. Soil atmosphere. *In* E. W. Carson (ed.) The plant root and its environment. Charlottesville, Virginia. Univ. Press of Virginia.

Trudinger, P. A. 1979. The biological sulfur cycle. p. 293–368. *In* P. A. Trudinger and D. J. Swaine (ed.) Biogeochemical cycling of mineral forming elements. Elsevier Science Publishing Co., N.Y.

Turner, F. T., and J. W. Gilliam. 1976a. Effect of moisture and oxidation status of alkaline rice soils on the adsorption of soil phosphorus by anion resin. Plant Soil 45:353–363.

----, and ----. 1976b. Increased P diffusion as an explanation of increased P availability in flooded rice soils. Plant Soil 45:365–377.

----, and W. H. Patrick, Jr. 1968. Chemical changes in waterlogged soils as a result of oxygen depletion. Trans. 9th Int. Congr. Soil Sci. 4:53.

Van Doren, D. M., and A. E. Erickson. 1966. Factors affecting the platinum microelectrode method for measuring the rate of oxygen diffusion through the soil solution. Soil Sci. 102:23–28.

Van Schaik, J. C., W. D. Kemper, and S. R. Olsen. 1966. Contribution of adsorbed cations to diffusion in a clay water system. Soil Sci. Soc. Am. Proc. 31:17–22.

Warncke, D. D., and S. A. Barber. 1972. Diffusion of Zn in soil: Influence of soil moisture. Soil Sci. Soc. Am. Proc. 36:35–39.

West, D. W., and J. D. F. Black. 1978. Irrigation timing—its influence on the effects of salinity and waterlogging stresses in tobacco plants. Soil Sci. 125:367–376.

Chapter 3

Physical Processes Influencing Water and Solute Transport in Soils[1]

J. M. DAVIDSON, P. S. C. RAO, AND P. NKEDI-KIZZA[2]

The mass transport of water and solutes in soils takes place through a complex three-dimensional inter-connected network of pores or voids of nonuniform sizes and shapes. A multitude of pore structure models has been devised, varying from simple bundles of capillary tubes to sophisticated pore network models. Dullien (1979) has grouped pore structure models into two broad categories: (i) those that are based on the arrangement of soil (or solid) particles, which in most cases are assumed to be spherical in shape but of different sizes and (ii) those that are based on shapes, sizes, and arrangements of pores or voids. Because of the complexity and intractability of soil pores that contain water and solutes, a continuum approach rather than a geometric approach is generally used to describe mathematically water and solute transport in soils. The continuum approach (macroscopic scale) is adequate for the practitioner interested only in a phenomenological description of water and solute transport processes in soils. The continuum approach fails, however, to provide the insight required in many instances to explain observations that depend on the properties of the pores and the behavior of water and solutes at a molecular scale. Attempts have been made to explain the transport of water and solutes at a microscopic scale, but these ap-

---

[1] This paper was presented at a Symposium "Chemical Mobility and Reactivity in Soils Systems" held during the American Society of Agronomy Annual Meetings in Atlanta, Georgia, 29 Nov.–4 Dec. 1981. Contribution from Soil Sci. Dep., Univ. of Florida. Approved for publication as Florida Agric. Exp. Stn. J. Series No. 4322.

[2] Professor, associate professor, and research associate, respectively, Soil Science Dep., Institute of Food and Agricultural Sciences, Univ. of Florida, Gainesville, FL 32611.

Copyright © 1983 ASA, SSSA, 677 South Segoe Road, Madison, WI 53711. *Chemical Mobility and Reactivity in Soil Systems.*

proaches have not been completely satisfactory, primarily because physical and chemical measurements within the individual soil pores or voids are not feasible. However, each effort has provided a valuable insight into what remains a complex problem.

The major purpose of this paper is to review the pertinent literature on water and solute transport in soils and the mathematical-physical relationships that have been developed and are amenable to analysis. Where appropriate, limitations in the theory will be identified. A discussion of the physical processes involved in the transport of water and solutes is emphasized here since the chemical and microbial aspects of transport of specific solutes (e.g., fertilizers, salts, pesticides, etc.) are the subjects of other chapters in this book.

TRANSPORT OF WATER

The equation for isothermal, steady-state water flow in homogeneous, isotropic, and water-saturated porous media was formulated by Henri Darcy as a result of his studies on water flow through sand filters in the City of Dijon, France (Darcy, 1856). Darcy's equation, though originally conceived for saturated water flow, was later extended by Richards (1931) to describe unsaturated water flow:

$$q = -K(h)\nabla H, \qquad [1]$$

where q is the soil-water flux (LT^{-1}), K(h) is the soil hydraulic conductivity (LT^{-1}) which is a function of the matric potential or soil-water pressure head h(L), ∇ is the standard differential operator of vector notation, $H = (h-z)$ is the total potential expressed as the total head (L), and z is the gravitation head (L) with the vertical space coordinate taken as positive downward. Note that all other components of soil-water potential have been neglected.

If a water-saturated soil could be considered as a bundle of straight and smooth capillary tubes, Poiseuille's equation and Darcy's equation would be analogous:

$$q = \frac{-R^2 \varrho g}{8\eta} \nabla H \qquad [2]$$

where R is the radius of a tube or pore (L), ϱ is the density of water (ML^{-3}), g is the gravitational acceleration (LT^{-2}), and η is the viscosity of water ($MT^{-1} L^{-1}$). In comparing Eqs. [1] and [2], note that K(h) is proportional to R^2, where R may be thought of as some "average effective" pore size of the soil.

During water flow under saturated conditions, all of the pores in the soil are filled with water; thus, a hydraulic continuity exists and the soil hydraulic conductivity is at its maximum. When the soil desaturates, as it does during the redistribution of water following an infiltration event, some pores become filled with air (soil-water pressure head decreases), and the cross sectional area of soil available to conduct water decreases. The first pores to empty during desaturation of a soil are the largest pores, which offer the least resistance to water flow; thus, during further redistribution, water flow occurs progressively in the smaller pores which offer a greater resistance to water flow (Eq. [2]). For this reason, the transition from a saturated to an unsaturated soil entails a sharp drop in the soil-water hydraulic conductivity. Decreases of three to four orders of magnitude in the hydraulic conductivity are not uncommon in field soils during redistribution of water following a thorough wetting of the upper portion of a deep well-drained soil profile (Davidson et al., 1969). It is this sharp decrease in soil hydraulic conductivity that produces what is commonly referred to as field capacity or that condition in the soil profile where changes in soil-water flux and soil-water content (or potential) at a given soil depth with time are too small to be detected over short time intervals (1 to 2 d).

Equations [1] and [2] imply that the soil-water flux vs. hydraulic gradient relationship is linear. This is only true for laminar flow conditions. At high fluxes when turbulent flow occurs and also at very low hydraulic gradients when the water may act as a non-Newtonian fluid, nonlinear relationship between q and H may be observed (Miller and Low, 1963; Nerpin et al., 1966). However, the hydraulic gradients and fluxes encountered in dealing with water flow in most soils are such that laminar flow is the norm rather than the exception and Eq. [1] is applicable.

The quantitative application of the unsaturated water flow theory to field or laboratory soil systems requires a knowledge of soil hydraulic conductivity [K(h)] and soil-water characteristic [θ(h)] relationships. Klute (1972) reviewed various methods for measuring soil hydraulic conductivity and discussed the advantages and disadvantages of each method; methods he reviewed include steady-state, unsteady-state, instantaneous profile methods, and those involving the use of the soil-water characteristic curves to estimate K(h) relationships. Combining Eq. [1] with the equation of continuity:

$$\frac{\partial \theta}{\partial t} = - \nabla q, \qquad [3]$$

where θ is volumetric soil-water content ($L^3\ L^{-3}$), t is time (T), and other parameters are as defined earlier. Substitution of Eq. [1] for q in Eq. [3] yields:

$$\frac{\partial \theta}{\partial t} = \nabla [K(h) \nabla H]. \qquad [4]$$

Equation [4] is appropriate for n-dimensional water flow in a heterogeneous anisotropic soil. Considering only vertical one-dimensional water flow, Eq. [4] simplifies to:

$$\frac{\partial \theta}{\partial t} = \frac{\partial}{\partial z} \left[K(h) \frac{\partial h}{\partial z} \right] - \frac{\partial [K(h)]}{\partial z} \qquad [5]$$

Assuming θ to be a unique function of h during wetting and drying cycles (i.e., no hysteresis), an alternate form of Eq. [5] can be derived (Childs and Collis-George, 1950) by defining a new variable, the soil-water deffusivity, $D(\theta) = [K(\theta) (dh/d\theta)]$ such that:

$$\frac{\partial \theta}{\partial t} = \frac{\partial}{\partial z} \left[D(\theta) \frac{\partial \theta}{\partial z} \right] - \left[\frac{\partial}{\partial z} K(\theta) \right]. \qquad [6]$$

Equations [5] and [6] provide the basis for predicting soil-water transport for both steady-state and transient conditions in soils. The dependence of K and D upon θ (or h) make both Eqs. [5] and [6] nonlinear, as such these equations are more difficult to solve than the classical linear equations for describing the flow of heat and electricity. Thus, analytical solutions of Eqs. [5] and [6] are not available except for simple initial and boundary conditions. Numerical and analytical solutions to Eqs. [5] and [6] for some of these cases are presented in Kirkham and Powers (1972). Refinements to existing techniques for solving Eqs. [5] and [6] numerically are presented by Haverkamp et al. (1977), Perrens and Watson (1977), Hayhoe (1978a), and Zaradny (1978). Use of the finite element method has been described by Gray and Pinder (1976) and Hayhoe (1978b). Solutions to Eqs. [5] and [6] using perturbation, iterative, and optimization techniques have been presented by Brutsaert (1976) and Parlange and Babu (1977).

Several attempts have been made during the past few years to predict the movement of water in laboratory soil columns (Nielsen et al., 1962; Kirda et al., 1973; Wood and Davidson, 1975; Elrick et al., 1979) as well as in field soil profiles (LaRue et al., 1968; Watts, 1975[3]; Warrick et al., 1977; Bresler et al., 1979). The success of these efforts has depended, to a large extent, upon whether or not the $K(\theta)$ and $\theta(h)$ relationships required to solve Eqs. [5] or [6] were measured within or external to the column or field site being simulated.

Based upon research to date, it is apparent that soil spatial variability confounds the problem of estimating soil-water flux and soil-water content profile distributions with time in field soils. Following Philip (1980), two types of field soil heterogeneity can be identified. Deterministic heterogeneity exists when soil properties vary spatially (and possibly temporally) in a known and well-defined manner. This type of heterogeneity can generally be handled by extending the mathematical analysis to include it. Stochastic heterogeneity is said to exist when the spatial variation of soil properties is irregular and is imperfectly known, i.e., it is essentially random.

Scaling offers an approach for handling the inherent stochastic spatial variability of specific soil-water microhydrologic characteristics $[K(\theta)$ and $\theta(h)]$. Early contributions to the scaling theory include work by Miller and Miller (1955, 1956), Klute and Wilkinson (1958), Wilkinson

---

[3] Watts, D. G. 1975. A Soil-water-nitrogen-plant model for irrigating corn on coarse textured soils. Ph.D. Diss. Logan, Utah. Utah State Univ. Publ. No. 76-6353. (Diss. Abstr. 36:4593B).

and Klute (1959), Elrick et al. (1959), and Philip (1967). The scaling approach has also been used by Peck et al. (1977), to model the water balance for a watershed as well as by Warrick et al. (1977) and Simmons et al. (1979) to analyze field data using $\theta(h)$ and $K(\theta)$ curves.

The major advantage offered by scaling is that it permits a generalized formulation of Eq. [6] for a field-site, as illustrated by Warrick and Amoozegar-Fard (1979):

$$\frac{\partial S}{\partial \tau} = \frac{\partial}{\partial \zeta} \left[\hat{D} \frac{\partial S}{\partial \zeta} \right] - \frac{\partial \hat{K}}{\partial \zeta}. \qquad [7]$$

The scaled variables used in the above equation are defined as follows:

$$\zeta = \alpha_r z; \quad \tau = (\alpha_r^3/\theta_s)t; \quad S = (\theta/\theta_s) \qquad [8]$$

and

$$\hat{h} = h_r \alpha_r; \quad \hat{K} = K_r \alpha_r^{-2}; \quad \hat{D} = \hat{K} (\partial \hat{h}/\partial S), \qquad [9]$$

where θ_s is saturated volumetric water content ($L^3 L^{-3}$), $\hat{h}$, $\hat{K}$, and $\hat{D}$ are the *scaled* soil-water suction, soil hydraulic conductivity, and soil-water diffusivity, respectively, and α_r is the scaling parameter for a particular location within the field site. Warrick and Amoozegar-Fard (1979), using the "average" $\hat{K}(\theta)$ and $S(\hat{h})$ functions obtained by scaling field data and the analytical solution to Eq. [7] outlined by Philip (1969), were able to calculate infiltration rates and soil-water content vs. depth profiles. Although the scaling procedure does not allow one to determine the actual values of θ, h, or q at a specific location, it is possible to compute the probability of a specific θ, q, or h value occurring at a location or the probability of a specific portion of the field having a given θ, q, or h (Sisson and Wierenga, 1981). The concept of geometrically similar media at similar states, the foundation of scaling theory, has been shown to be approximately valid for selected field soils, but general use of theory may be limited (Warrick et al., 1977; Sharma et al., 1980; Rao et al., 1983).

TRANSPORT OF SOLUTES

As water moves through the soil, dissolved solutes are carried along in the convective stream. During water flow through soil pores, some solutes may be subject to various interactions (e.g., sorption on soil surfaces, precipitation, plant uptake, microbial and chemical transformations, etc.). Because this paper deals only with the physical aspects of solute flow, both the soil and solutes are considered to be ideal, i.e., the soil-solute and solute-solute interactions are assumed negligible and the solute is considered to be conservative (i.e., no losses or gains). Solutes move not only with the soil water, but also within it in response to solute concentration gradients. An understanding of the simultaneous movement of water and solutes in soils is essential for improved efficiency of soil-applied agrochemicals, prevention of groundwater contamination, waste disposal site management, and wastewater renovation using soils.

The mass flow of water through a soil results in a concurrent convective flux (J_c) of solutes:

$$J_c = q\,C = v\theta C, \qquad [10]$$

where q is the Darcy flux discussed in the section on transport of water, $v = (q/\theta)$ is average pore-water velocity (LT^{-1}), C is solute concentration in the soil-water (ML^{-3}), and J_c ($ML^{-2}\,T^{-1}$) is the mass of solute flowing across a unit cross-sectional area of soil per unit time. To estimate the average distance a solute will travel by convection, the following relationship is generally used:

$$Z = vt, \qquad [11]$$

where Z is the distance a solute front travels in time t.

Solute concentration gradients which exist within the soil pore network are reduced with time through molecular diffusion. The diffusive solute flux (J_D) in a saturated or an unsaturated porous medium is given by Fick's first law:

$$J_D = -D_e\,\theta\,\nabla C \qquad [12]$$

where J_D is the diffusive solute flux per unit cross-sectional area ($ML^{-2}\,T^{-1}$), and D_e is the effective molecular diffusion coefficient ($L^2\,T^{-1}$). In Eq. [12], $D_e = D_o\gamma$, where D_o is the molecular diffusion coefficient ($L^2\,T^{-1}$) in bulk water and γ is the complexity or formation factor which accounts for sinuous and tortuous nature of the soil pore sequences in which diffusion occurs (Porter et al., 1960; Olsen and Kemper, 1968). Note that $0 \le \gamma \le 1$.

Equation [10] disregards the geometric complexity of pores through which the solute passes. Also, the average pore-water velocity concept does not account for the wide range in solute velocities within and between pore domains. It is probable that certain pore domains may be nonconducting (or stagnant) and that these regions act as diffusive sinks/sources for solutes in the conducting (or mobile) pore domains (Coats and Smith, 1964; Philip, 1968; Skopp and Warrick, 1974; van Genuchten and Wierenga, 1976, 1977; Rao et al., 1980). Thus, the use of Eqs. [10] and [12] for describing solute transport in soils represents a gross over-simplification of the microscopic-scale transport phenomenon.

Three physical processes contribute to solute dispersion in a porous medium: (i) molecular diffusion within a pore domain, (ii) mechanical dispersion owing to pore-water velocity variations between pores within the conducting or mobile pore domains, and (iii) sink/source effects due to diffusive solute transfer between conducting (mobile) and nonconducting (stagnant) pore domains. Because of the similarity in effect (though not in mechanism) between molecular diffusion, mechanical dispersion and sink/source effects, it is frequently assumed that these three processes are additive. Thus, the coefficients for diffusion, mechanical dispersion, and sink/source effects are combined into a single term called hydrodynamic dispersion, D_h, defined as (Passioura, 1971; Rao et al., 1980):

$$D_h = [D_e + D_M + D_s]. \qquad [13]$$

In Eq. [13], D_e is the *effective molecular diffusion coefficient* and D_M is the *mechanical dispersion coefficient* which is a function of the average pore-velocity (Bresler, 1972; Biggar and Nielsen, 1976):

$$D_M = mv^n. \qquad [14]$$

When $n = 1$ and m is a constant, Eq. [14] reduces to a linear relationship as proposed by Bresler (1972), such that,

$$D_M = mv. \qquad [15]$$

The term D_s in Eq. [13] accounts for the dispersion caused by diffusive solute transfer between stagnant and mobile pore domains. Based on recent research (Passioura, 1971; DeSmedt, 1979[4]; Rao et al., 1980) it may be proposed that:

$$D_s = f[\phi, (\ell^2/D_e), (L/v)....], \qquad [16]$$

where ϕ is the fraction of pore-water residing in the conducting pore domains, ℓ is the diffusion path length, D_e is the effective diffusion coefficient within the stagnant domains, L is column length, and v is average pore-water velocity. Given a specific geometry of the mobile and the stagnant pore domains, exact analytic expressions for Eq. [16] can be derived. For example, Passioura (1971) and Rao et al. (1980) have proposed for spherical aggregates that:

$$D_s = \frac{v^2 a^2 (1 - \phi)}{15 D_e}, \qquad [17]$$

where a is the aggregate radius, and other parameters are as defined earlier.

Combining the appropriate terms, Eq. [13] can now be restated as (for a soil with spherical aggregates):

$$D_h = \left[(D_o \gamma) + (mv^n) + \frac{v^2 a^2 (1 - \phi)}{15 D_e}\right], \qquad [18]$$

Equation [18] suggests that as $v \to 0$, the second and third terms on the right hand side can be neglected,

$$\lim_{v \to 0} D_h \cong D_o \gamma \qquad [19]$$

---

[4] DeSmedt, F. 1979. Theoretical and experimental study of solute movement through porous media with mobile and immobile water. Doctoral Diss. Vrije Universiteit. Brussels, Belgium.

resulting simply in molecular diffusion in a porous medium. However, for a large v, the molecular diffusion term becomes negligible and the two "dispersion" terms in Eq. [18] become more important.

Equations [16] to [18] were derived based on the principal assumption that during flow, "near-equilibrium" conditions exist for solute transfer between the mobile and stagnant pore-water domains. For example, for spherical aggregates Rao et al. (1980) have shown that the near-equilibrium assumption is valid only when:

$$\frac{D_e L (1 - \phi)}{a^2 v^2 0.3} \geq 1. \quad [20]$$

When Eq. [20] is not satisfied, the diffusive solute transfer into stagnant domains must be explicitly defined (van Genuchten and Wierenga, 1976; Rao et al., 1980).

Given that Eq. [13] is valid for defining D_h and that hydrodynamic dispersion is analogous to diffusion in effect, but not in mechanism, Eq. [12] can be restated as:

$$J_{D_h} = - D_h \theta \nabla C. \quad [21]$$

The convective-diffusive-dispersive solute flux (J_s) is the sum of solute fluxes owing to convection (J_c) and hydrodynamic dispersion (J_{D_h}). Thus, $J_s = (J_c + J_{D_h})$.

Combining Eqs. [10] and [21] with the continuity equation yields for one-dimensional flow:

$$\frac{\partial}{\partial t}(\theta C) = \frac{\partial}{\partial z}\left[D_h \theta \frac{\partial C}{\partial z} - v\theta C\right]. \quad [22]$$

In order to describe the simultaneous movement of water and solute during transient, one-dimensional flow, Eqs. [6] and [22] must be solved simultaneously. For steady one-dimensional water flow ($\partial \theta/\partial t = 0$) in homogeneous soils, Eq. [22] simplifies to:

$$\frac{\partial C}{\partial t} = D_h \frac{\partial^2 C}{\partial z^2} - v \frac{\partial C}{\partial z}. \quad [23]$$

Equation [23] has been solved analytically for a variety of initial and boundary conditions (van Genuchten and Alves, 1982). However, for many cases of practical interest, closed-form analytical solutions to Eq. [23] are unavailable and numerical methods have to be used (Fried and Combarnous, 1971).

The early work of Danckwerts (1953), Day (1956), and Nielsen and Biggar (1961, 1962) involving the miscible displacement of nonreactive solutes illustrated the applicability of Eq. [23] for simple porous media such as generally exists in laboratory columns packed with glass beads, sands, and sieved loam soils. Equation [23] subject to Eq. [20] has also been successfully used to describe solute transport in aggregated soils (Passioura, 1971; Rao et al., 1976, 1980; and Nkedi-Kizza et al., 1983). However, Eqs. [22] and [23] have not been satisfactory for describing solute transport in strongly aggregated or structured soils or in unsaturated soils (Biggar and Nielsen, 1962; Green et al., 1972; Gupta et al., 1973; Rao et al., 1974; Quisenberry and Phillips, 1976, 1978; van Genuchten and Wierenga, 1976, 1977; Gaudet et al., 1977) where the condition specified in Eq. [20] was not valid. Excellent predictions, however, can be obtained by explicitly accounting for the nonequilibrium diffusive solute transfer between the mobile and stagnant regions (van Genuchten and Wierenga, 1976, 1977; Rao et al., 1980, Nkedi-Kizza et al., 1982).

Recent reviews of the literature on hydrodynamic dispersion (Fried, 1975; Cherry et al., 1975; Bredehoeft et al., 1976; Anderson, 1979; Biggar and Nielsen, 1980) have pointed out two general trends: (i) D_h values measured in laboratory columns are much smaller than those obtained in undisturbed soil cores or field plots and (ii) D_h values may vary both spatially and temporally within a given field site. The first trend may be interpreted by examining the three components of D_h as specified in Eq. [18]. Most laboratory soil column studies have involved sieved soils (a $\leq$ 0.1 cm) and fairly low pore-water velocities (v $\leq$ 15 cm/h). The process of interest in undisturbed cores and field soils is, of course, diffusion into large structural units (i.e., preferential flow along biochannels or cracks with simultaneous diffusion into primary or secondary peds). While the contribution of the D_s term to D_h may be small in laboratory columns, D_s may be the primary term with relatively minor contributions from D_e and D_M under field conditions. Spatial and temporal variations in D_h may also be explained by Eq. [18] or some other more generalized expression. Because soil structure (and pore geometry) frequently change spatially and temporally at a given site, factors such as v, d, a, etc. also vary, all of which could lead to variations in the effective D_h.

The condition specified in Eq. [20] and required for Eq. [18] to be valid may, unfortunately, not be met in most field sites. Field soils are not composed of spherical aggregates and even if the structural units could be approximated as such, their radii are generally too large to assume near-equilibrium solute transfer between the mobile and stagnant regions as required by Eq. [20]. Thus eventhough Eq. [18] may provide a physical explanation for the large and frequently variable dispersion coefficient observed under field conditions, it is not intended to provide estimate of dispersion coefficients in aggregated or well-structured field soils.

During the past few years, solutions to Eq. [21] have been used in an effort to describe the mass transport of nonadsorbed solutes through field soils and groundwater aquifers. Application has required a scaling-up of dispersion coefficients from the small values observed in laboratory columns to the larger values required to describe field data. Evidence now exists to suggest that the use of a large dispersion coefficient in association with uniform hydraulic characteristics may be inappropriate for describing solute transport in field and geological systems (Anderson, 1979; Smith and Schwartz, 1980). The primary issue centers around the fact that modeling field-scale solute dispersion using the convective-dispersion equation (Eq. [21]) is an unsatisfactory approximation to what is a more complex phenomenon.

Bresler and Dagan (1979), Dagan and Bresler (1979), Smith and Schwartz (1980), and Russo and Bresler (1981a, 1981b) have used a stochastic modeling concept that considers macroscopic dispersion to be a process related to the spatial heterogeneity of hydraulic conductivity and soil-water characteristic relationships. This approach represents a realistic analog of the actual field situation and does not require the use of large dispersion coefficients to describe the observed macroscopic dispersion. Jury (1982), beginning with the "point of view that many causes of spatial variability of water and solute transport render measurement of hydraulic and retention parameters of a field soil all but impossible," proposed a simple transfer function model as an alternate to the deterministic approaches for describing solute transport in field soils. In his model, with only piston flow (Eq. [11]) being considered, solute dispersion was ascribed to spatially-varying (log-normally distributed) pore-water velocities. The unknown velocity (or solute travel time) distribution was calibrated using measured solute concentration histories at a single depth and then this information was used to predict solute concentration profiles at other times or depths.

SUMMARY

The continuum approach to describe mathematically water and solute transport in laboratory soil columns has been reasonably successful. Recent experiments have further elucidated the physical processes responsible for the large dispersion coefficients associated with unsaturated and/or aggregated field soils. However physically-based models now appear to be inadequate and unnecessarily complex for describing water and solute transport for large field sites exhibiting spatial variability with regard to soil hydraulic conductivity and other soil-water characteristics. In systems of this nature solute dispersion occurs primarily because of macroscale spatial changes in direction and magnitude of water flow. Once dispersion is accounted for in this manner, it may not be necessary to use the large dispersion coefficients presently required to account for the macroscopic dispersion observed under field conditions. Jury et al. (1982) showed that it is possible in a field study to collect the necessary information to apply the above modeling approach. It is important to test similar simple models with data from a number of field experiments.

LITERATURE CITED

Anderson, M. P. 1979. Using models to simulate the movement of contaminants through ground water flow systems. Chemical Rubber Company Critical Reviews in Environmental Control 9:97–156.

Biggar, J. W., and D. R. Nielsen. 1962. Miscible displacement: II. Behavior of tracers. Soil Sci. Soc. Am. Proc. 26:125–128.

----, and ----. 1976. Spatial variability of the leaching characteristics of a field soil. Water Resour. Res. 12:78-84.

----, and ----. 1980. Mechanisms of chemical movement in soils. p. 213-227. In A. Banin and U. Kafkafi (ed.) Agrochemicals in soil. Pergamon Press, N.Y.

Bredehoeft, J. D., H. B. Counts, S. G. Robson, and J. B. Robertson. 1976. Solute transport in ground water systems. p. 229-256. In J. C. Rodda et al. (ed.) Facets of hydrology. John Wiley and Sons, N.Y.

Bresler, E. 1972. Control of soil salinity. p. 102-132. In D. Hillel (ed.) Optimizing the soil physical environment toward greater crop yields. Academic Press, N.Y.

----, H. Bielorai, and A. Laufer. 1979. Field test of solution flow models in a heterogeneous irrigated soil. Water Resour. Res. 15:645-652.

----, and G. Dagan. 1979. Solute dispersion in unsaturated heterogeneous soil at field scale: II. Applications. Soil Sci. Soc. Am. J. 43:367-472.

Brutsaert, W. 1976. The concise formulation of diffusive sorption of water in a dry soil. Water Resour. Res. 12:1118-1124.

Cherry, J. A., R. W. Gillham, and J. F. Pickens. 1975. Contaminant hydrogeology: I. Physical processes. Geosci. Can. 2:76-83.

Childs, E. C., and C. N. Collis-George. 1950. The permeability of porous materials. Proc. R. Soc. 201A:392-405.

Coats, K. H., and B. D. Smith. 1964. Dead-end pore volume and dispersion in porous media. Am. Inst. Mech. Eng. Trans. 231:73-84.

Dagan, G., and E. Bresler. 1979. Solute dispersion in unsaturated heterogeneous soil at field scale: I. Theory. Soil Sci. Soc. Am. J. 43:461-467.

Danckwerts, P. V. 1953. Continuous flow systems: Distribution of residence times. Chem. Eng. Sci. 2:1-13.

Darcy, H. 1856. Les Fontaines Publique de la Ville de Dijon. p. 570, 590-594. Victor Dalmont, Paris.

Davidson, J. M., L. R. Stone, D. R. Nielsen, and M. E. LaRue. 1969. Field measurement and use of soil water properties. Water Resour. Res. 5:1312-1321.

Day, P. R. 1956. Dispersion of a moving salt water boundary advancing through a saturated sand. Trans. Am. Geophys. Union 37:595-601.

Dullien, F. A. L. 1979. Porous Media—Fluid Transport and Pore Structure. Academic Press, N.Y.

Elrick, D. E., J. H. Scandrett, and E. E. Miller. 1959. Tests of capillary flow scaling. Soil Sci. Soc. Am. Proc. 23:329-333.

----, K. B. Laryea, and P. H. Groenevelt. 1979. Hydrodynamic dispersion during infiltration of water into soils. Soil Sci. Soc. Am. J. 43:856-865.

Fried, J. J. 1975. Groundwater pollution. Elsevier Science Publishing Co., Amsterdam, The Netherlands.

----, and M. A. Combarnous. 1971. Dispersion in porous media. Adv. Hydrosci. 7:169-282.

Gaudet, J. P., H. Jegat, G. Vachaud, and P. J. Wierenga. 1977. Solute transfer, with exchange between mobile and stagnant water, through unsaturated sand. Soil Sci. Soc. Am. J. 41:665-671.

Gray, W. G., and G. F. Pinder. 1976. An analysis of the numerical solution of the transport equation. Water Resour. Res. 12:547-555.

Green, R. E., P. S. C. Rao, and J. C. Corey. 1972. Solute transport in aggregated soils: Tracer zone shape in relation to porevelocity distribution and absorption. In Proc. 2nd Symp. on Fundamentals of Transport Phenomena in Porous Media. IAHR-ISSS. Guelph, Canada. 2:732-752.

Gupta, R. K., R. J. Millington, and A. Klute. 1973. Hydrodynamic dispersion in unsaturated porous media: II. The stagnant zone concept and the dispersion coefficient. J. Indian Soc. Soil Sci. 21:121-128.

Haverkamp, R., M. Vauclin, J. Touma, P. J. Wierenga, and G. Vachaud. 1977. A comparison of numerical simulation models for one dimensional infiltration. Soil Sci. Soc. Am. J. 41:285-294.

Hayhoe, H. N. 1978a. Numerical study of quasi-analytic and finite difference solutions of the soil-water transfer equation. Soil Sci. 125:68–74.

——. 1978b. Study of the relative efficiency of finite difference and Galerkin techniques for modeling soil-water transfer. Water Resour. Res. 14:97–102.

Jury, W. A. 1982. Simulation of solute transport using a transfer function model. Water Resour. Res. 18:363–368.

——, L. H. Stolzy, and P. Shouse. 1982. A field test of the transfer function model for predicting solute transport. Water Resour. Res. 18:369–375.

Kirda, C., D. R. Nielsen, and J. W. Biggar. 1973. Simultaneous transport of chloride and water during infiltration. Soil Sci. Soc. Am. Proc. 37:339–345.

Kirkham, D., and W. F. Powers. 1972. Advanced soil physics. Wiley-Interscience, N.Y.

Klute, A. 1972. The determination of the hydraulic conductivity and diffusivity of unsaturated soils. Soil Sci. 113:264–276.

——, and G. E. Wilkinson. 1958. Some tests of the similar media concept of capillary flow: I. Reduced capillary conductivity and moisture characteristic data. Soil Sci. Soc. Am. Proc. 22:273–281.

LaRue, M. E., D. R. Nielsen, and R. M. Hagan. 1968. Soil water flux below a ryegrass root zone. Agron. J. 60:625–629.

Miller, E. E., and R. D. Miller. 1956. Physical theory for capillary flow phenomena. J. Appl. Phys. 27:324–332.

Miller, R. D., and E. E. Miller. 1955. Theory of capillary flow: I. Practical implications. Soil Sci. Soc. Am. Proc. 19:267–271.

Miller, R. J., and D. F. Low. 1963. Threshold gradient for water flow in clay systems. Soil Sci. Soc. Am. Proc. 27:605–609.

Nerpin, S., S. Pashkina, and N. Bondareko. 1966. The evaporation from bare soil and the way of its reduction. Symp. Water Unsaturated Zone. Wageningen, The Netherlands.

Nielsen, D. R., and J. W. Biggar. 1961. Miscible displacement in soils: I. Experimental information. Soil Sci. Soc. Am. Proc. 25:1–5.

——, and ——. 1962. Miscible displacement: III. Theoretical considerations. Soil Sci. Soc. Am. Proc. 26:216–221.

——, ——, and J. M. Davidson. 1962. Experimental consideration of diffusion analysis in unsaturated flow problems. Soil Sci. Soc. Am. Proc. 26:107–111.

Nkedi-Kizza, P., J. W. Biggar, M. Th. van Genuchten, P. J. Wierenga, H. M. Selim, J. M. Davidson, and D. R. Nielsen. 1983. Modeling tritium and chloride-36 transport through a porous medium (Oxisol) exhibiting physical non-equilibrium. Water Resour. Res. (In press).

——, P. S. C. Rao, R. E. Jessup, and J. M. Davidson. 1982. Ion-exchange and diffusive mass transfer during miscible displacement through an aggregated Oxisol. Soil Sci. Soc. Am. J. 46:471–476.

Olsen, S. R., and W. D. Kemper. 1968. Movement of nutrients to plant roots. Adv. Agron. 20:91–151.

Parlange, J-Y., and D. K. Babu. 1977. On solving nonlinear diffusion equations: A comparison of perturbation, iterative, and optimal techniques for an arbitrary diffusivity. Water Resour. Res. 12:213–214.

Passioura, J. B. 1971. Hydrodynamic dispersion in aggregated media: I. Theory. Soil Sci. 111:339–344.

Peck, A. J., R. J. Luxmoore, and J. L. Stolzy. 1977. Effects of spatial variability of soil hydraulic properties in water budget modeling. Water Resour. Res. 13:348–354.

Perrens, S. J., and K. K. Watson. 1977. Numerical analysis of two-dimensional infiltration and redistribution. Water Resour. Res. 13:781–790.

Philip, J. R. 1964. Sorption and infiltration in heterogeneous media. Aust. J. Soil Res. 5:1–10.

——. 1967. Sorption and infiltration in heterogeneous media. Aust. J. Soil Res. 5:1–10.

——. 1968. Diffusion dead-end pores, and linearized absorption in aggregated media. Aust. J. Soil Res. 6:31–39.

——. 1969. Theory of infiltration. Adv. Hydrosci. 5:215–305.

----. 1980. Field heterogeneity: Some basic issues. Water Resour. Res. 16:443–448.
Porter, L. K., W. D. Kemper, R. J. Jackson, and B. A. Stewart. 1960. Chloride diffusion in soils as influenced by moisture content. Soil Sci. Soc. Am. Proc. 24:460–463.
Quisenberry, V. L., and R. E. Phillips. 1976. Percolation of surface-applied water in the field. Soil Sci. Soc. Am. J. 40:484–489.
----, and ----. 1978. Displacement of soil water by simulated rainfall. Soil Sci. Soc. Am. J. 43:675–679.
Rao, P. S. C., R. E. Green, V. Balasubramanian, and Y. Kanehiro. 1974. Field study of solute movement in a highly aggregated Oxisol with intermittent flooding: II. Picloram. J. Environ. Qual. 3:197–202.
----, ----, L. R. Ahuja, and J. M. Davidson. 1976. Evaluation of a capillary bundle model for describing solute dispersion in aggregated soils. Soil Sci. Soc. Am. J. 40:815–820.
----, ----, A. G. Hornsby, D. K. Cassel, and W. A. Pollans. 1983. Scaling soil microhydrological properties of Lakeland and Konawa soils using similar media concepts. Agric. Water Manage. (In press).
----, D. E. Rolston, R. E. Jessup, and J. M. Davidson. 1980. Solute transport in aggregated porous media: Theoretical and experimental evaluation. Soil Sci. Soc. Am. J. 44:1139–1146.
Richards, L. A. 1931. Capillary conductivity of liquids through porous media. Physics 1:318–333.
Russo, D., and E. Bresler. 1981a. Effect of field variability in soil hydraulic properties on solutions of unsaturated water and salt flows. Soil Sci. Soc. Am. J. 45:675–681.
----, and ----. 1981b. Soil hydraulic properties as stochastic processes: I. An analysis of field spatial variability. Soil Sci. Soc. Am. J. 45:682–687.
Sharma, M. L., G. A. Grander, and C. G. Hunt. 1980. Spatial variability of infiltration in a watershed. J. Hydrol. 45:101–122.
Simmons, C. S., D. R. Nielsen, and J. W. Biggar. 1979. Scaling of field-measured soil properties. Hilgardia 47:77–154.
Sisson, J. B., and P. J. Wierenga. 1981. Spatial variability of steady-state infiltration rates as a stochastic process. Soil Sci. Soc. Am. J. 45:699–704.
Skopp, J., and A. W. Warrick. 1974. A two-phase model for the miscible displacement of reactive solutes through soils. Soil Sci. Soc. Am. Proc. 38:545–550.
Smith, L., and F. W. Schwartz. 1980. Mass transport: I. A stochastic analysis of macroscopic dispersion. Water Resour. Res. 16:303–313.
van Genuchten, M. Th., and W. J. Alves. 1982. A compendium of available analytical solutions of the one-dimensional convective-dispersive solute transport equation. USDA-SEA Technical Bull. 1661. 149 p.
----, and P. J. Wierenga. 1976. Mass transfer studies in sorbing porous media: I. Analytical solutions. Soil Sci. Soc. Am. J. 40:473–480.
----, and ----. 1977. Mass transfer studies in sorbing porous media: II. Experimental evaluation with tritium (^3H_2O). Soil Sci. Soc. Am. J. 41:272–278.
Warrick, A. W., and A. Amoozegar-Fard. 1979. Infiltration and drainage calculations using spatially scaled hydraulic properties. Water Resour. Res. 15:1116–1120.
----, G. J. Mullen, and D. R. Nielsen. 1977. Scaling field-measured soil hydraulic properties using a similar media concept. Water Resour. Res. 13:355–362.
Wilkinson, G. E., and A. Klute. 1959. Some tests of the similar media concept of capillary flow: II. Flow systems data. Soil Sci. Soc. Am. Proc. 23:434–437.
Wood, A. L., and J. M. Davidson. 1975. Fluometuron and water distributions during infiltration: Measured and calculated. Soil Sci. Soc. Proc. 39:820–825.
Zaradny, H. 1978. Boundary conditions in modeling water flow in unsaturated soils. Soil Sci. 125:75–82.

Chapter 4

Chemical Transport Modeling: Current Approaches and Unresolved Problems

WILLIAM A. JURY[1]

Prediction of downward movement of chemicals below land operations is one of the major tasks facing soil scientists today. These operations, ranging from farming inputs of fertilizer and pesticide, to sludge pond disposal of heavy metals, to shallow burial of nuclear waste, all pose a potential hazard to underlying or adjacent water supplies. For those chemicals which are hazardous even at trace concentrations, the task of prediction must extend to "worst case" behavior, such as might occur through soil cracks or other preferential flow channels, in order to give a realistic estimate of the risk involved in such a land use.

Unfortunately, worst case behavior such as flow through channels and cracks, may not be limited to highly heterogeneous soil profiles. There is an increasing body of experimental information to suggest that vertical water and chemical transport in the vadose zone can vary significantly from place to place in the horizontal direction in natural soil profiles, even when the input of material at the surface is uniform (Jury et al., 1976; Biggar and Nielsen, 1976; Van de Pol et al., 1977; Starr et al., 1978). Models are needed which can simulate this variability while at the same time requiring only such calibration as may be reasonably performed in situ.

At present, few if any of our transport models are appropriate for this sort of problem. Rather, most water and solute transport models used in the vadose zone are an outgrowth of laboratory models developed to simulate flow in uniformly packed porous media. As such they are either one-dimensional models which assume vertical flow with no lateral vari-

---

[1] Professor, Dep. of Soil and Environmental Sci., Univ. of California, Riverside, CA 92521.

Copyright © 1983 ASA, SSSA, 677 South Segoe Road, Madison, WI 53711. *Chemical Mobility and Reactivity in Soil Systems.*

ability (Rose et al., 1982) or are formal three-dimensional models with the transport coefficients represented as tensors which are impossible to calibrate (Dugiud and Reeves, 1976). Only in recent years have soil scientists turned to the more difficult problem of simulating flow under conditions of natural variability. In this chapter I will briefly review some recent solute transport field experiments and identify the major models used currently in the field, and also show how such models may give widely differing interpretations when applied to fields of prespecified variability.

REVIEW OF SOLUTE TRANSPORT VARIABILITY STUDIES

There have been several significant studies performed in recent years which demonstrate the extreme lateral variability of solute applied to the unsaturated zone of soil. In 1976, Biggar and Nielsen conducted a steady state ponded flow experiment studying the movement of a solute pulse on 20 subplots of a 150-ha field. They measured water velocity by determining the arrival time lag between adjacent solution sampler depths from the surface to 180 cm. The resulting measurements of water velocity were lognormally distributed with a mean, median, and mode value of 44.2, 20.3, 4.3 cm/d, respectively. This variability was much higher than the corresponding variability of the saturated hydraulic conductivity measurements on the same field.

There also have been studies conducted where the input of water at the surface was uniform across the field. In 1976, Jury et al. added a tracer dye to the irrigation water on a sprinkler-irrigated potato (*Solanum tuberosum* L.) field and observed a highly nonuniform water infiltration pattern underneath the hill, shoulder, and furrow surfaces of the field even though the infiltration capacity of the sandy soil was very high. Van de Pol et al. (1977) studied the movement of a pulse of chloride added by a trickle irrigation system with a high application uniformity, and measured a lognormal water velocity distribution even though the system was in steady state and the trickle emitters were only 30 cm apart. Jury et al. (1982) studied movement of a pulse of bromide applied uniformly to the surface and leached by rainfall and again found lognormally distributed velocities. Table 1 shows a summary of these four experiments with each observed velocity distribution fitted to a lognormal distribution. The third and fourth columns give the expectation and variance of the log of the water velocities. As expected, the variance of the ponded experiment is quite high, reflecting not only variability in the infiltration rate from point to point, but also variability in water transport within the soil. The other studies, however, show substantial variability even though the input of water to the soil was spatially uniform. It is clear from these studies that regardless of the uniformity of the input, a distribution of water output velocities in the horizontal direction is to be expected.

Table 1. Measured solute velocity distributions in field experiments using mobile chemicals, fitted to lognormal distributions.

Experiment	Soil type	E(V)	Var(V)	E(lnV)	Var(lnV)	Literature citation
		cm/d	cm²/d²			
1. Tracer movement under sprinkler irrigated potato crop	sand	14.4	24.0	2.61	0.11	(7)
2. Steady state Cl⁻ ponded bare soil plots over 150 ha	loam-clay loam	44.3	7362.1	2.01	1.56	(2)
3. Steady state Cl⁻ flow under 2 cm/day trickle irrigation	clay	3.9	5.7	1.20	0.32	(13)
4. Leaching of Br⁻ pulse by rainfall	sand-loamy sand	1.0†	0.4	−0.15†	0.31	(8)

† Average velocity over 30 days.

REVIEW OF FIELD MODELS OF SOLUTE TRANSPORT

There have been a number of comprehensive reviews of solute transport in soil published in the recent literature (Biggar and Nielsen, 1976; Boast, 1973), including an excellent review of all models potentially applicable to waste disposal modeling (van Genuchten, 1978). The vast majority of all field solute transport models used in the unsaturated zone are based on the convection-dispersion equation, which developed from laboratory simulations and experiments (Biggar and Nielsen, 1967).

$$\frac{\partial C_T}{\partial t} + R = \underline{\nabla} \cdot \underline{\underline{D}} \cdot \underline{\nabla} C_L - \nabla \cdot \underline{J}_w C_L, \qquad [1]$$

where

$C_T = \varrho_b C_S + \Theta C_L$ is total solute concentration (mass of solute/soil volume)
C_L is dissolved solute concentration (mass of solute/volume of solution)
C_S is adsorbed solute concentration (mass of solute/mass of dry soil)
ϱ_b is soil dry bulk density (mass/volume)
Θ is volumetric water content
$\underline{J}_w$ is water flux (volume flow/area/time)
$\underline{\underline{D}}$ is the diffusion-dispersion tensor (area/time)
R is a generalized sink term modeled for specific reactions (mass of solute/volume/time)

This equation has been thoroughly tested under laboratory conditions both for steady state and transient water flow. Its applicability to the field, however, has been hampered by an inability to predict multidimensional water flows using the water flow equation, and a corresponding inability to measure or model the diffusion-dispersion tensor. In fact, the form and size of $\underline{D}$ is a function of the precision of the representation used for the water flux J_w. For example, if only a one-dimensional average water flux is used in a situation where lateral variations in water flux are present, then solute spreading caused by the variable convective flux will have to be modeled as though it arose from dispersion. For this case, a large and complex form for $\underline{D}$ would be needed to simulate any observed solute movement.

As an alternative to the deterministic approach which uses the convection-dispersion equation and a tensor representation of the transport coefficients, several researchers have begun to look at stochastic representations of water movement. One such approach has been to use scaling theory to relate various regions of a field soil. In this method, heterogeneous regions of the field are imagined to be magnified versions of a reference region, related to each other by a macroscopic length parameter or scaling factor. If the assumptions of scaling theory are met in a field, then knowledge of the scaling factor distribution would allow calculations to be made of the transport coefficient distribution from a standard set of measurements of a transport coefficient at the reference location. Once the distribution of transport coefficients is obtained, laterally variable water and/or solute transport during, e.g., infiltration, could be modeled by assuming the field to consist of a parallel non-interacting network of soil columns, each with its characteristic scaling factor and corresponding transport properties. Computer simulation of such processes could be simplified through dimensionless calculations (Peck et al., 1977; Warrick and Amoozegar-Fard, 1979). The few field tests of the scaling factor model of variability have yielded mixed results (Warrick et al., 1977; Sharma et al., 1980) when used to compare transport coefficients or processes measured at different locations in a field.

A novel variation in solute transport modeling was achieved by Dagan and Bresler (1979), who combined a piston flow model of downward solute flow under the influence of gravity at a given point in the field, with a stochastic model for predicting the probability of solute movement to a given depth. They defined the hydraulic transport properties of any point in the field through a probability distribution for scaling factors. Finally they combined the soil transport model with spatial variations in water application rate defined through a frequency function.

A recent model for solute transport in the field (Jury, 1982) used a transfer function model to predict solute movement through the unsaturated zone, assuming the transport of solute to a given depth was a stochastic function of the net water applied at the surface. Unlike the model of Dagan and Bresler (1979), this approach did not use a specific model for the transport mechanism and the entire transport process was defined through a travel time probability density which could be obtained through field measurements.

Although the solute transport models mentioned above differ widely in their working assumptions, they may be broadly summarized into categories which use the convection-dispersion equation without horizontal variations in velocity while requiring the dispersion coefficient to explain all solute spreading, and a second group of models which explicitly include horizontal variations in water velocity.

Although there are many models currently used for simulation of solute transport in studies projecting the environmental impact of a given land use operation, they have almost exclusively used the convection-dispersion equation (CDE) without modeling water velocity variations. To contrast the two approaches outlined above, in the remainder of this chapter a comparison will be made between the predictions of the prototype CDE (Eq. [1]) and the predictions of the transfer function model (TFM) of Jury (1982) when applied to several typical transport problems.

ILLUSTRATIVE SOLUTE TRANSPORT CALCULATIONS

Transfer Function Model

The transfer function model has no mechanisms and represents the transport from the surface to a depth L as a stochastic function of the net cumulative amount of water I applied at the surface (i.e., rainfall minus evaporation). Thus, the probability P that a tracer injected at the surface will reach a depth L after a net quantity of water I has entered the soil surface is

$$P_L(I) = \int_0^I f_L(I')dI' \qquad [2]$$

where $f_L(I)$ is the probability density function given the frequency of net applied water volume I to reach $z = L$. Since for any application rate i, $I = it$, $f_L(I)$ is equivalent to a travel time distribution as well if the rate of water input is specified, or to a velocity distribution $V = L/t$. The method of measuring $f_L(I)$ and a successful field test of the TFM are given in Jury at al. (1982).

By superimposing all of the different travel times or breakthrough volumes and their probability of occurrence together with the phase lagged inlet concentrations, an average concentration at $z = L$ is obtained.

$$C_L(I) = C_L(it) = \int_0^\infty C_{IN}(I-I') f_L(I')dI' \qquad [3]$$

Notice that in this model no mixing or dispersion occurs between stream tubes and solute entering a tube with a breakthrough volume I' arrives unchanged at the outlet as $C_{IN}(I-I')$. Thus, all of the solute mixing which arises from spatially variable transport properties is represented by travel time variations in this model.

Although in principle each depth z requires its own transfer function model and thus its own $f_z(I)$, in many cases of interest the distribution of factors giving rise to variable breakthrough volumes (textural variations, cracks, etc.) may be reasonably unchanged over a large depth interval. In this case, the frequency functions $f_z(I)$ for different z may be highly enough correlated with each other to be represented in terms of the reference distribution $f_L(I)$. Thus in this case, the probability of reaching a depth z not equal to L when an amount $I = I_o$ of water has entered the surface, is equal to the probability of reaching $z = L$ when $I = LI_o/z$ has entered the surface which implies (Jury, 1982)

$$P_z(I) = \int_o^I f_z(I')dI' = P_L\left(\frac{LI_o}{z}\right) = \int_o^{LI_{o/z}} f_L(I')dI' \qquad [4]$$

or

$$f_z(I) = \frac{L}{Z} f_L(IL/z). \qquad [5]$$

Thus, the average concentration at any z and t may be calculated as

$$C(z,t) = \int_o^\infty C_{in}(t-t') f_z(i_o t) i_o dt', \qquad [6]$$

where i_o is the average net input flux rate and $I = i_o t$ has been substituted for cumulative water input.

Ultimately, the correlations which produce Eq. [5] must decrease due to lateral mixing. However, no experimental information is available to indicate at what depth this may occur. Thus the TFM, which is based on the assumption that Eq. [5] is valid, has a domain of validity verified experimentally near the surface (Jury et al., 1982), and projections outside this domain of validity to greater depths are only specualtive.

Movement of Nonadsorbed Chemicals

In these calculations it will be assumed that a pulse of mobile chemical (e.g., chloride) is introduced for a time period Δt with a steady state water flow application rate i applied to a large field. The field will be assumed to have a log-normally distributed population of water velocities with mean and variance as measured in the experiment of Jury et al. (1982) (Table 1). The measured distribution in this experiment of net applied water I required to displace chemical past 30 cm is shown in Fig. 1, along with normal and lognormal fits to the data. As mentioned earlier, the velocity is related to I by $V = Li_o/I$. The convection-dispersion equation CDE will be fitted to the predicted output of the Jury (1982) transfer function model TFM at one depth and the calibrated CDE will then be compared with the predictions of the TFM for other depths and for other circumstances. The dispersion coefficient will be assumed constant $D = D_o$ or will be assumed to be a linear function of velocity $D = \epsilon V$ where ϵ is the dispersivity. The following initial and boundary conditions are assumed in the simulations

$$C(z,o) = 0 \qquad 0 < z < \infty$$
$$-\Theta D \frac{\partial C}{\partial z} + iC = iC_o \qquad z = 0 \qquad 0 < t < \Delta t \qquad [7]$$
$$t > \Delta$$
$$C = 0 \qquad z \to \infty,$$

where Θ is water content, and C_o is pulse concentration at the surface. Since the TFM does not have dispersion, its surface condition is equivalent to a step function change in concentration. The solutions to Eq. [7] for the two solute transport models (Eq. [1] and [6]) are given below:

$$C(z,t) = C_o [H(z,t) - H(z,t - \Delta t)], \quad [8]$$

where for the CDE

$$H(z,t) = \frac{1}{2}\text{erfc}\left[\frac{z - Vt}{2\sqrt{Dt}}\right] + V\sqrt{\frac{t}{\pi D}}\exp\left[-\frac{(z - Vt)^2}{4Dt}\right]$$
$$- \frac{1}{2}\left(1 + \frac{Vz}{D} + \frac{V^2 t}{D}\right) \exp\left(\frac{Vz}{D}\right) \text{erfc}\left[\frac{z + Vt}{2\sqrt{Dt}}\right], \quad [9]$$

and for the TFM

$$H(z,t) = \frac{1}{2}\left[1 + \text{erf}\left[\frac{\ln(iLt/z) - \mu}{\sqrt{2}\,\sigma}\right]\right], \quad [10]$$

where μ, σ^2 are the log mean and variance of the net applied water distribution of $f_L(I)$ to reach depth L from the surface when the input flux rate is uniform.

Figure 2 shows simulated breakthrough curves for a pulse of concentration C_o applied over 5 days and leached to various depths all at a uniform surface input of i = 1 cm day. The solution to the CDE model Eq.

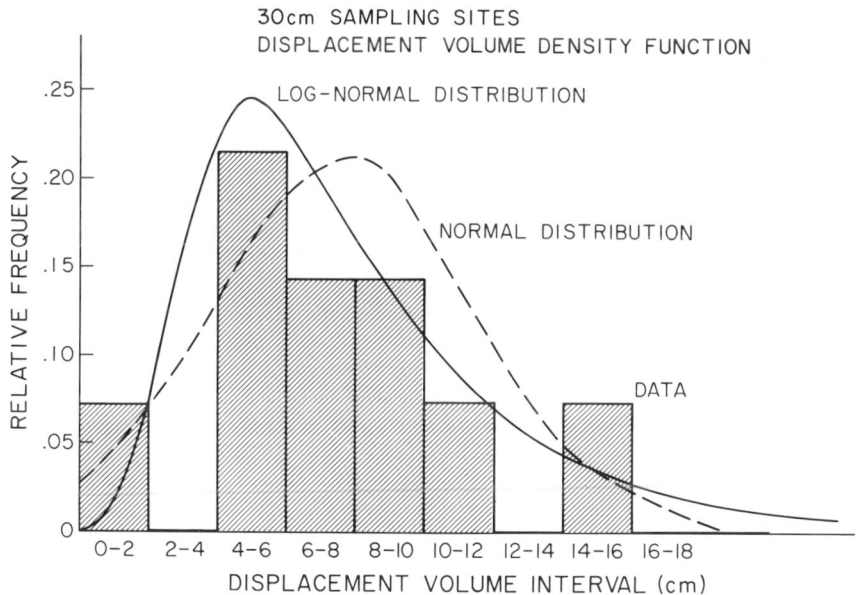

Fig. 1. Measured distribution of net applied water I required to move bromide pulse past 30 cm, in experiment of Jury et al. (1982) fitted to a log-normal and normal distribution.

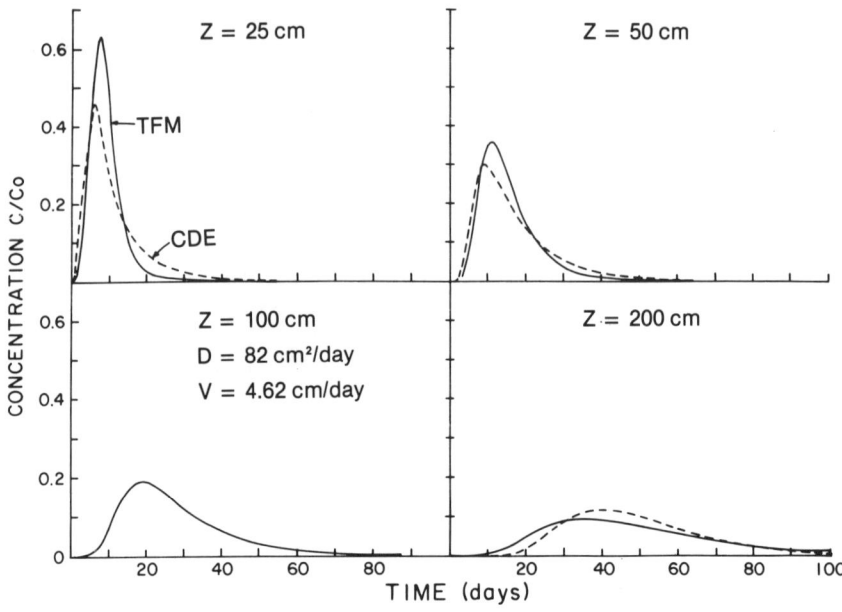

Fig. 2. Simulated solute breakthrough curves at various depths for a pulse of concentration C_o applied with 5 cm of water using TFM and CDE models calibrated at $Z = 100$ cm.

[3] was fitted to the TFM model at $z = 100$ cm with essentially perfect agreement between models. This resulted in a dispersion coefficient of $D_o = 82$ cm^2/d and a velocity of 4.62 cm/d so that $\epsilon = 17.8$ cm. The two models were then compared at depths of 25, 50, and 200 cm for the same simulation at the same water input rate. As seen, the CDE and TFM models agree only where the calibration was made at $z = 100$ cm. When the water input flux is changed (Fig. 3) and outputs compared at $z = 100$ cm, the constant dispersion coefficient CDE model using the calibrated value of $D_o = 82$ cm^2/d will match values predicted by the TFM model only at the calibration flux of $i = 1$ cm/day. In this case the dispersivity CDE model $D = \epsilon V$ agrees with the TFM model at all flux rates. However, neither CDE model of the dispersion coefficient agrees with the transfer function model when different depths and water input fluxes are used (Fig. 4). In this figure the ratio i/z is constant so that the TFM model predicts the same outflow shape for all simulations (See Eq. [10]). The other two models give different predictions for other depths and input rates. For these cases as well as the one in Fig. 3, the dispersivity model of the dispersion coefficient agrees more closely with the TFM than does the constant dispersion coefficient model. However, when a simulation is made of transport under a smaller input flux rate $i = 0.25$ to a greater depth $z = 500$ cm (Fig. 5), which might be a typical example of a long-term simulation of chemical movement for environmental impact assessment, the dispersivity model disagrees greatly with the predictions of the TFM, whereas in this case, the constant dispersion coefficient model gives similar estimates to the TFM.

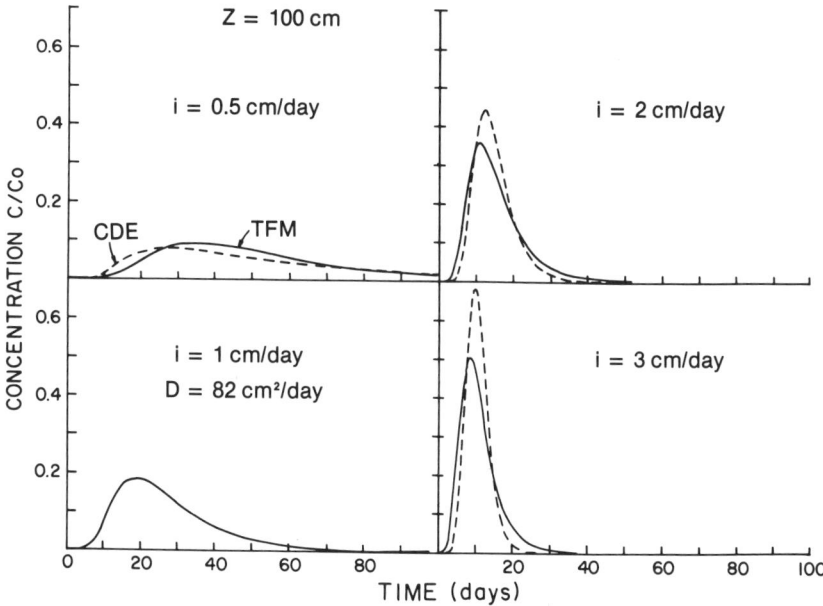

Fig. 3. Simulated solute breakthrough curves at Z = 100 cm for various input flux rates i for a pulse of concentration C_o applied with 5 cm of water using TFM and CDE models (constant D) calibrated at i = 1 cm/d.

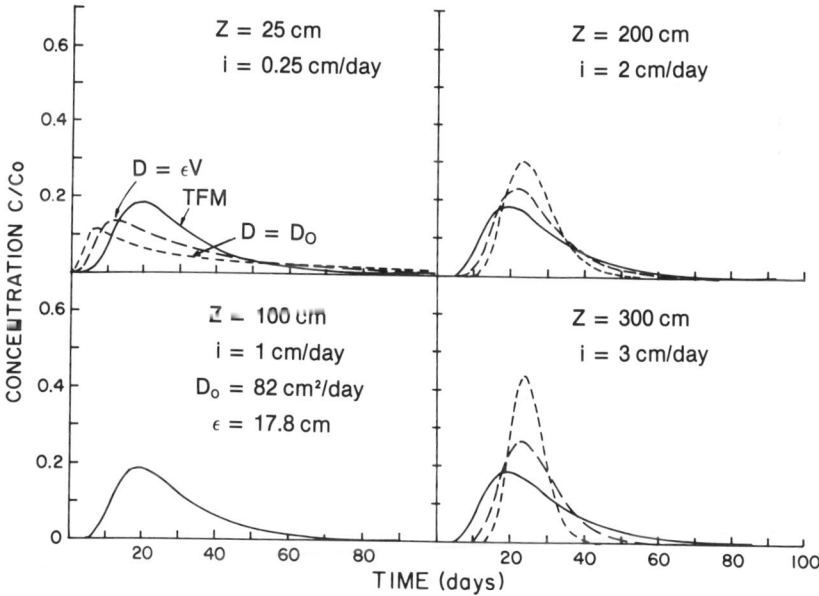

Fig. 4. Simulated solute breakthrough curves for various depths and input flux rates for a pulse of concentration C_o applied with 5 cm of water using TFM and CDE models (constant D and D = ϵV) calibrated at Z = 100, i = 1 cm/d.

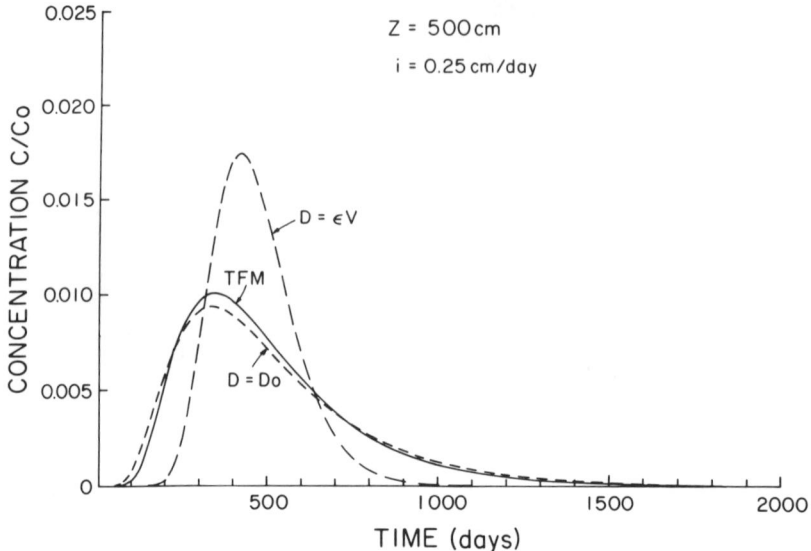

Fig. 5. Simulated solute breakthrough curve for $i = 0.25$ cm/d and $Z = 500$ cm for a pulse of concentration C_o applied with 5 cm of water using TFM and CDE models calibrated at $i = 1$ cm/d, $Z = 100$ cm.

The essential point made in these simulations is that the predictions of solute transport made by the convection-dispersion equation for either a constant dispersion coefficient or for a linear relationship between dispersion and velocity will agree with the variable velocity simulation of solute transport only under conditions similar to the ones where and when the calibration was obtained. Therefore, these models, if calibrated on the same data, will disagree when used to simulate other processes. Unfortunately, there is no experimental evidence to suggest which model is appropriate for simulations of downward movement to great depth. The experiment of Jury et al. (1982) as well as the earlier mentioned experiments where velocity variations were found, would give credance to the hypothesis that a variable velocity model is most appropriate near the soil surface. However, whether this model may be extended to depths of the orders of tens of meters or not remains to be discovered.

Movement of Adsorbed Chemicals

The majority of chemicals which are judged to be environmentally harmful do not move freely with water through the soil but are adsorbed on mineral surfaces or organic matter. A common hypothesis, particularly for trace organics and pesticides, is to assume a linear equilibrium adsorption isotherm

$$C_S = RC_L \qquad [11]$$

where C_S is adsorbed concentration, C_L is solution concentration, and R is the slope of the adsorption isotherm. It can be shown that the travel time for an adsorbed chemical t_A which obeys Eq. [11] will relate to the travel time for a non-adsorbed or mobile chemical t_M by

$$t_A = (1 + \varrho_b R/\Theta) t_M \equiv K t_M, \qquad [12]$$

where ϱ_b is the bulk density and K is the partition or distribution coefficient. If water content and bulk density are constant, then K will also be. In undisturbed field soil one might expect the partition coefficient K to have a variety of values at different locations which depend on local values of organic matter concentration, exchange surfaces, soil bulk density, etc. However, since this detailed information is usually not available, simulation models of adsorbed chemical transport commonly assume a constant value of the partition coefficient K and use the appropriate solution of the convection-dispersion Eq. [1] to predict movement of the solute in question.

If the partition coefficient has a distribution of values rather than being constant the TFM may be used to estimate the adsorbed chemical movement if the distribution of adsorbed travel times is known. This distribution function $f_L(t_A)$ will be a function of the distribution of both K and t_M as well as being a function of the degree of correlation between them. For example, if physical properties of soil which tend to cause short travel times or high water velocity for non-adsorbed chemical, e.g., cracks, holes, coarse-textured soil, also tend to possess small partition coefficients K then the individual values of K and t_M will be correlated and the distribution of $t_A = K t_M$ will reflect this correlation. In such a circumstance, use of the mean values of these properties will not give accurate estimates and the extreme behavior, i.e., the shortest arrival times of adsorbed chemicals, will not be adequately reflected.

To calculate the probability that an adsorbed chemical injected at t = 0 and z = 0 will reach z = L between t_A and $t_A + dt$, we multiply the probability $f_K(K)dK$ of obtaining a given value of K by the probability $f_M(t_A/K) d(t_A/K)$ of obtaining a given value of $t_M = t_A/K$ and integrate over all possible values of K, or (assuming zero or perfect correlation)

$$f_A(t_A) dt = \int_{-\infty}^{\infty} f_K(K) f_M(t_A/K) \frac{dt}{K} dK, \qquad [13]$$

where f_A, f_K, f_M are the probability density functions for t_A, K, and t_M, respectively.

Two cases of interest will be examined in the calculations below: (i) K is perfectly correlated with t_M and (ii) K is uncorrelated with t_M. In the first case, the correlation will be expressed through the functional representation

$$K = K_o (t_M/\tau)^\beta \qquad [14]$$

where τ is the median of the t_M distribution and β is a constant to be given values of 0, 0.5, and 1. Equation [14] represents a perfect correlation, such that given a value of t_M, K is completely determined. This idealization is put in for mathematical simplicity to represent an extreme case to contrast with the assumption of no correlation or complete independence. The choice $\beta = 0.5$ is consistent with the capillary bundle model of soil water flow, if K is assumed proportional to surface area and water velocities are calculated by Poiseuille's Law. The choice $\beta = 1.0$ is consistent with the hypothesis that travel time or vertical water velocity variations are due to path length variations and that K is proportional to surface area. If t_M is lognormally distributed with $E(\ln t_M) = \mu$ and $\text{Var}(\ln t_M) = \sigma^2$, it follows from Eq. [14] that K is lognormally distributed with

$$E(\ln K) = \ln K_o \quad [15]$$

$$\text{Var}(\ln K) = \beta^2 \sigma^2$$

since $\tau = \exp(\mu)$. Figure 6 shows the probability distributions for K for the three values of β and assuming $K_o = 10$. The value of $\sigma^2 = 0.31$ was taken from the study of Jury et al. (1982). The same distributions will be used when case ii, K uncorrelated with t_M is studied.

If Eq. [14] is valid, then the probability density function $f_K(K)$ in Eq. [14] is replaced by the single-valued delta function distribution $f_K(K) = \delta(K - K_o t_M^\beta/\tau^\beta)$. If K is independent of t_M, $f_K(K)$ is given by the lognormal distribution with mean and variance in Eq. [15]. In either case, however, evaluation of Eq. [14] shows that the probability density function $f_A(t_A)$ will be lognormally distributed, with

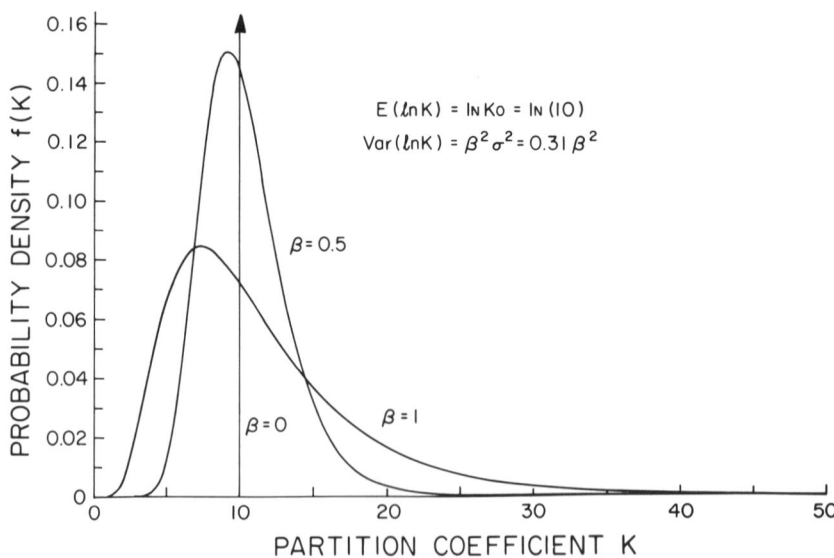

Fig. 6. The probability distribution for partition coefficient K, represented as a log-normal distribution with $E(\ln (K)) = \ln K_o$, $\text{var}(\ln (K)) = \beta^2\sigma^2$ for $\beta = 0, 0.5, 1.0$.

CHEMICAL TRANSPORT MODELING

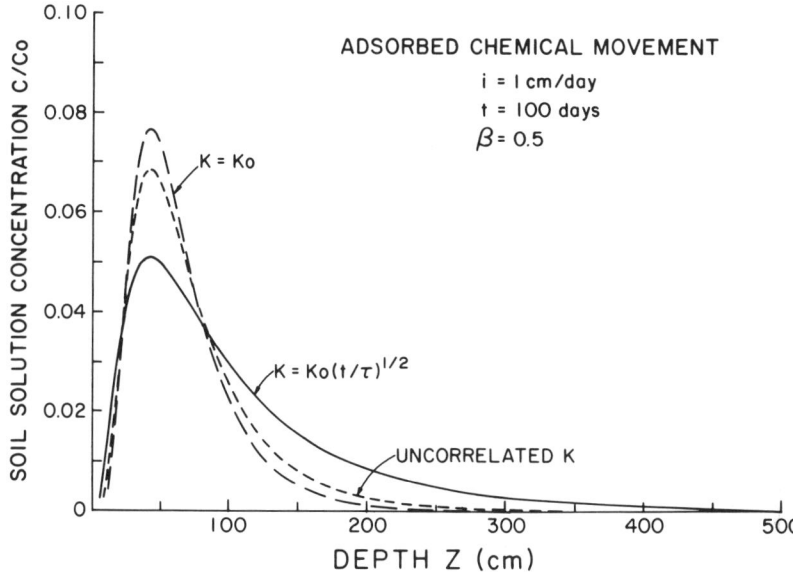

Fig. 7. Predicted solute concentrations vs. depth obtained at t = 100 d for an adsorbed chemical of concentration C_o added with 5 cm of irrigation water for a steady rate I = 1 cm/d. K is obtained from Fig. 6 with $\beta = 0.5$ or 0.

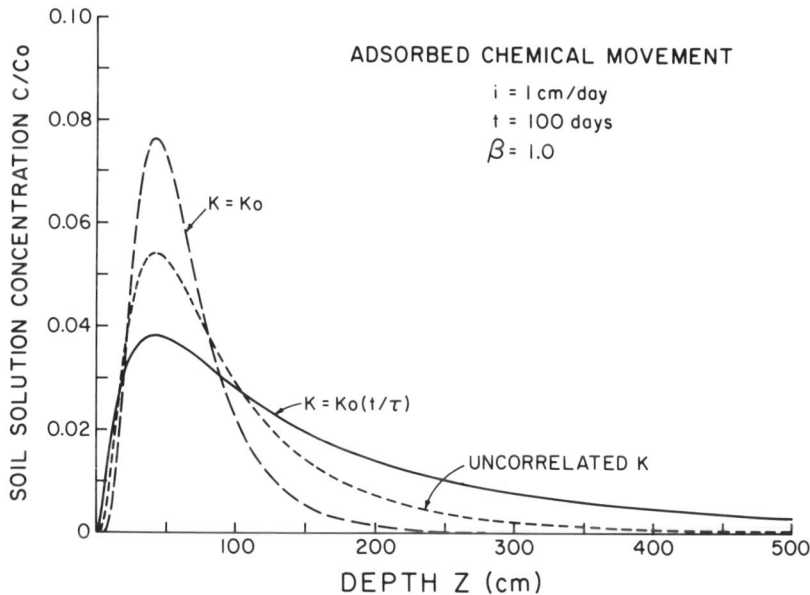

Fig. 8. Predicted solute concentrations vs. depth obtained at t = 100 d for an adsorbed chemical of concentration C_o added with 5 cm of irrigation water for a steady rate I = 1 cm/d. K is obtained from Fig. 6 with $\beta = 1.0$ or 0.

Table 2. Net amount of applied water I (cm) required to leach P % of an adsorbed chemical past Z = 500 cm, assuming either $K = 10 \, (\tau_M/\tau)^\beta$ (correlated K), or K independent of t_M and lognormally distributed (uncorrelated K).

	K correlated with t_M			K uncorrelated with t_M		
	P=0.1%	P=1.0%	P=5.0%	P=0.1%	P=1.0%	P=5.0%
$\beta = 0$	207	312	455	207	312	455
$\beta = 1/2$	89	165	290	141	234	372
$\beta = 1$	38	87	184	103	183	313

$$E(\ln t_A) = E(\ln t_M) + E(\ln K) = \mu + \ln K_o \qquad [16]$$

$$\text{Var}(\ln t_A) = (1 + \beta^2) \, \sigma^2 \text{ if K is uncorrelated with } t_M$$

$$\text{or Var}(\ln t_A) = (1 + \beta)^2 \, \sigma^2 \text{ if Eq. [14] is valid.}$$

All calculations for adsorbed chemical transport are now identical to the nonadsorbed chemical calculations above, with the adsorbed travel time distribution log mean and variance (Eq. [16]) substituted for μ and σ^2 in Eq. [10]. The prediction of the CDE for adsorbed chemical is obtained by substituting t/K_o for t in Eq. [10].

Figure 7 shows the predicted total concentration vs. depth distribution at t = 100 days of an adsorbed chemical added as a pulse with 5 cm of irrigation water at a steady rate of i = 1 cm/d, for $\beta = 0.5$. The $K = K_o$ curve corresponds to the TFM model with $t_A = K_o t_M$ and thus represents the prediction of a CDE calibrated to a variable velocity field. The other two curves are the predicted concentrations for the case where a distribution of K values with median K_o is found in the field. The uncorrelated K simulation is very similar to the constant K simulation, but the correlated K simulation shows a much greater spread of concentration. The differences between the three simulations are even more pronounced in Fig. 8 where $\beta = 1$. Now substantial differences exist between the constant K and the uncorrelated K case and the correlated K concentrations have reached well below 500 cm.

The extreme differences to be found between predictions of the models described above are shown in Table 2 which gives the predicted net amount of water in centimeters required to leach 0.1, 1.0, or 5% of an adsorbed chemical past z = 500 cm, using the three presumed models for adsorption. The 0.1% probability represents the worst case behavior discussed above in that it represents the earliest arrival time of a potentially dangerous chemical. It is obvious by looking at this table that significant differences are found between models calibrated on the same data. For example, if one found substantial variability in the values of the distribution coefficient, K, depending on whether one averaged these values, considered them to be randomly varying but independent of the water flow regime, or whether one considered them to be correlated with the water flow variations, one would make estimates ranging from 38 to 207 cm of water. Such uncertainty is clearly unsuitable for a predictive model.

SUMMARY AND CONCLUSIONS

The simulations above have shown the fragile foundation of simulation models of chemical movement through the unsaturated zone when applied to natural fields where substantial lateral variations in vertical water velocity are found. Preliminary testing indicates that convective models such as the transfer function model which take velocity variations into account, are able to describe the spatial variability of chemical movement near the surface. What is unknown, however, is to what depth the correlations inherent in the assumption of the transfer function model will still hold. Below this depth where correlations die out, perhaps models such as the convection-dispersion equation using asymptotic and average values will be more appropriate. Unfortunately, at the present time, such information is lacking. In the future it is hoped that experimentation will merge more closely with modeling under actual field conditions to study in greater detail the interaction between variations in the water velocity field and variations in soil and chemical properties.

ACKNOWLEDGMENT

The author wishes to thank the Southern California Edison Company for financial assistance on this project.

LITERATURE CITED

Biggar, J. W., and D. R. Nielsen. 1967. Miscible displacement and leaching phenomenon. Agron. Monogr. 11:254–274.

----, and ----. 1976. Spatial variability of the leaching characteristics of a field soil. Water Resour. Res. 12:78–84.

Boast, C. W. 1973. Modeling the movement of chemicals in soil by water. Soil Sci. 115:224–230.

Dagan, G., and E. Bresler. 1979. Solute dispersion in unsaturated soil at field scale. Soil Sci. Soc. Am. J. 43:461–466.

Duguid, J. O., and M. Reeves. 1976. Material transport through porous media. A finite-element galerkin model. Oak Ridge National Laboratory ORNL 4928.

Jury, W. A. 1982. Simulation of solute transport using a transfer function model. Water Resour. Res. 18:363–368.

----, W. R. Gardner, P. G. Saffigna, and C. B. Tanner. 1976. Model for predicting movement of nitrate and water through a loamy sand. Soil Sci. 122:36–43.

----, L. H. Stolzy, and P. Shouse. 1982. A field test of the transfer function model for predicting solute transport. Water Resour. Res. 18:369–374.

Peck, A. J., R. J. Luxmoore, and J. L. Stolzy. 1977. Effects of spatial variability of soil hydraulic properties in water budget modeling. Water Resour. Res. 13:348–354.

Rose, C. W., F. W. Chichester, J. R. Williams, and J. T. Ritchie. 1982. A contribution to simplified models for solute transport. J. Environ. Qual. 11:146–150.

Sharma, M. L., G. A. Gander, and C. G. Hunt. 1980. Spatial variability of infiltration in a watershed. J. Hydrol. 45:101–122.

Starr, J. L., H. C. DeRoo, C. R. Frink, and J. Y. Parlange. 1978. Leaching characteristics of a layered field soil. Soil Sci. Soc. Am. J. 42:386–391.

Van de Pol, R. m., P. J. Wierenga, and D. R. Nielsen. 1977. Solute movement in a field soil. Soil Sci. Soc. Am. J. 41:10–13.

van Genuchten, M. 1978. Land disposal of hazardous wastes. Proc. of 4th Annual Res. Symp. EPA-600/9-78-016.

Warrick, A. W., and A. Amoozegar-Fard. 1979. Infiltration and drainage calculations using spatially scaled hydraulic properties. Water Resour. Res. 15:1116–1120.

----, G. W. Mullen, and D. R. Nielsen. 1977. Scaling field-measured soil hydraulic properties using a similar media concept. Water Resour. Res. 13:355–362.

Chapter 5

Spatial Soil Variability and Mass Transfers from Agricultural Soils

D. R. NIELSEN, P. J. WIERENGA, AND J. W. BIGGAR[1]

The need to improve our understanding of chemical mobility and reactivity in soils and soil materials has increased markedly within the last decade owing primarily to state and federal legislation to improve environmental quality. These legislative mandates which have accelerated regulatory programs through old and newly created environmental action agencies have intensified research efforts on mass transfers within soil profiles. These programs coupled with rapidly increasing costs of agrochemicals and the disposal or containment of solid and liquid wastes have focused our attention further on broader classes of problems that impinge directly on the quality of surface and groundwaters. Today, soil and crop scientists, irrigation and drainage engineers, hydrologists, ecologists and others, including resource economists and analysts, are now heavily engaged in laboratory and field experiments, pilot demonstration projects, and countless feasibility studies to monitor and control the retention and transport of solutes. Concern over potential and existing contamination has stimulated researchers to develop a variety of new techniques and models to analyze and predict the transport of contaminants to and in groundwater systems. A fundamental difficulty with these techniques and models is how to incorporate the complex natural variations in the soil which occur in both space and time. This difficulty is confounded by the convention of describing water flow in saturated and unsaturated soils

---

[1] Contribution from the Dep. of Land, Air and Water Resources, Univ. of California, Davis; and the Agron. Dep., New Mexico State Univ., Las Cruces. First and third authors are professors, Univ. of California, and the second author is professor, New Mexico State Univ.

Copyright © 1983 ASA, SSSA, 677 South Segoe Road, Madison, WI 53711. *Chemical Mobility and Reactivity in Soil Systems.*

using macroscopic considerations assuming a Darcian continuum which have not yet been satisfactorily reconciled with microscopic or megascopic velocities of water and solutes in soils. The lack of reconciliation of water and solute behavior using these conceptual definitions has been manifested clearly by incongruous results of experimentalists trying to monitor, model or predict water and solute transport in agricultural soils (e.g., Van de Pol et al., 1977; Quisenberry and Phillips, 1976; Starr et al., 1978), the vadose zone and groundwater aquifers (e.g., Pickens and Grisak, 1981). Earlier, Nielsen and Biggar (1962) working with laboratory soil columns and Elrick and French (1966) working with soil cores noted the importance of the pore size distribution by examining solute displacement at different soil water contents and flow velocities. Many soil scientists now suggest that because of macropores, leaching of solutes takes place through a limited number of preferential paths (Bouma, 1981; Thomas and Phillips, 1979), and hence, the convective-diffusion equation is inappropriate (van Genuchten and Wierenga, 1976). Other investigators are looking for a theoretical basis or for criteria for determining the appropriate number and/or frequency in time and space of soil solution samples (Warrick and Amoozegar-Fard, 1979; Biggar and Nielsen, 1976). Still others have been testing the merits of different kinds of porous devices to sample the soil solution (Silkworth and Grigal, 1981; Talsma et al., 1979; Hansen and Harris, 1975) rather than looking at the spatial and temporal variance structure of the soil solution. And, in general, the verification and calibration of almost any model to describe the mobility and retention of solutes in the field remain incomplete regardless of the extent or intensity of the effort spent.

It is our intention to review briefly a few of the fundamental concepts of flow in porous media, point out issues that remain obscure or unaddressed by current-day investigators, and suggest research alternatives that we believe will advance technology for a greater understanding and a more comprehensive analysis of solute transfer.

Let us reexamine the concept of a representative elementary volume (REV) of a porous material discussed in detail by Hubbert (1956) and later adapted by Bear (1972) and Freeze and Cherry (1979). We consider here a volume scale much greater than that of the molecular level commensurate with either a single soil particle or a great number of particles. We wish to assign a value to the porosity f at a location P within the porous material, as a function of the size of the elemental bulk volume ΔV. If location P happens to reside inside a soil pore, we obtain f as a function of ΔV as shown in Fig. 1. As the value of ΔV increases from zero to REV, the value of f fluctuates in a manner consistent with the microscopic distribution of solid particles within the porous medium. At REV and beyond, if indeed REV exists, the value of f appears in a statistical sense as a constant f_{REV}. As ΔV increases still further, the value of f may continue to remain invariant or depart from f_{REV} depending upon the heterogeneity of the soil being investigated. For example, as values of the volume ΔV become larger, the value of f may suddenly increase if the soil manifests cracks between the soil peds. For the application of Darcy's law, groundwater hydrologists consider the size of REV to be smaller than

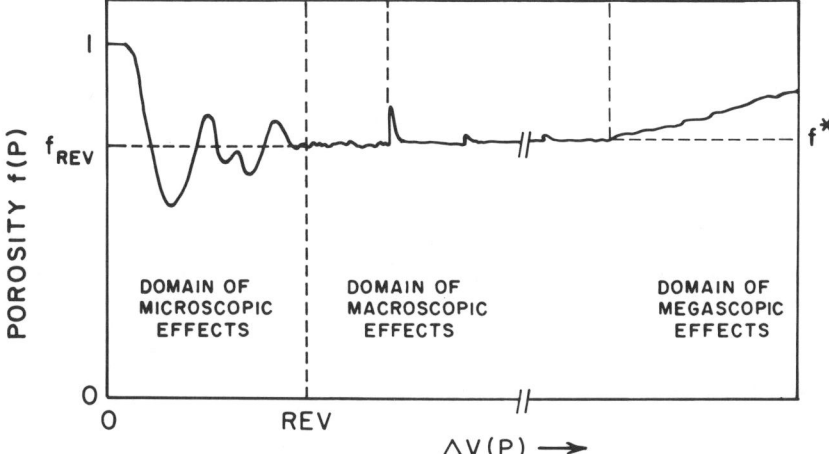

Fig. 1. Definition of porosity and representative elementary volume REV. Adapted from Hubbert (1956), Bear (1972), and Freeze and Cherry (1979).

the volume of soil removed with a soil core sampler (Freeze and Cherry, 1979). Taking progressively larger and larger values of ΔV eventually yields the value of f^*—the porosity of an entire field had it been sampled in its entirety. At still greater sample volumes or radii as sampling transcends over more than one soil mapping unit, the porosity f departs from f^*.

Although the above definition of REV is illustrated for soil porosity, similar considerations of soil water content, air-filled porosity in an unsaturated soil, soil water viscosity, soil water flux, etc. would not necessarily yield equal values of REV for each parameter. To date, it has been tacitly assumed that experimental observations and theoretical considerations of each of the terms in Darcy's law and Richards' equation are applicable to the same REV. This assumption has been given credence by efforts to derive the macroscopically applicable Darcy's law from the microscopically applicable Stokes-Navier equation of motion of viscous fluids (e.g., Hubbert, 1956). The two concepts are linked with a proportionality factor whose magnitude is a function of the geometry of the void space through which the fluid flows—the hydraulic conductivity. Inasmuch as the hydraulic conductivity is only obtained experimentally, its existence as a unique parameter independent of the particular experimental techniques to measure the flux and the hydraulic gradient (or that they pertain to the same REV) has seldom been examined. Under practical conditions, the parameter $(K(\theta)$ is usually adjusted to describe a particular set of soil water observations without explicit verification that the experimental methods were identifying equivalent REV's for each parameter.

Values of REV for chemical and biological parameters would differ and each would also be different than that for other physical or soil water parameters. For example, the concentration of the soil solution c replacing f in Fig. 1 would fluctuate in the microscopic domain depending upon the size of the soil pore within which P is situated owing to the fact that even with chemical equilibrium, the concentration and distribution of the ionic species depend upon the relative size of the soil pore. Similarly, the distribution of soil microbes or plant roots would yield different values of REV. On the other hand, as the value of ΔV approaches that of an entire field, field average values (associated with r of order 10^4 cm) of f, c, etc., are conceptually consistent even though experimentally difficult or impossible to ascertain exactly.

In the case of describing and observing water and solute movement through field soils, it is not surprising that investigators are perplexed by seemingly anomalous results. They have discovered once again (Lawes et al., 1882; Means and Holmes, 1901) that water and solutes move faster through the larger water-filled pore sequences. The same is true for laboratory soil columns. Such events signal inadequate attention given to the construction of mathematical models that are based upon consistent and experimentally appropriate REVs for both water and solutes—a point to be discussed below.

The concept illustrated in Fig. 1 addressed the size of a REV at a point P in the soil. The size of an area (or volume) characterized by a given soil sample is another important concept. Closely linked to this concept is the identification of the appropriate size and spacing of experimental observations. The autocorrelogram, a measure of the strength of the linear association between pairs of observations, is useful in defining the separation distance between samples beyond which there is no correlation between pairs of values. The autocorrelation coefficient ϱ is a function of the separation distance h defined as

$$\varrho(h) = \frac{\text{autocovar}(f_x, f_{xth})}{\text{var}(f_x)} \qquad [1]$$

is idealized in Fig. 2. The figure shows that as the distance between pairs of samples increases, the correlation between the observations decreases. The value l, sometimes referred to as the scale of observation, defines the

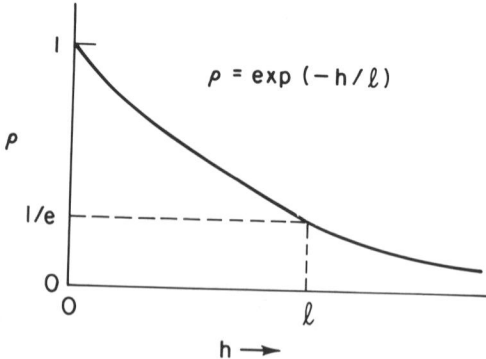

Fig. 2. Idealized autocorrelogram defining the scale of observation l for a series of measurements separated by distances h.

distance beyond which the correlation is equal to or less than 1/e. In general, there are several ways to defining scales of observations (Lumley and Panofsky, 1964) as well as ways of interpreting autocorrelograms (Davis, 1973). Our purpose here is to merely point out that Fig. 2 defines the distance over which observations are correlated to some degree. In other words, a sample observed at h = 0 characterizes a spatial region of radius l.

The above considers the spacing between centers of observations without necessarily considering the size of the observations. In Fig. 3, an over-simplified view is presented of salt being leached through a soil profile with some large water-filled pore sequences. The soil is being sampled with two different sizes of soil cores or soil suction probes. For the small samples, the autocorrelogram shows that the observations are spatially independent with the salt passing through a sample only infrequently and with no salt passing through the majority of samples. The small samples would manifest a large variance which yields a large uncertainty for the sample mean concentration to be anywhere near the population mean concentration at that depth in the field—a common experimental result. On the other hand, the larger samples yield an autocorrelogram that defines a large scale of observation. Because the scale of observation for the larger samples is several times larger than the diameter of the sample, fewer observations would be needed to adequately estimate the mean salt concentration (Sisson and Wierenga, 1981). An examination of the relations between sample size and sample spacing through the use of autocorrelograms for similar soil mapping units would help optimize the sampling process and/or soil management schemes (Burgess and Webster, 1980).

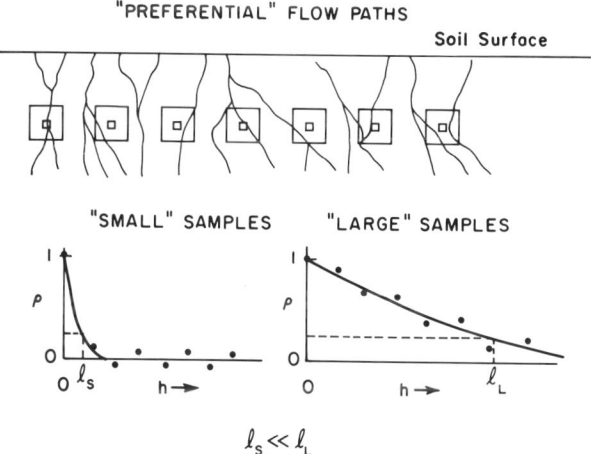

Fig. 3. Scales of observation l_s and l_L from idealized autocorrelograms for "small" and "large" samples taken to observe solute concentrations in a soil profile manifesting "preferential" flow paths.

How do the above spatial concepts apply in still another way to solute mobility and retention? Let us assume that a simple relationship exists between the solute associated with the solid phase s and the solute associated with the solution phase c represented by the expression

$$s = Rc, \qquad [2]$$

where R is the distribution coefficient, a parameter of great importance in many modeling efforts and obtainable only by experimental observation. Assume that samples of soil removed from a field in a regular spacing were analyzed for a solute concentration in terms of both s and c. Further, assume that their autocorrelograms were those shown in Fig. 4. Figure 4 shows that the observations of c represent an areal radius of l_c, while those of s represent an areal radius l_s several times larger. In other words, the concentration c represents a different volume of soil than that of the concentration s, and hence, a linear correlation between s and c will *not* necessarily give a meaningful value of the parameter R. Such a linear correlation will be appropriate when the domains represented by s and c are identical.

A simple extension of the linear correlation is the spatial cross-correlation for two series of observations (say, s and c) illustrated in Fig. 5A and expressed as

$$\varrho(m) = \frac{\text{covar}(s,c)}{\sqrt{\text{var }s}\sqrt{\text{var }c}}. \qquad [3]$$

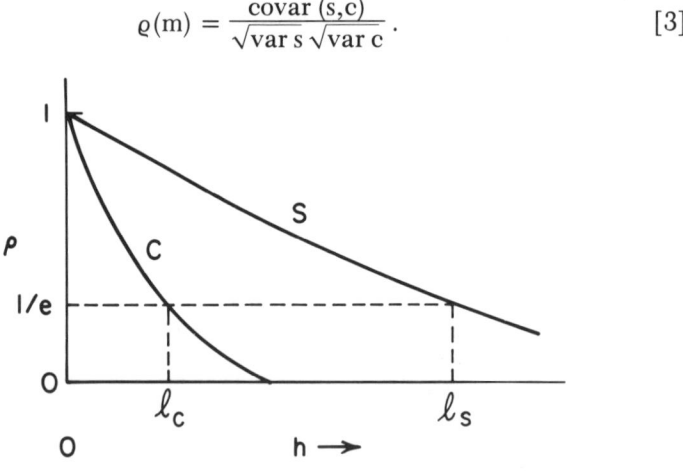

Fig. 4. Idealized autocorrelograms for observations of the solute associated with the solid phase s and the solute associated with the solution phase c.

$\varrho(m)$ is calculated for increasing distances over which the two series overlap. Figure 5B illustrates the change in cross-correlation coefficient ϱ by increasing the horizontal translation of the two curves. The area under the curve or the range of m over which the value of ϱ remains near unity is an indication of the spatial area over which the linear correlation between s and c exists. For the solid curve, the correlation is appropriate for a radial distance of about 5 m. For the broken curve, the correlation holds for a radial distance less than 1 m. Larger experimental samples would be necessary to increase the latter distance.

Similarly, when dealing with Darcy's law, the hydraulic conductivity can only be meaningful when observations of the flux of water and the hydraulic gradient pertain to the same size spatial domain. And still

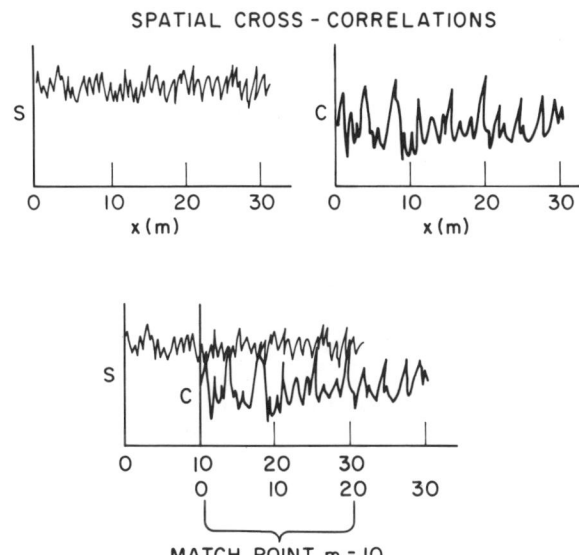

Fig. 5A. Observations of s and c as a function of horizontal distance × illustrating match point m for the calculation of a cross-correlation shown in Fig. 5B.

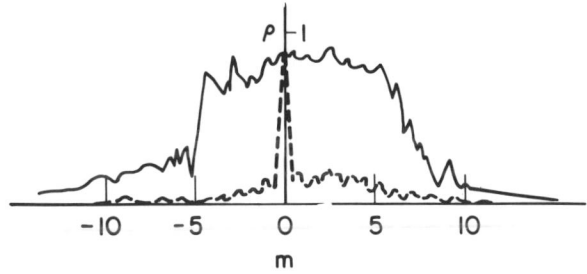

Fig. 5B. Cross-correlograms between two sets of observations of s and c. The solid line indicates a set of observations cross-correlated to match points as great as ±5 m while the broken line indicates a set of observations correlated to a distance less than 1 m.

further, the convective flux of solute, equal to the flux of water J_w, multiplied by the solute concentration c, will be meaningful only when the values of J_w and c apply to the same domain. Because there has been no systematic attempt to describe the appropriate size or spacing of samples used to quantify terms used in even the simplest of models, it is not unreasonable that investigators present seemingly incongruous experimental results in relation to mass balances, solute fluxes, and solute retentions in field soils.

Depending upon the genetic nature of the soil profile, the distribution of the larger pore sequences (and, hence, the zones of higher concentration as illustrated in Fig. 3) may be somewhat cyclic for each soil depth as shown schematically in Fig. 6. Excavating soil profiles or the removal of soil cores to observe the distribution of water soluble dyes and other tracers observed in field soils or from soil cores has revealed such patterns (e.g., Bouma, 1981). A means of quantifying these cyclic variations is to calculate their spectral density function. This calculation is illustrated with the data c(x) in Fig. 6, from which an autocorrelogram $\varrho(h)$ is constructed in Fig. 7. And from $\varrho(h)$, the spectral density function S(f) in Fig. 7 is calculated by

$$S(f) = 2 \int_0^\infty \varrho(h) \cos(2\pi fh) dh, \qquad [4]$$

where f is the frequency equal to 1/p where p is the period. The spectral density function is useful in isolating periodicities or the spacings associated with repetitious observations or properties such as the larger pore sequences in Fig. 6. It should be noted that had either the sample size or the sample spacing been larger in Fig. 6, the data would have revealed a smaller frequency (larger period) as illustrated by a simple sine curve in Fig. 8. The solid line was obtained by frequent, small samples while the broken curve (solid circle) was obtained by less frequent samples of the

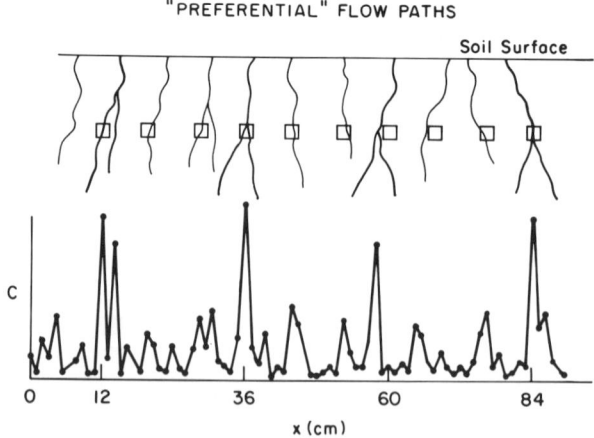

Fig. 6. Schematic cross-section of soil profile sampled at regular intervals illustrating variations in solute concentration associated with "preferential" flow paths.

same size. Had the size of the samples taken been progressively larger (open circles), in the limit, there would have been no periodicity manifested. Hence, autocorrelations and their spectral densities of solute concentrations provide information on scales of observation, spatial dependence, and periodicities. If such parameters were known and cataloged in soil mapping units, then strategies for sampling soil profiles would be less

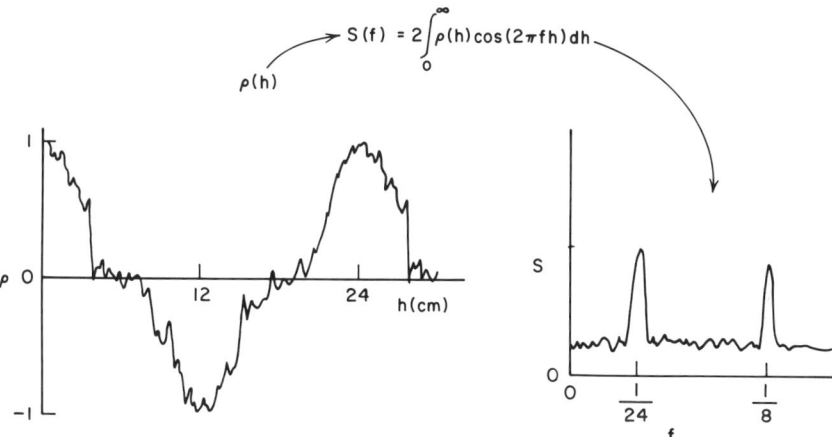

Fig. 7. Idealized autocorrelogram of c(x) given in Fig. 6 and the resulting spectral density function S(f or 1/p).

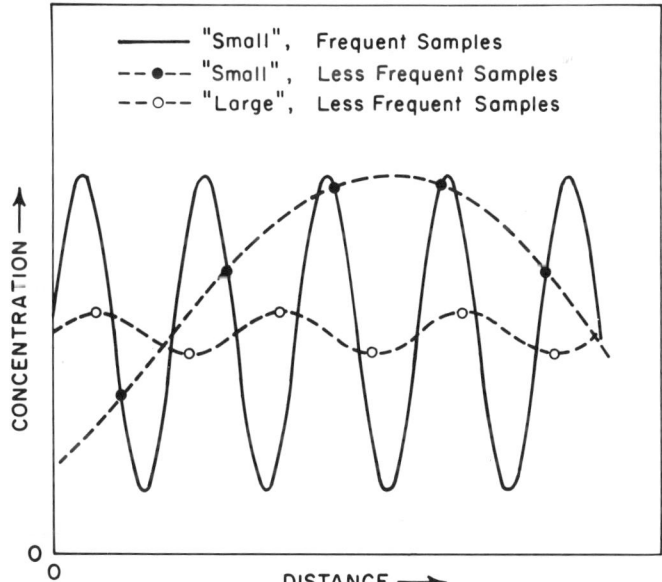

Fig. 8. Idealized sinusoidal solute concentration as a function of horizontal distance from small and larger samples taken frequently and less frequently.

ambiguous and would be more easily analyzed for a particular mathematical model.

Let us now consider the utility of several solute transport and retention models in light of the foregoing discussions. The now common convective-diffusion equation is based upon deterministic flux and mass balance considerations yielding

$$\frac{\partial \delta s}{\partial t} + \frac{\partial \theta c}{\partial t} = D \frac{\partial^2 c}{\partial z^2} - \frac{\partial J_w c}{\partial x} + \phi, \qquad [5]$$

where δ is the soil bulk density, t time, θ the soil water content, D the apparent diffusion coefficient, J_w the flux of water, ϕ an irreversible source or sink term and z the vertical coordinate. Each of the terms, δ, s, θ, c, J_w, and ϕ are based implicitly on the identification of a REV, and yet few investigators have defined the dimensions of a REV whose size exists simultaneously for each of the above terms so that the equation is conceptually correct. For example, J_w is often defined in terms of Darcy's law, a macroscopic term which allows a flux of water averaged over a cross-section of an unknown distribution of microscopic pore water velocities, while the dispersion of the solute and thus the value of D is most often based upon microscopic considerations. These include (i) flow rates within and between soil-water pores, (ii) molecular diffusion, and (iii) water-solute-soil particle interactions. Hence, investigators have had to resort to defining empirical relations between the magnitude of D, essentially a microscopic term, and J_w, a macroscopic term. An alternative strategy has been to consider microscopic distributions of pore water velocity and solute concentration while losing sight of the macroscopic REV for the other terms. Whether the investigator has in mind preferential flow paths around soil aggregates, through cracks or around soil peds, or is considering macroscopic average flows at different soil water contents, almost without exception studies would be more informative and more generally applicable to other studies if the size and spacing of the experimental observations used in the mathematical model were chosen on the basis of their spatial variance structures.

A second approach to solute transfer involves lumped parameters without having to explicitly integrate the macroscopic solute fluxes in Eq. [5]. An example of such a model is given by Rose et al. (1979) describing the movement of solute through a very slowly permeable soil whose hydraulic properties are extremely difficult to ascertain. They expressed the time rate of change of the spatial mean solute concentration $\bar{s}$ within the soil profile to depth z as

$$\frac{d\bar{s}}{dt} = (Ic - Ls_z)/z\bar{\theta}_s, \qquad [6]$$

where I and L are annual rates of irrigation and leaching, respectively, c is the solute concentration in the irrigation water, s_z the soil solute concentration at depth z, and $\bar{\theta}_s$ the spatial mean saturated soil water content. Equation [6] was integrated assuming that the initial mean soil solute concentration s_o was known and that the s_z followed the relation

$$s_z = \lambda \bar{s}, \qquad [7]$$

where λ is a proportionality constant. Each of the terms $\bar{s}$, I, c, L, s_z, and $\bar{\theta}_s$ in Eq. [6] are spatial averages and, hence, are subject to spatial considerations in relation to REV and scales of observation. Equation [7] is similar to Eq. [2] as regards spatial cross-correlation considerations.

The above two models are considered deterministic. With the intent of coping more effectively with the apparent spatial variability of soils, other investigators have considered some of the parameters or terms in Eq. [5] to be stochastic. Proponents of statistical approaches suggest that it is not possible to give an exact mathematical description of the solute path and, hence, its location can be determined only in a probabilistic view (e.g., Scheidegger, 1954; Saffman, 1959). Theoretical representations on a field scale or alternative formulations based upon scaling theory (e.g., Peck et al., 1977; Warrick and Amoozegar-Fard, 1979) rely on measured scaling factors while other probability models (e.g., Dagan and Bresler, 1979; Amoozegar-Fard et al., 1982) rely on probability density functions estimated from a limited set of experimental observations.

Solutions to stochastic differential equations have shown their usefulness in relating variations in hydraulic conductivity to the dispersion of solutes. The work of Naff (1978)[2], Gelhar et al. (1979), and Smith and Schwartz (1980) are good examples that need to be explored and extended further through a greater effort on experimental observations in the field for unsaturated water flow conditions, particularly where spatial variations in chemical and microbiological reactions impinge on solute transport.

All of the above models, including those which contained random or stochastic functions, were based upon deterministic or mechanistic concepts that describe processes occurring within the soil profile. There exists another large class of stochastic formulations, transfer function models (Box and Jenkins, 1976), that are not based on a knowledge of mechanisms within the profile but merely treats the transformation of inputs into outputs. Those models, usually used by electrical engineers and hydrologists, have been examined recently by Jury (1982) to investigate the transfer of solutes added to the soil surface to a greater depth in the soil profile. The average and extreme behavior of solute moving through the soil depends upon the measurement of the distribution of solute travel times. With these distributions known for a given soil domain (Fig. 9), there is no need to measure apparent diffusion or other reaction term coefficients as used in Eq. [5]. On the other hand, travel time observations—taking repetitive samples of solute concentrations at specified depths and time intervals—are not necessarily unique and, in fact, depend directly upon the size and the 3-dimensional spacing of samples. Also, the depth to

---

[2] Naff, R. L. 1978. A continuum approach to the study and determination of field longitudinal dispersion coefficients. Ph.D. Diss. Socorro, NM: New Mexico Institute of Mining and Technology. 176 p.

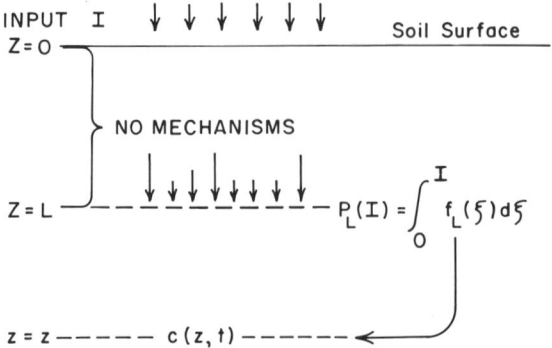

Fig. 9. An illustration of the utility of a probability density function of the solute travel times to depth L for an infiltration input I at the soil surface to simulate the average concentration $\bar{c}$ at depth z as a function of time t.

which predictions of solute movement can be made based upon a calibration at one depth, depends upon the same probability density function being applicable over all soil depths of interest. While in a deterministic model all appropriate, significant physical, chemical and biological reactions should be included, this is not the case for a transfer function model. For the latter, these reactions are implicitly included through the probability of the amount and rate at which the solute arrives at a greater soil depth.

In conclusion, a fully developed field technology will depend upon a better understanding of the spatial and temporal variability of samples taken from the field, the description of this variability and its inclusion into the derivation of appropriate mathematical models. Furthermore, such technology requires that measuring devices be designed and spaced such that they indeed indicate values averaged over the appropriate REV of the soil profile. We encourage the use of both deterministic and stochastic models as well as the use of models that are mechanistic or are transfer functions without explicit mechanisms. It should not be argued, in general, that one is better than another inasmuch as each is designed for particular sets of conditions.

LITERATURE CITED

Amoozegar-Fard, A., D. R. Nielsen, and A. W. Warrick. 1982. Soil solute concentration distributions for spatially varying pore water velocities and apparent diffusion coefficients. Soil Sci. Soc. Am. J. 46:3–9.

Bear, J. 1972. Dynamics of fluids in porous media. American Elsevier, N.Y.

Biggar, J. W., and D. R. Nielsen. 1976. Spatial variability of the leaching characteristics of a field soil. Water Resour. Res. 12:78–82.

Bouma, J. 1981. Soil morphology and preferential flow along macropores. Agric. Water Manage. 3:235–250.

Box, G. E. P., and G. M. Jenkins. 1976. Time series analysis. Holden-Day, Inc., San Francisco.

Burgess, T. M., and R. Webster. 1980. Optimal interpolation isarithmic mapping of soil properties: I. The semivariogram and puntual kriging. J. Soil Sci. 13:315–331.

Dagan, G., and E. Bresler. 1979. Solute dispersion in unsaturated soil at field scale. I. Theory. Soil Sci. Soc. Am. J. 43:461–466.

Davis, J. C. 1973. Statistics and data analysis in geology. John Wiley and Sons, N.Y. 550 p.

Elrick, D. E., and L. K. French. 1966. Miscible displacement patterns on disturbed and undisturbed soil cores. Soil Sci. Soc. Am. Proc. 30:153–156.

Freeze, R. A., and J. A. Cherry. 1979. Groundwater. Prentice-Hall, Inc., N.J.

Gelhar, L. W., A. L. Gutjahr, and R. L. Naff. 1979. Stochastic analysis of macrodispersion in a stratified aquifer. Water Resour. Res. 15:1387–1397.

Hansen, E. A., and A. R. Harris. 1975. Validity of soil-water samples collected with porous ceramic cups. Soil Sci. Soc. Am. Proc. 39:528–536.

Hubbert, M. K. 1956. Darcy's law and the field equations of the flow of underground fluids. Trans. Am. Inst. Min. Met. Eng. 207:222–239.

Jury, W. A. 1982. Simulation of solute transport using a transfer function model. Water Resour. Res. 18:363–369.

Lawes, J. B., J. H. Gilbert, and R. Warington. 1882. On the amount and composition of the rain and drainage waters collected at Rothamsted. William Clowes and Sons, Ltd, London.

Lumley, J. L., and H. A. Panofsky. 1964. The structure of atmospheric turbulence. John Wiley and Sons, N.Y.

Means, T. H., and J. G. Holmes. 1901. Soil survey around Fresno, California. p. 333–384. In Field operations of the division of soils, House of Representatives, Document No. 526, U.S. Government Printing Office.

Nielsen, D. R., and J. W. Biggar. 1962. Miscible displacement. III. Theoretical considerations. Soil Sci. Soc. Am. Proc. 26:216–221.

Peck, A. J., R. J. Luxmoore, and J. L. Stolzy. 1977. Effects of spatial variability of soil hydraulic properties in water budget modeling. Water Resour. Res. 13:348–354.

Pickens, J. F., and C. E. Griak. 1981. Scale dependent dispersion in a stratified granular aquifer. Water Resour. Res. 17:1191–1211.

Quisenberry, V. L., and R. E. Phillips. 1976. Percolation of surface applied water in the field. Soil Sci. Soc. Am. J. 40:484–489.

Rose, C. W., P. W. A. Dayananda, D. R. Nielsen, and J. W. Biggar. 1979. Long-term solute dynamics and hydrology in irrigated slowly permeable soils. Irrig. Sci. 1:77–87.

Saffman, P. G. 1959. A theory of dispersion in porous medium. J. Fluid Mech. 6:321–349.

Scheidegger, A. E. 1954. Statistical hydrodynamics in porous media. J. Appl. Physics 25:994–1001.

Silkworth, D. R., and D. F. Grigal. 1981. Field comparison of soil solution samplers. Soil Sci. Soc. Am. J. 45:440–442.

Sisson, J. B., and P. J. Wierenga. 1981. Spatial variability of steady-state infiltration rates as a stochastic process. Soil Sci. Soc. Am. J. 45:699–704.

Smith, L., and F. W. Schwartz. 1980. Mass transport. 1. A stochastic analysis of macrodispersion. Water Resour. Res. 16:303–313.

Starr, J. L., H. C. DeRoo, C. R. Frink, and J. Y. Parlange. 1978. Leaching characteristics of a layered field soil. Soil Sci. Soc. Am. J. 42:386–391.

Talsma, T., P. A. Hallam, and R. J. Mansell. 1979. Evaluation of porous cup soil-water extractors: Physical factors. Austr. J. Soil Res. 17:417–422.

Thomas, G. W., and R. E. Phillips. 1979. Consequences of water movement in macropores. J. Environ. Qual. 8:149–152.

Van de Pol, R. M., P. J. Wierenga, and D. R. Nielsen. 1977. Solute movement in a layered field soil. Soil Sci. Soc. Am. J. 41:10–13.

van Genuchten, M. Th., and P. J. Wierenga. 1976. Mass transfer studies in sorbing porous media. I. Analytical solution. Soil Sci. Soc. Am. Proc. 40:473–480.

Warrick, A. W., and A. Amoozegar-Fard. 1979. Infiltration and drainage calculations using spatially scaled hydraulic properties. Water Resour. Res. 15:1116–1120.

Chapter 6

Assessing Nitrogen Movement in the Field

J. L. STARR[1]

Nitrogen is one of the most actively studied elements in natural systems. This activity is due to a combination of several factors, e.g., the comparatively large amounts that are required for plant growth, the many organic and inorganic forms in which it naturally occurs, and its potential for becoming a pollutant in the environment. These studies have provided the basis for many symposia, reports and reviews, including those on the movement of N through soils (e.g., Gardner, 1965; Beek and Frissel, 1973; Nielsen and MacDonald, 1978; Pratt, 1979; Frissel and van Veen, 1981; Nielsen et al., 1982; Rao et al., 1982; Tanji, 1982). The purpose of this paper is to supplement rather than duplicate these recent publications, though some duplication cannot be avoided. In supplementing these works, as well as related chapters in this volume, I will illustrate some of the reasons for continued and even increased research on this topic. It is hoped that this paper will serve as a stimulus for researchers to take a new look at the assumptions and methods that are commonly used in research on the fate of N in soils and may thereby play a small role in the advancement of knowledge and management of N in soils.

---

[1] Research soil scientist, Soil Nitrogen and Environmental Chemistry Lab., Agric. Environmental Quality Institute, USDA-ARS, NE Region, Beltsville Agric. Res. Center, Beltsville, MD 20705.

Copyright © 1983 ASA, SSSA, 677 South Segoe Road, Madison, WI 53711. *Chemical Mobility and Reactivity in Soil Systems.*

DETERMINISTIC EQUATIONS

Owing to the many sources and sinks of mobile N in soils quantitative description of its movement necessitates the consideration of the many physical, chemical, and biological reactions that are involved. Our understanding of the physicochemical factors influencing N movement in soils (e.g., water flux and reversible exchange reactions) is much more advanced than it is of the same processes when they are coupled with microbiological and plant physiological reacitons (e.g., mineralization, nitrification, denitrification, immobilization, N uptake by plants, etc.). These and other processes are presented schematically in many references (e.g., Stevenson, 1982; Frissel and van Veen, 1981). The general equation used to describe one-dimensional flow with reversible ion exchange reactions and irreversible biologically mediated reactions is

$$\frac{\partial}{\partial t}(s_i + \Theta c_i) = \frac{\partial}{\partial z}\left[D_i(v_i,\Theta)\frac{\partial c}{\partial z} - \frac{\partial (qc_i)}{\partial z}\right] \pm \sum_{j=1}^{n} \phi_{ij}, \qquad [1]$$

where s_i is the concentration of the ith N species in association with the adsorbed phase (mg cm^{-3} soil), Θ the volumetric soil water content (cm^3 cm^{-3} soil), c_i the concentration of the ith N species in the soil solution (g cm^{-3}) not identified with the adsorbed phase, t the time (h), z the downward space coordinate (cm), D_i the hydrodynamic dispersion coefficient, (cm^2 h^{-1}), v_i the average interstitial flow velocity (cm h^{-1}), q the volumetric flux of soil water (cm h^{-1}), ϕ_{ij} the net time rate at which the mass of the ith N species is being produced (+) or consumed (−) due to the jth N transformation (g cm^{-3} h^{-1}) involving c_i. Examples of j include microbiological transformations, immobilization by soil organisms, and uptake by plants.

The theoretical basis, common assumptions and applications associated with Eq. [1], where $\phi_{ij} = 0$, are presented elsewhere in this volume (Wagenet, 1983). To evaluate the transformation term ϕ_{ij} in Eq. [1], the first four terms of the equation must first be measured in the absence of irreversible reactions, i.e., $\phi_{ij} = 0$, or determined for a tracer of c_i that has similar physicochemical reaction and transport properties but for which there are no sources or sinks, i.e., $\phi = 0$. Analytical and numerical solutions to Eq. [1] are available for several initial and boundary conditions involving one or more of the processes and reactions of N in soils (e.g., McLaren, 1969; Cho, 1971; Starr et al., 1974; Misra et al., 1974; Starr and Parlange, 1975; Wagenet et al., 1976; Davidson et al., 1978; Kanwar et al., 1980; Parlange et al., 1982; van Genuchten and Alves, 1982). Most of the solutions of Eq. [1] apply to N movement through soils without plants, and under steady state conditions with respect to Θ and v. The kinetics for nitrification and denitrification are commonly assumed to be either first-order, i.e., $\phi_i = -k_i c_i$ where k_i is the rate constant of the first-order reaction (e.g., McLaren, 1969; Cho, 1971; Kirda et al., 1974; Misra et al., 1974; Starr et al., 1974; Stanford et al., 1975); or zero-order kinetics, i.e., $\phi_i = k_i$ (e.g., Doner et al., 1974; McLaren, 1976; Phillips et al., 1978; Kanwar et al., 1980; Parlange et al., 1982). Good agreement between data and theory is not in itself an adequate basis for choosing the kinetics involved in the reactions since an equally good agreement can often be obtained with either a zero or first-order reaction model (Starr and Parlange, 1976). The rate constants associated with ϕ_{ij} in soils are a function of microbial activity, which in turn are a function of several additional soil factors, e.g., temperature, substrate concentration, aeration, pH, etc. Hence, it is important to have at least a conceptual model for microbial growth, maintenance, and decay associated with each ith source or sink of N in the soil. McLaren (1973) proposed the following equation for modeling the rate that a given substrate c_i is utilized by microorganisms in soil by,

$$\frac{\partial c_i}{\partial t} = -A\frac{dm}{dt} - \alpha m - \beta m, \qquad [2]$$

where m is biomass of the microbes and A, α and β are, coefficients related to growth, maintenance and waste, respectively (see also Hattori (1973) p. 127–129, 373ff). The equation states that microbiological transformations such as the oxidation or reduction of N compounds stem from the metabolism rate per microbe whose numbers and kinds within a soil element are not fixed. If a steady state is assumed and the biomass m is constant, then for a particular substrate c_i Eq. [2] reduces to the Michaelis-Menten rate (McLaren, 1973),

$$\partial c_i/\partial t = -\frac{kmc_i}{K + c_i}; \qquad [3]$$

where k and K are constants. In application, Eq. [3] is inserted into Eq. [1] with $\phi_i = \partial c_i/\partial t$, and km is commonly lumped together into a new constant since the rate limiting substrates for growth and maintenance are not easily identified for different soil depths and time. Hence for many situations, either first- or zero-order kinetics is considered to be appropriate, i.e., $K \ll c_i$ or $K \gg c_i$, respectively (e.g., Starr et al., 1974; Misra et al., 1974; Ardakani et al., 1973; Kanwar et al., 1980). It is also apparent that as the substrate c_i is consumed that at some point in time or position the zero-order rate will usually reduce to the Michaelis-Menten rate (Eq. [3]). Numerical solutions of Eqs. [1] and [3] for steady flow conditions have been presented by Davidson et al. (1978), and analytical approximations have been presented by Parlange et al., 1982.

FIELD VERIFICATION

Equation [1] provides a mechanistic description of the N reactions and transport in soils and can be used to identify the most significant factors affecting the fate of N under specified conditions. However much refinement in modeling N movement can be accomplished by including terms for other important transformations involving N (e.g., mineralization of N and C, immobilization). A major difficulty is that of independently verifying each component in ϕ_{ij}. There is a risk, however, of including so many parameters in the model that "the theory can easily be adapted to fit any set of experimental data, without necessarily proving anything or providing substantative guidance for future research or technology" (Nielsen et al., 1979).

The difficulty of field verification of reaction and transport models is the principle reason that we are still essentially unable to describe quantitatively the movement of N through field soils, or to extend our predictions to other field sites with any reasonable degree of certainty. As model verification moves from the laboratory to the field, variations in space and time requires the inclusion of stochastic parameters in the model. Soil and plant scientists have commonly assumed that their experimental observations were spatially independent and normally distributed. These assumptions seem to be largely a product of the experimental techniques and statistical tools available. In recent years mathematical and statistical procedures have been developed which do not require these assumptions and also allow for spatial analyses. Elsewhere in this volume, Nielsen et al. (1983) demonstrate the use of the autocorrelogram to assess the strength of the linear association between pairs of observations and in defining the distance at which observations are no longer correlated. Another structural analysis that can be a powerful tool in assessing spatial variability is that of the variogram, which comes out of the theory of regionalized variables (Matheron, 1971). The variogram coefficient γ, is called the semivariance and is defined as one-half the mean variance of the differences between samples, or

$$\gamma(h) = \frac{1}{2n} \sum_{i=1}^{n} (f_i - f_{i+h})^2, \qquad [4]$$

where the lag h, is the distance between observations, n the number of pairs of observations, and f_i and f_{i+h} are two values of the observation at two locations a distance h, apart (e.g., Olea, 1977; Clark, 1979; Delhomme, 1978; Journel and Huijbregts, 1978; Nielsen et al., 1982). The shape of the variogram near the origin indicates the strength of the spatial dependence of the variable, whereas the shape of the variogram at large distances indicates whether the sampling area is bounded or unboudned, i.e., whether γ stabilizes around some limiting value or continually increases as h increases. Figure 1 illustrates the variograms that might be observed for two different variables (e.g., total-N and inorganic-N concentration). The dashed line represents a purely random set of observations, i.e., they are spatially independent beyond the shortest lag measured. The solid line represents a strong spatial dependence up to a distance of α lags, at which point the sample variance approaches the total variance, b, and beyond which the observation are spatially independent. The value of h at which the semivariance approaches to within an arbitrarily selected small value ϵ of the total semivariance, b, is called the

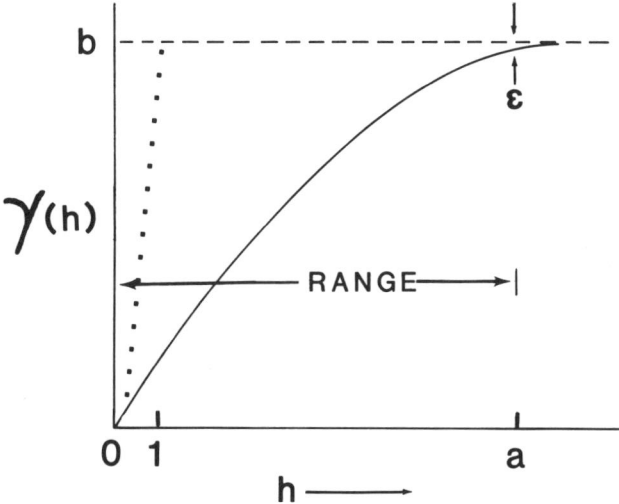

Fig. 1. Idealized variograms for spatially independent, ----, and spatially dependent, ——, observations.

range, beyond which the values of the parameter are considered to be independent of one another. For spatially independent observations, field data can be characterized by classical Fisher statistics (e.g., Steel and Torrie, 1980). For the spatially dependent observations, e.g., the solid line of Fig. 1, the spatial dependence should be considered in future sampling as well as in the experimental design. For a known spatially dependent structure of the observations additional techniques, derived from the theory of regionalized variables, are available which not only estimate the values of the variable for all points in the region, but also provide a means of computing the probable error associated with the estimates (e.g., Delhomme, 1978; Davis, 1973, p. 381–407; Journel and Huijbregts, 1978, p. 303–443). In this way field variability can be separated into spatially dependent and independent observations and the cause and effect relationships can more accurately be defined.

FIELD STUDIES

The following studies, conducted under a variety of field conditions and experimental techniques, are presented to illustrate the "success" commonly obtained in quantifying N movement in soils and to provide a basis for an assessment of future research needs.

Nitrate-N movement through a sandy loam soil and losses to ground water under shade tobacco (*Nicotiana tabacum* L.) were studied for 3 years by Starr and DeRoo (Barker, 1980). Nitrate-N movement was monitored via duplicate suction probe samples obtained from incremental soil depths, ranging from 30 to 300 cm. Time and depth distribution of NO_3-N

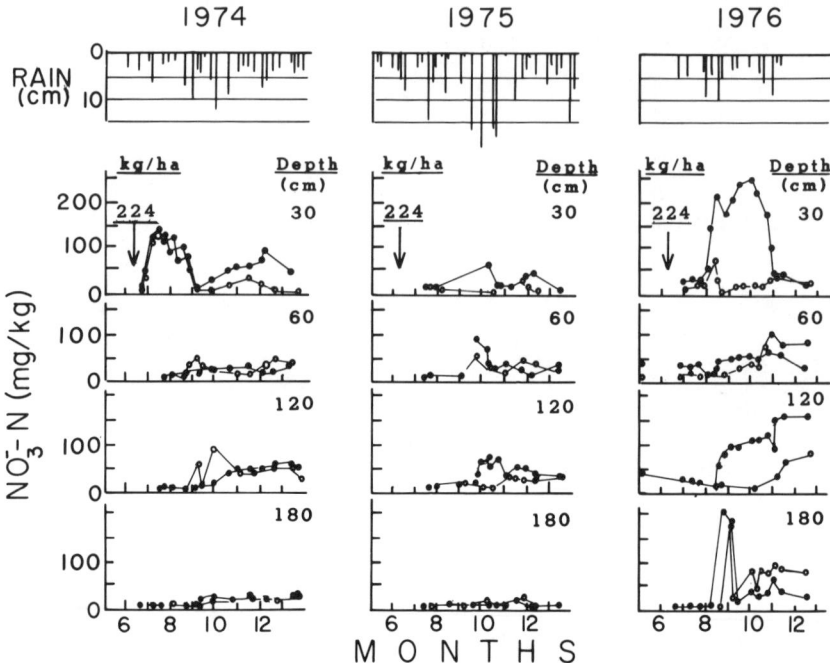

Fig. 2. Rainfall distribution with time and NO_3^--N distributions with soil depth and time beneath shade tobacco. Open and closed circles represent duplicate observations at each soil depth following application of organic-N fertilizer (240 kg ha^{-1}) each year. Derived from Barker, 1980.

stemming from annual applications of 224 kg ha^{-1} of organic-N fertilizer (primarily cotton seed meal) in each of 3 years is shown in Fig. 2. Rainfall (cm) distribution for each year is shown across the top of the figure. Open and closed circles represent the duplicate suction probe samples at each depth. These data illustrate the variability in time (within and between years) and space (horizontally and vertically) that is commonly observed in field measurements of NO_3-N movement in soils. Much of the temporal variability can be correlated with the observed pattern of rainfall. However, the wide spatial variability that was observed in duplicate probes cannot be easily reconciled with other parameters. Hence, with this experimental design and data base, it is extremely difficult to describe quantitatively the mass movement of N through the soil. Nevertheless, on these tobacco plots, it is apparent that NO_3-N moved readily in this soil in response to rainfall. In contrast, a similar study conducted on nearby plots but with a different crop, Kentucky bluegrass (*Poa pratensis* L.) rather than tobacco, resulted in very little movement of N below the 30 cm depth (Starr and DeRoo, 1981). Although several unknown factors may also have been involved, these two experiments on nearby plots indicate that different crops can have a large effect on the movement of N in soils. The specific and quantitative ways that different crops affect the fate of N in soils is not known. Clearly, there is a need to identify the time and space effects of different crops on the transformations and immobilization of soil N.

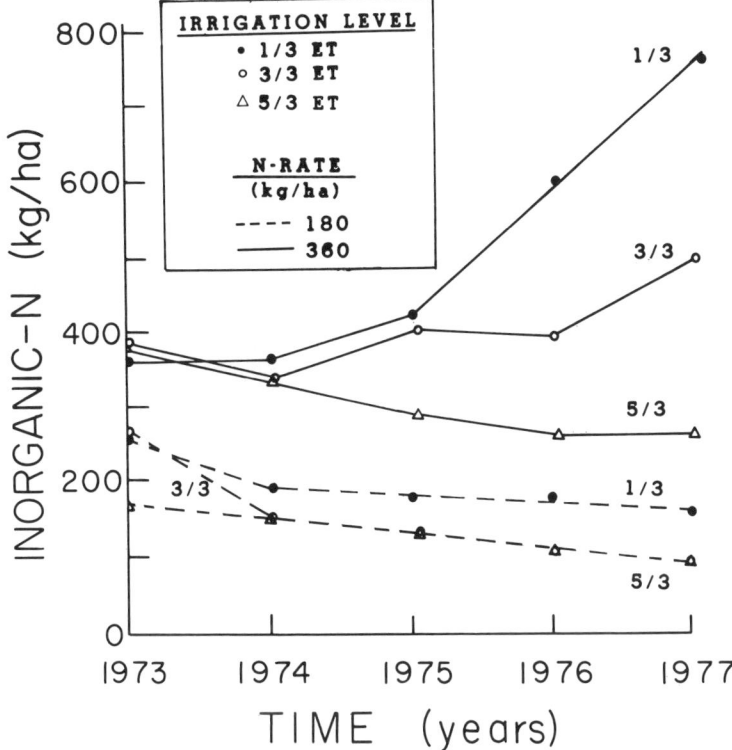

Fig. 3. Inorganic-N in 0–3 m depth of soil after crop harvest at three irrigation levels, where ——— and ---- represents the 360 and 180 kg ha^{-1} rates of fertilizer-N application, respectively. Derived from Pratt, 1979.

A multifaceted study undertaken at the Univ. of California to assess the movement of NO_3-N from irrigated lands (Pratt, 1979) overcomes some of the above problems associated with quantifying N movement in soils. One portion of this study was conducted on Yolo fine sandy loam (fine-silty, mixed, nonacid, thermic Typic Xerorthents) in which 48 field plots were instrumented to obtain in situ samples of soil water, soil water pressure, and soil water content with space and time. These plots were then subjected to four fertilizer-N rates, using ^{15}N depleted N fertilizers as a tracer for the added N, and three levels of irrigation, applied to corn (*Zea mays* L.) for 6 consecutive years. Irrigation quantities during the growing season were adjusted to apply 5/3, 3/3, and 1/3 of the estimated water loss by evapotranspiration (ET). Annual water inputs during the winter rainy season varied from 10 to 30 cm over the experimental period. Total amounts of inorganic-N in the soil profile, 0 to 3 m, at the end of each harvest year are shown in Fig. 3. These data represent the average value from eight soil cores, with coefficients of variation (CV) ranging from 4 to 50%. The crop appears to have been an efficient sink for much of the inorganic-N applied at the 180 kg ha^{-1} rate, resulting in only a slow linear decrease of inorganic-N remaining in the profile in the presumed absence of leaching, i.e., 1/3 ET irrigation rate. At the same deficit levels

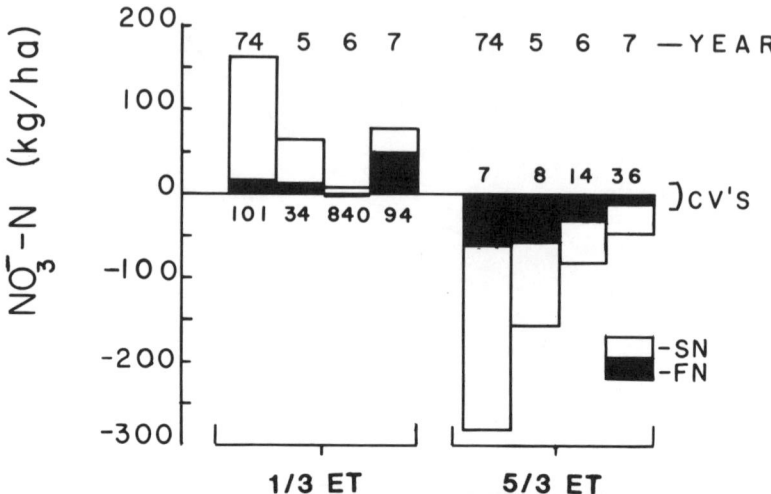

Fig. 4. Net annual inorganic-N flux past the 3 m depth suction samplers for plots with 360 kg ha^{-1} N applied, where SN and FN are soil- and fertilizer-N, respectively. Derived from Pratt, 1979.

of water addition, but at the 360 kg ha^{-1} N-rate, the inorganic-N increased linearly with time from 1975 to 1977 at about 180 kg ha^{-1} year^{-1}. In contrast, the 5/3 ET irrigation treatment for the high N rate addition, inorganic-N levels linearly decreased with time, approaching a quasi-constant level of inorganic-N in the soil profile from 1976–1977 that was about 180 kg/ha greater than that for the 180 kg ha^{-1} rate. The absence of net accumulation from year to year for the 5/3 ET and 360 kg ha^{-1} N-rate treatments suggests that this "excess" 180 kg/ha^{-1} was lost from the soil during the subsequent rainy season. However, N loss by leaching is not supported by the N flux data. The annual mass flux of NO$_3$-N past the 3-m depth N$_f$, was calculated by integrating daily averages of neutron probe, tensiometer and soil solution concentration data as (Pratt, 1979, p. 489):

$$N_f = \int_o^t c_i J_w dt, \qquad [5]$$

where J_w is the Darcy flux of soil water (cm^3 cm^{-2} h^{-1}) and c_i the concentration of nitrate-N in the soil solution.

A summary of the results of those calculations for the 360 kg ha^{-1} N rate are shown in Fig. 4. On the ordinate is represented the yearly net inorganic-N movement, past the 3 m depth [i.e., soil-N (SN) plus fertilizer-N (FN) for the 1/3 and 5/3 ET irrigation rates]. The sign of the ordinate indicates net upward movement as positive and net downward movement as negative. The variability of this data is shown by the CV for the total inorganic-N printed with each bar graph, which range in value from 7 to 840%. Note that the 840% value reflects large proportionate concentration changes which occurred in the soil solution from both downward and upward movement, past the 3 m depth, but with the net result being nearly zero—as shown here. In general, it may be seen that the net NO_3-N movement was always upward into the relatively dry soil profile of the 1/3 ET treatments and always downward for 5/3 ET treatment. The large fluxes observed for 1974 appear to be the result of residual SN which had accumulated in the soil profile prior to the initiation of the experiment. The net upward movement of FN for the 1/3 ET treatment may result from undetected N movement past the probes. This undetected movement could have occurred between sampling dates or as a result of large pore water movement past the samplers following saturated flow conditions at the soil surface (Bouma et al., 1977), followed by the slower upward movement in the smaller pores during the subsequent growing season. It may also reflect the errors associated with integrating products of calculation derived from measurements of soil solution and soil water content, Eq. [5], separated in both time and space.

The set of bar graphs associated with the 5/3 ET and the 360 kg ha^{-1} N-rate treatments is of particular interest. Recall that the inorganic-N for the 3-m profile (Fig. 3) showed that there was about 180 kg ha^{-1} year^{-1} of inorganic-N available for leaching during the winter rainy season. Yet Fig. 4 represents a logarithmic decrease in inorganic-N movement past the 3 m depth with increasing time. It is difficult to imagine a set of conditions that would result in a logarithmically increasing sink from year to year. Perhaps this too reflects the difficulty of sampling large pore water movement with suction probes as well as the separation in time and space of the various measurements needed to calculate N fluxes in soils.

Recent research by USDA and Univ. of Maryland researchers exemplifies some of the problems associated with gravimetric sampling techniques to quantify N movement in field soils. The history of these experiments has been reported by Bandel et al. (1975). Briefly, long-term field experiments were established to study the effects of two tillage systems on the fate of N in soils, using ^{15}N depleted fertilizer as a tracer for the nitrogenous fertilizer. The treatments for this study consisted of two tillage methods, four N rates, replicated four times at each of three field sites.

The primary movement of N through soils in this climatic region typically occurs during the fall-winter-spring period. To assess the magnitude of the N movement in these plots, soil samples were obtained in October, December, and May of 1980–1981. The soil sampling program was chosen to reflect most of the previous years sampling procedures, i.e., compositing four cores per plot. However, these cores were not composited in order to assess the variance structure associated with the sampling. The samples were analyzed for inorganic-N and for Cl$^-$ remaining in the soil from the NH_4NO_3 and KCl fertilizers broadcast in May 1980. The results and conclusions of this study are yet to be published.

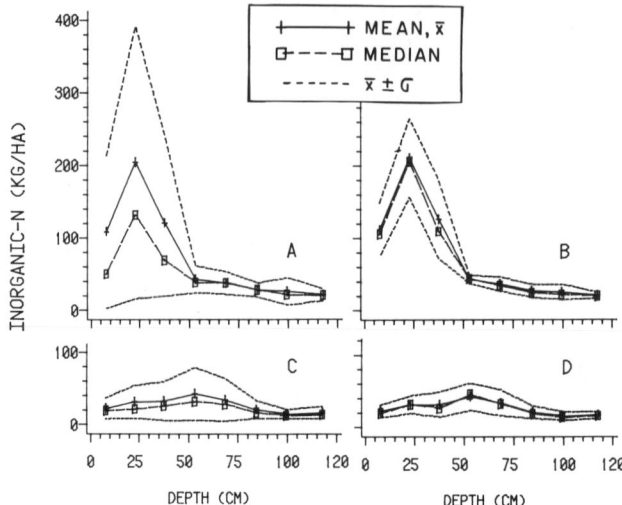

Fig. 5. Quantities of inorganic-N in successive 15 cm layers of soil under corn in December 1980, A and B, and in May 1981, C and D. The left half is derived from 16 uncomposited cores and the right half is derived from composited cores and four replicates.

However, partial results are shown in Fig. 5, in which the mass of inorganic-N in each 15-cm layer is plotted with soil depth. These data represent the results from the conventional till plots that received 270 kg ha^{-1} FN and is illustrative of the NO_3-N distribution in the soil profile from all plots receiving ≥ 135 kg ha^{-1} N. The four parts of this figure illustrates the effect of time of soil sampling (A, B = December; C, D = May), and sample handling (A, C = uncomposited samples; B, D = composited samples) on the apparent total amount of inorganic-N. Samples were mathematically composited by taking the mean value of the four cores in each plot before subsequent calculations.

Owing to the effect of compositing on the apparent frequency distribution and the standard deviation, compositing samples (Figs. 5A and 5B) has a direct effect on the conclusions that may be drawn. The composited samples not only show a much lower variability, but more importantly, the true frequency distribution of the raw data has been distorted (e.g., mean vs. median). It should also be noted that both effects (variability and skewing) are greatly reduced by the subsequent spring (Fig. 5C and 5D). Hence, no serious loss of information would have occurred from compositing the spring samples. It is apparent from this data that much information may be lost by the common practice of compositing soil samples before performing chemical analyses.

The characteristically high CV observed in Fig. 2, 3, 4, and 5 relate to the problem of determining the number of samples that are required to measure variables in the field within specified confidence limits. Standard techniques are available to determine the number of samples required for spatially independent and normally distributed observations (e.g., Harris et al., 1948). However, as noted above, the assumptions associated with these techniques, i.e., spatial independence and normal population distributions, are often invalid for field scale observation. Hence, for field experiments, these assumptions should be tested and verified before applying standard experimental methods and statistical procedures.

An experiment that illustrates an approach to testing these assumptions was conducted at the Univ. of California at Davis (Waynick, 1918). A fallow field site was chosen for its uniformity in texture, color, and topography. Soil samples were obtained from the 0 to 15 cm and 15 to 60 cm soil layer on transects in four directions on a 1.5 m interval spacing (Fig. 6). Analyses included NO_3^-N concentrations which had an overall CV of 26% for the 0 to 15 cm layer and 51% for the 15 to 60 cm layer. With the statistical analysis available at the time, Waynik was not able to assess the spatial dependence in his data. Hence, analysis of the variance structure as a function of distance (Eq. [4]), in each of the four directions was applied to his data as shown in Fig. 7. The variograms for the 0 to 15

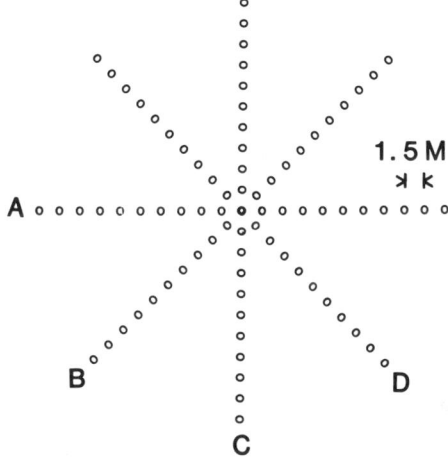

Fig. 6. Sampling pattern used to study spatial variation in nitrate distribution in a fallow soil (Waynick, 1918).

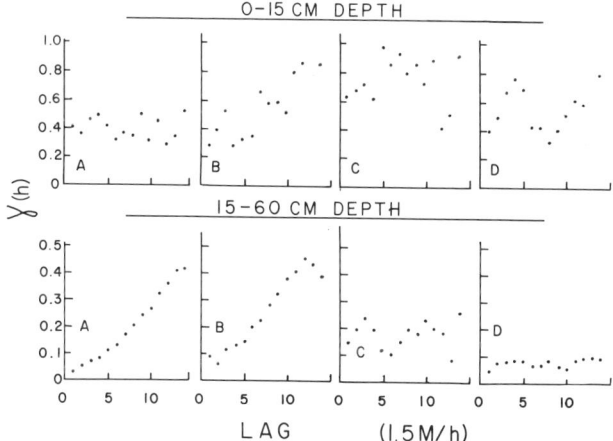

Fig. 7. Semivariograms for each soil depth and sampling transect. Derived from Waynick, 1918.

cm layer show little or no spatial dependence for this scale of observation (1.5 m) and length of transects (30 m). In contrast there is a strong spatial dependence for two of the transects in the 15 to 60 cm layer, up to a distance of 12 lags (about 18 m). The cause of the strong spatial dependence cannot be assessed from this analysis, even though it is clearly evident that the concentration of N in the subsoil is spatially and directionally dependent. Hence, this fact should be taken into account in subsequent experimental designs, sampling procedures and data analysis associated with this research site.

CONCLUSIONS

In summary, deterministic models provide a conceptual interpretation of N transformations and movement through soils. A major factor limiting the use of such models in the field is the lack of data which allow for stochastic inputs which can account for time and space variations in the parameters which are associated with the reaction-transport processes. Future improvements in understanding and predictive capability of the movement of solutes in general and N compounds in particular in soils will result from: (i) the development and application of better field sampling technology, especially for measuring interdependent variables that vary both in space and time (e.g., What is the number, size, and spacing of samples required for each variable of interest? Under what conditions is it appropriate to composite samples? How can interdependent variables be measured at precisely the same space/time coordinates? and How may solute movement in large and small soil pores be quantitatively measured?); (ii) testing for and verification of common a priori assumptions regarding the kinds and causes of variation that are observed in field studies (e.g., frequency distributions and spatial independency); (iii) an applied realization that understanding of natural systems requires interdisciplinary approaches to research; and (iv) a greater deversity in the use of analytical tools to assess observations that vary both in time and space, such as regionalized variable analysis.

LITERATURE CITED

Ardakani, M. S., J. T. Rehbock, and A. D. McLaren. 1973. Oxidation of nitrite to nitrate in a soil column. Soil Sci. Soc. Am. Proc. 37:53–56.

Bandel, V. A., Stanislaw Dzienia, George Stanford, and J. O. Legg. 1975. I. First-year results using unlabeled N fertilizer. Agron. J. 67:782–786.

Barker, A. V. (ed.) 1980. Efficient use of nitrogen on crop land in the northeast. Northeast Regional Research Pub. Project NE-39. Bull. 792. The Conn. Agric. Exp. Stn. New Haven.

Beek, J., and M. J. Frissel. 1973. Simulation of nitrogen behaviour in soils. Wageningen: Pudoc, 67 p.

Bouma, J., A. Jongerius, O. Boersma, A. Jager, and D. Schoonderbeek. 1977. The function of different types of macropores during saturated flow through four swelling soil horizons. J. Soil Sci. Soc. Am. 41:945–950.

Cho, C. M. 1971. Convective transport of ammonium with nitrification in soil. Can. J. Soil Sci. 51:339–350.

Clark, Isobel. 1979. Practical geostatistics. Applied Science Pub. LTD. London. 129 p.

Davidson, J. M., D. A. Greetz, P. Suresh, C. Rao, and H. Magdi Selim. 1978. Simulation of nitrogen movement, transformation, and uptake in plant root zone. U.S. Dep. of Commerce. National Technical Information Service Pub. No. PB-280712.

Davis, J. C. 1973. Statistics and data analysis in geology. John Wiley and Sons, N.Y. 550 p.

Delhomme, J. P. 1978. Kriging in the hydrosciences. Adv. Water Res. I:251–266.

Doner, H. E., M. G. Volz, and A. D. McLaren. 1974. Column studies of denitrification in soil. Soil Biol. Biochem. 6:341–346.

Frissel, M. J., and J. A. van Veen (ed.) 1981. Simulation of nitrogen behaviour of soil-plant systems. Wageningen: Pudoc 227 p.

Gardner, W. R. 1965. Movement of nitrogen in soil. *In* W. V. Bartholomew and F. E. Clark (ed.) Soil nitrogen. Agronomy 10:550–572.

Harris, Marilyn, D. G. Horvitz, and A. M. Mood. 1948. On the determination of sample sizes in designing experiments. J. Am. Stat. Assoc. 43:391–402.

Hattori, T. 1973. Microbiol life in the soil: An introduction. Marcel Dekker, Inc., N.Y. 427 p.

Journel, A. G., and Ch. J. Huijbregts. 1978. Mining geostatistics. Academic Press, London.

Kanwar, R. S., J. L. Baker, H. P. Johnson, and D. Kirkham. 1980. Nitrate movement with zero-order denitrification in a soil profile. Soil Sci. Soc. Am. J. 44:898–902.

Kirda, C., J. L. Starr, C. Misra, J. W. Biggar, and D. R. Nielsen. 1974. Nitrification and denitrification during miscible displacement in unsaturated soil. Soil Sci. Soc. Am. Proc. 38:772–776.

Matheron, G. 1971. The theory of regionalized variables and its applications. Les Cahiers du Centre de Morphologie Mathematique, Fasc. 5, CG Fontaine-bleau. 212 p.

McLaren, A. D. 1969. Steady state studies of nitrification in soil: Theoretical considerations. Soil Sci. Soc. Am. Proc. 33:273–275.

----. 1973. A need for counting microorganisms in soil mineral cycles. Environ. Lett. 5:143–154.

----. 1976. Rate constants for nitrification and denitrification in soils. Radiat. Environ. Biophys. 13:294–299.

Misra, C., D. R. Nielsen, and J. W. Biggar. 1974. Nitrogen transformation in soil during leaching: I. Theoretical considerations. Soil Sci. Soc. Am. Proc. 38:289–293.

Nielsen, D. R., J. W. Biggar, and Y. Barrada. 1979. Water and solute movement in field soils. p. 165–183. *In* Isotopes and radiation in research on soil-plant relationships. International Atomic Energy Agency, Vienna.

----, and J. G. MacDonald (ed.) 1978. Nitrogen in the environment. Vol. 1. Academic Press, N.Y. 526 p.

----, P. J. Wierenga, and J. W. Biggar. 1982. Nitrogen transport processes in soil. *In* F. J. Stevenson (ed.) Nitrogen in agricultural soils. Agronomy 22:423–448. Am. Soc. of Agron., Soil Sci. Soc. of Am., and Crop Sci. Soc. of Am., Madison, Wis.

----, P. J. Wierenga, and J. W. Biggar. 1983. Spatial soil variation and mass transfers from agricultural soils. p. 65–78. *In* D. W. Nelson, K. K. Tanji, and D. E. Elrick (ed.) Chemical mobility and reactivity in soil systems. Am. Soc. of Agron. and Soil Sci. Soc. of Am. Spec. Pub. No. 11, Am. Soc. of Agron., Madison, Wis.

Olea, R. A., 1977. Measuring spatial dependance with semivariograms: Kansas Geol. Survey, Series on Spatial Analysis No. 3, Univ. Kansas, Lawrence, Kans. 29 p.

Parlange, J.-Y., J. L. Starr, D. A. Barry, and R. D. Braddock. 1982. A theoretical study of the inclusion of dispersion in boundary conditions and transport equations for zero-order kinetics. Soil Sci. Soc. Am. J. 46:701–704.

Phillips, R. E., K. R. Reddy, and W. H. Patrick. 1978. The role of nitrogen diffusion in determining the order and rate of denitrification in flooded soil: II. Theoretical analysis and interpretation. Soil Sci. Soc. Am. J. 42:272–278.

Pratt, P. F. 1979. Nitrate in effluents from irrigated land. Final report to National Science Foundation, Univ. of Calif. 822 p.

Rao, P. S. C., R. E. Jessup, and A. G. Hornsby. 1982. Simulation of nitrogen in agro-ecosystems: criteria for model selection. Plant Soil 67:35–43.

Stanford, G., R. A. VanderPol and S. Dzienia. 1975. Denitrification rates in relation to total and extractable soil carbon. Soil Sci. Soc. Am. Proc. 39:284–289.

Starr, J. L., F. E. Broadbent, and D. R. Nielsen. 1974. Nitrogen transformations during continuous leaching. Soil Sci. Soc. Am. Proc. 38:283–289.

----, and H. C. DeRoo. 1981. The fate of nitrogen fertilizer applied to turfgrass. Crop Sci. Soc. Am. 21:531–536.

----, and J.-Y. Parlange. 1975. Nonlinear denitrification kinetics with continuous flow in soil columns. Proc. Soil Sci. Soc. Am. 39:875–880.

----, and ----. 1976. Relation between the kinetics of nitrogen transformation and biomass distribution in a soil column during continuous leaching. Soil Sci. Soc. Am. J. 40:458–460.

Steel, R. G. D., and J. H. Torrie. 1980. Principles and Procedures of Statistics. McGraw-Hill Book Co., N.Y. 666 p.

Stevenson, F. J. (ed.) 1982. Nitrogen in agricultural soils. Agronomy 22. Am. Soc. of Agron., Soil Sci. Soc. of Am., and Crop Sci. Soc. of Am., Madison, Wis.

Tanji, K. K. 1982. Modeling of the soil nitrogen cycle. In F. J. Stevenson (ed.) Nitrogen in agricultural soils. Agronomy 22:721–772. Am. Soc. of Agron., Soil Sci. Soc. of Am., Crop Sci. Soc. of Am., Madison, Wis.

van Genuchten, M. Th., and W. J. Alves. 1982. Analytical solutions of the one-dimensional convective-dispersive solute transport equation. Technical Bull. 1661. USDA, ARS. 149 p.

Wagenet, R. J. 1983. Principles of salt movement in soils. p. 123–140. In D. W. Nelson, K. K. Tanji, and D. E. Elrick (ed.) Chemical mobility and reactivity in soil systems. Am. Soc. of Agron. and Soil Sci. Soc. of Am. Spec. Pub. No. 11, Am. Soc. of Agron., Madison, Wis.

----, J. W. Biggar, and D. R. Nielsen. 1976. Analytical solutions of miscible displacement equations describing the sequential microbiological transformations of urea, ammonium and nitrate. Water Science and Engineering Papers No. 6001. Univ. Calif., Davis. 53 p.

Waynick, D. D. 1918. Variability in soils and its signifiance to past and future soil investigations. I. A statistical study of nitrification in soils. Agric. Sci. 3:243–270.

Chapter 7

The Movement of Phosphorus in Soil[1]

CARL G. ENFIELD AND ROSCOE ELLIS, JR[2]

Eutrophication of lakes and streams is a natural process which can be accelerated by the movement of nutrient rich water into waterways. Phosphorus has been identified as the nutrient most likely limiting primary productivity in lakes and streams. A prerequisite to understanding the impact of land application wastewater treatment or the environmental impact of P fertilization is an assessment of the interaction of P and soil constituents.

The soil is a dynamic chemical and biological system in a state of constant flux. To describe this complex system, chemical forms of P and associated soil constituents are discussed first and then the kinetics of the reaction are considered. After describing the soil P interaction, P transport is compared to measured P activity in dynamic systems.

There are many forms of P which may be applied to the soil ranging from complex organics to simpler orthophosphates. Mineralization of many of the P compounds appear to be reasonably rapid (Blanchar and Hossner, 1969). The discussion presented here will be limited to the reaction and movement of orthophosphate.

---

[1] Although the research described in this article has been funded wholly or in part by the United States Environmental Protection Agency through in-house research, it has not been subjected to the Agency's required peer and policy review and therefore does not necessarily reflect the views of the Agency and no official endorsement should be inferred.

[2] U.S. Environmental Protection Agency, Ada, Okl.; deceased, respectively.

Copyright © 1983 ASA, SSSA, 677 South Segoe Road, Madison, WI 53711. *Chemical Mobility and Reactivity in Soil Systems.*

CHEMICAL STABILITY MODELS

Most inorganic phosphates found in the soil can be classified into three groups: (i) those containing calcium phosphates; (ii) those containing iron and aluminum phosphates; and (iii) those combining with silicate materials. The relative importance of these compounds can be roughly correlated to pH of the soil environment (Buckman and Brady, 1970). In acid soils, iron and aluminum phosphates are more common. In basic soils, calcium phosphates predominate.

By studying the solubility products of some of the more important iron, aluminum, and calcium compounds commonly found in soil, it is possible to quantitatively project stability diagrams for the equilibrium concentration of P in soil solution.

The equilibrium isotherms, plotted in Fig. 1 for selected P compounds, were developed from Gibbs free energies given in Table 1. The values selected for the Gibbs free energies were from single source (Sadiq and Lindsay, 1979) so that they would be internally consistent with each other. There is a wide range in the values reported in the literature, and the projected concentrations could range several orders of magnitude if the reported extremes were used. For this reason, using data that is internally consistent is important. Several assumptions were made in calculating Fig. 1: (i) calcium was assumed to be 0.01 molar; (ii) the ionic

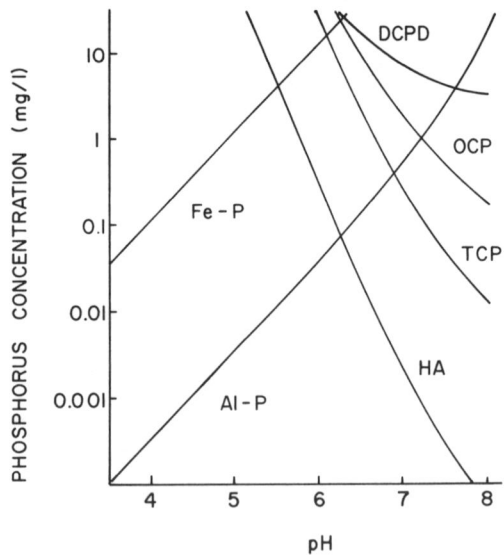

Fig. 1. Stability diagram for selected phosphate compounds. Fe-P is the stability line for strengite (FePO$_4$·2H$_2$O) in equilibrium with Fe(OH)$_3$. Al-P is the stability line for variscite (AlPO$_4$·2H$_2$O) in equilibrium with amorphous Al(OH)$_3$. The calcium phosphates dicalcium phosphate dyhydrate (DCPD) Ca(H$_2$PO$_4$)$_2$·H$_2$O, octacalcium phosphate (OCP) Ca$_8$H$_2$(PO$_4$)$_6$·5H$_2$O, tricalcium phosphate (TCP) Ca$_3$(PO$_4$)$_2$, and hydroxyapatite (HA) Ca$_{10}$(OH)$_2$(PO$_4$)$_6$ are calculated for 0.01 M Ca and an ionic strength of 0.026 mol L^{-1}.

strength was assumed to be 0.026 mol L^{-1} (ionic strength assumed equalled 0.013 times the electrical conductivity in mmho cm^{-1} Griffin, R. A. and J. J. Jurnak, 1973); (iii) variscite (AlPO$_4 \cdot$2H$_2$O) was assumed to be in equilibrium with amorphous aluminum hydroxide [Al(OH)$_3$]; (iv) strengite (FePO$_4 \cdot$2H$_2$O) was assumed to be in equilibrium with iron hydroxide [Fe(OH)$_3$]; and (v) in the pH range of interest (HPO$_4^{2-}$) and (H$_2$PO$_4^-$) were the only phosphate species significantly affecting the P activity. When possible, it would be better to know the activity of the iron, aluminum, calcium, and the oxidation state rather than estimate activity from an assumed associated compound. The stability model can then be used to estimate the equilibrium concentration of P in soil solution when pH and activities of Ca^{2+}, Fe^{2+}, Fe^{3+}, and Al^{3+} are known.

Table 1. Reported Gibbs free energy in Kcal mol^{-1} of selected compounds.†

	ΔG_F°
Al^{3+}	−117.33
AlOH^{2+}	−167.17
Al(OH)$_2^+$	−218.02
Al(OH)$_3^0$	−266.94
Al(OH)$_3$ (amorphous)	−274.21
α-Al(OH)$_3$ (bayerite)	−275.78
γ-Al(OH)$_3$ (gibbsite)	−276.43
Al(OH)$_3$ (nordstrandite)	−276.3
AlPO$_4$ (berlinite)	−388.50
AlPO$_4 \cdot$2H$_2$O (variscite)	−505.97
Ca^{2+}	−132.52
CaCO$_3^0$	−263.00
CaCO$_3$ (calcite)	−270.18
CaPO$_4^-$	−386.51
CaHPO$_4^0$	−398.29
CaH$_2$PO$_4^+$	−406.28
CaHPO$_4$ (monetite)	−403.96
CaHPO$_4 \cdot$2H$_2$O (brushite)	−516.89
Ca(H$_2$PO$_4$)$_2 \cdot$H$_2$O (c)	−734.48
α-Ca$_3$(PO$_4$)$_2$ (c)	−922.70
β-Ca$_3$(PO$_4$)$_2$ (tricalcium phosphate)	−927.37
Ca$_8$H$_2$(PO$_4$)$_6 \cdot$5H$_2$O (octacalcium phosphate)	−2942.62
Ca$_{10}$(OH)$_2$(PO$_4$)$_6$ (hydroxyapatite)	−3030.24
FeOH$^+$	−69.29
FeOH^{2+}	−57.72
Fe(OH)$_2$ (c)	−117.58
Fe(OH)$_3$ (amorphous)	−169.25
Fe(OH)$_3$ (soil)	−170.40
α-FeOOH (geothite)	−117.42
FePO$_4 \cdot$2H$_2$O (strengite)	−398.59
Fe$_3$(PO$_4$)$_2 \cdot$8H$_2$O (vivianite)	−1058.36
PO$_4^{3-}$	−245.18
HPO$_4^{2-}$	−262.03
H$_2$PO$_4^-$	−271.85
H$_3$PO$_4^0$	−274.78
H$_2$O (l)	−56.69

† From Sadiq and Lindsay (20).

SORPTION-DESORPTION MODELS

There is confusion as to the terminology used in the literature describing the loss of P from solution. Some choose to describe their data using solubility product theory, as above, (van Riemsdijk et al., 1975; Bennett and Adams, 1976; Subbarao and Ellis, 1977; Lindsay, 1979; and Hsu, 1979) assuming a precipitation type of process. Other researchers (Holford et al., 1974; Shayan and Davey, 1978; and Sibbesen, 1981) attempt to describe the process using sorption isotherms. The most commonly used sorption expression for P is the Langmuir equation (Langmuir, 1918) which can be written in its simplest form as

$$\frac{C}{S} = \frac{1}{bS_m} + \frac{C}{S_m},\qquad [1]$$

where:
- C = solution concentration of P (mg per liter),
- S = adsorbed phosphate at concentration C (micrograms P per gram soil),
- S_m = maximum adsorption which is equivalent to a monolayer of phosphate on the adsorbing surface (micrograms P per gram soil), and
- b = a constant.

The second most common equation is the Freundlich equation

$$S = KC^n,\qquad [2]$$

where:
- K = constant and
- n = constant.

The Freundlich equation has been shown to be equivalent to a multiple layer Langmuir equation (Brunauer and Copeland, 1967). Both of the methods have successfully described the disappearance of P in soil after a given time for equilibration (Berkheiser et al., 1980). Both of these equations, as generally written, assume P reactions are instantaneous and the soil is at equilibrium. If these equations alone were used to predict the movement of P in soil, one would expect P movement to appear as if it were moving in a chromatographic column. The chemical, when applied, would move through the soil as a slug. This type of description does not appear consistent with most reported column data in literature or batch sorption data which shows the influence of equilibrating time.

Consider the sorption data of Stuanes (personal communication, 1981) shown in Fig. 2. The data presented is after 24 h of equilibration. The soil suspension was then centrifuged and the equilibrating solution changed. The amount of sorption was calculated by measuring the reduction in P in the supernatant liquid. The process was repeated 10 times to evaluate both sorption and desorption. The data show a nearly irreversible reaction. Equations [1] and [2] are generally written as if the system remains in local equilibrium. One can hypothesize, from the data in Fig.

2, that either sorption isotherms, as they are generally written, are not adequate for describing P sorption or one must also include the kinetics of the reaction and assume the experimental data of Stuanes do not represent an equilibrium condition to have a useable model.

Stuanes' data do not conflict with equilibrium solubility models since the mass of sorption or desorption will be different depending upon initial conditions. However, Fig. 3 shows the timed response in a batch equilibration study to several applications of P. The system depicted in Fig. 3

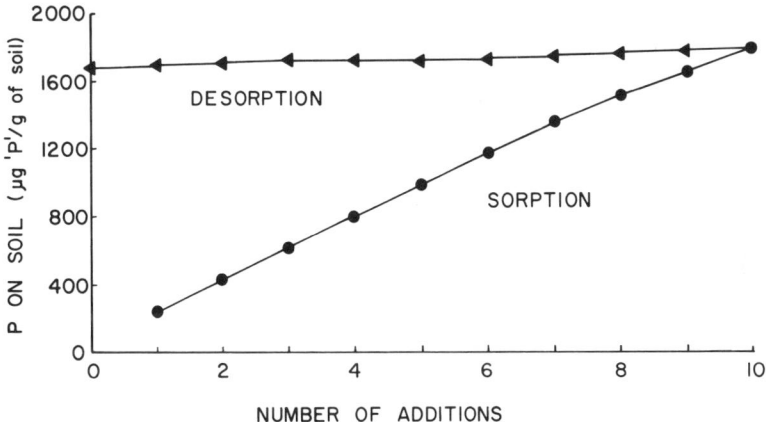

Fig. 2. Successive sorption, desorption of phosphorus by a single soil sample (personal communication Dr. Arne Stuanes, Norwegian Forest Research Institute, Aas, NLH 1981).

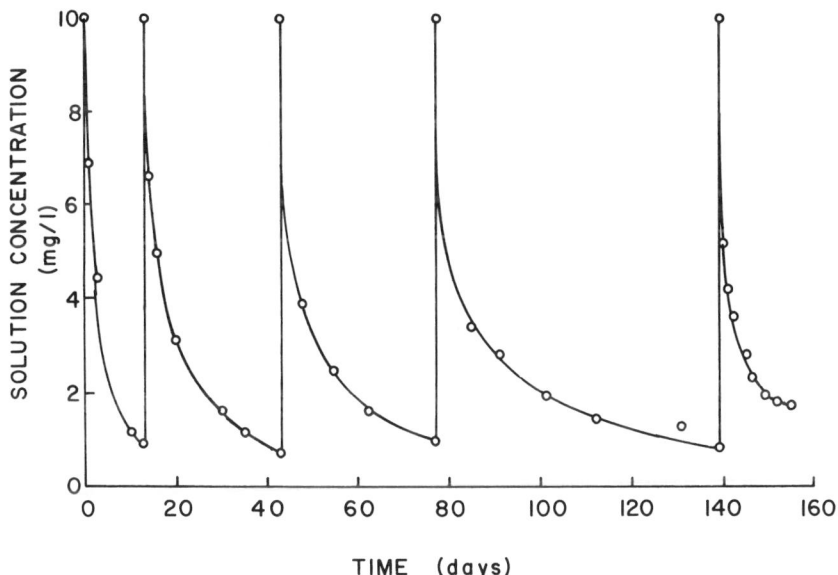

Fig. 3. Timed response for sorption of P by a single soil sample.

does not appear to be at equilibrium. This indicates the solubility product approach would also need to include a kinetic term to adequately describe P movement in soil profiles.

KINETIC MODELS

The disappearance of P from soil solution occurs as a fast initial reaction followed by a much slower reaction. Investigators believe sorption is the primary process during the initial rapid reaction, and precipitation to relatively insoluble phosphate controls the slow reaction (Sawhney, 1977). The quick response takes place in the first few hours followed by a gradual change for a long period of time.

Methods of describing this response varies widely among researchers and there appears to be little agreement as to the appropriate approach. One group will describe the response assuming sorption processes control the reaction; another group believes the process is controlled by the stability of the chemicals and the rate of nucleation or crystal growth. A third group attempts to describe the process empirically. Most researchers who use sorption isotherms base their description on the assumption that at least a portion of the sorption sites are not at equilibrium. De Camargo et al. (1979) describe their assumption as two types of sites: Type I sites are assumed to be in equilibrium following the equation

$$\frac{\partial S_1}{\partial t} = \frac{\Theta}{\varrho} K_1 C^n - K_2 S_1. \qquad [3]$$

Type II sites are assumed to not be at equilibrium following the equation

$$\frac{\partial S_2}{\partial t} = \frac{\Theta}{\varrho} K_3 C^n - K_4 S_2, \qquad [4]$$

where S_1 and S_2 the amount of P adsorbed at the respective sites, K_1 and K_2 are the forward and backward rate coefficients for Type I sites and K_3 and K_4 the forward and backward rate coefficients for Type II sites, Θ is the volumetric water content ($cm^3 cm^{-3}$), ϱ is the soil bulk density (g cm^{-3}), and t is time (h). To be at equilibrium, $\partial S_1/\partial t$ must equal zero. Equation [3] then becomes the same as a Freundlich equation $S_1 = KC^n$. The retardation of the pollutant as pointed out by de Camargo et al. (1979) will be $R = 1 + (\varrho n K C^{n-1}/\Theta)$. Overman et al. (1978) used a similar approach but assumed the reaction followed a linear equilibrium condition rather than a Freundlich equation.

Several investigators (Novak et al., 1975; Mansell et al., 1977a; and Enfield and Bledsoe, 1975) assumed equilibrium could be described by either the Langmuir equation or the Freundlich equation, but the system had not yet reached equilibrium. Novak et al. (1975) assumed the mass transfer is proportional to a concentration driving force $(C - C^*)$ where C^* is the equilibrium condition described by the Langmuir equation; the mass transfer rate is then described using first-order kinetics as

$$\frac{\partial S}{\partial t} = k(C - C^*) \qquad [6]$$

where k is the instantaneous rate coefficient (h^{-1}). Mansell et al. (1977b), on the other hand, used equilibrium described by a Freundlich equation. Enfield et al. (1976) assumed the process was diffusion limited following the equation

$$S_{avg} = F(C) - \frac{6 F(C)}{\pi^2} \sum_{j=1}^{n} \frac{1}{j^2} \exp(-\varkappa j^2 \pi^2 t/r^2), \qquad [7]$$

where:
  $F(C)$ = Langmuir or Freundlich equilibrium isotherm,
  S_{avg} = average concentration of P in the soil particle,
  $\varkappa$ = diffusion coefficient, and
  r = radius of assumed spherical soil particle where phosphorus is being sorbed.

Several of the approaches proposed by researchers (Novak et al., 1975; Sawhney, 1977; Mansell et al., 1977a; Mansell et al., 1977b; Overman et al., 1978; Fiskell et al., 1979; Enfield et al., 1975; and Enfield et al., 1976) may be adequate to describe the kinetics of the sorption process but do not appear to be adequate to describe the desorption process as presented in Fig. 2.

Enfield et al. (1981b) took the approach that sorption was an instantaneous reversible reaction following the Langmuir equation but the dominant factor controlling the reaction kinetics was the rate of nucleation and crystal growth considering solubility product theory. Their equation describing the kinetic reaction in soil was

$$S = \frac{S_m b C}{1 + b C} + \sum_{i=1}^{n} \int_0^t a_i (C - CE_i) \, dt, \qquad [8]$$

where:
  a_i = rate coefficients,
  CE_i = equilibrium concentration for a specific compound, and
  i = an index for the numerous compounds which might form.

As the equation is written, several compounds could be forming simultaneously each following first-order kinetics. Formation of several compounds at a given time has been identified recently by Freemand and Rowell (1981). As the equilibrating solution concentration goes down, the number of compounds forming would be reduced with an associated disassociation of the more soluble compounds. The equation would project the concentration to fall in steps; for example, going from DCPD-OCP-TCP-HA where the solution would appear to stay near the DCPD line if the sum of the rates of formation of OCP, TCP, and HA were less than the rate of dissolution of DCPD. This type of response has been reported by Subbarao and Ellis (1977).

TRANSPORT MODELS

The general equation describing the process of simultaneous diffusion, hydrodynamic dispersion mass flow, and sorption of a reactive solute in a soil in one dimension is

$$\frac{\partial C}{\partial t} = D \frac{\partial^2 C}{\partial X^2} - V \frac{\partial C}{\partial X} - \frac{\varrho}{\Theta} \frac{\partial S}{\partial t}, \qquad [9]$$

where C is the concentration of P in the soil solution (mg L^{-1}), t is time (h), D is a coefficient combining diffusion and hydrodynamic dispersion (cm^2 h^{-1}), V is the interstitial velocity (cm h^{-1}) which is the flux of solution divided by the volumetric water content Θ (cm^3 cm^{-3}), ϱ is the soil bulk density (g cm^{-3}), S is the amount of P sorbed (refers to the total amount adsorbed and/or precipitated) per unit mass of soil (μg g^{-1}), and X is the coordinate along the flow path (cm). There have been numerous solutions to this equation for a variety of boundary conditions and rate terms. Three typical solutions will be examined here. Assumptions for the first solution are:

(1) Θ, ϱ, D, and temperature remain constant during the experiment;
(2) the sorption-desorption processes remain in local equilibrium following the linear relationship $S_1 = k_d C$;
(3) an irreversible precipitation process takes place only when it is in the presence of the solid phase following first-order kinetics with respect to concentration;

$$S_2 = \int_0^t a\, C\, dt, \qquad [10]$$

where a is the instantaneous rate coefficient (h^{-1})
(4) the soil column behaves as if it were infinitely long with constant water velocity; and
(5) the total sorption is the sum of the reversibly sorbed and irreversibly precipitated following the equation

$$S = S_1 + S_2.$$

The initial and boundary conditions for the models considered are:

$C(x = 0, t) = g(t)$
$C(x = 0, t) = 0$
$C(x = 0, t = 0) = 0$
$C(x = \infty, t) = 0$

One solution has been proposed as

$$\frac{C}{C_T} = \frac{1}{2}\left(\text{erf}\, \frac{X + X_o - Vt/R}{2\sqrt{D/R}} - \text{erf}\, \frac{X - Vt/R}{2\sqrt{D/R}}\right), \qquad [11]$$

where:
- erf = error function,
- $C_T = C_o \exp(-\varrho ta/\Theta R)$,
- C_o = initial concentration (mg L^{-1}),
- R = retardation factor = $1 + \varrho/\Theta\, k_d$,
- X_o = depth of an equivalent length of P describing the applied chemical $X_o = t_o V/R$, and
- t_o = time P applied at surface (h).

The second approach is a numerical solution to Eq. [9] developed with appropriate boundary conditions by Mansell et al. (1977b) where:

$$\frac{\partial S}{\partial t} = K_r (K\, C^N - S), \qquad [12]$$

where $K = \dfrac{\Theta}{\varrho}\dfrac{k_a}{k_d}$.

The parameters k_a and k_d are the forward and backward rate coefficients (h^{-1}). The third model considered follows the kinetic Eq. [8] which was solved using the method of characteristics (Enfield et al., 1981a).

RESULTS AND DISCUSSION

The models were tested using data already in the literature. Model coefficients were developed where necessary when they were different from those presented in the literature. Figure 4 is the sorption data pre-

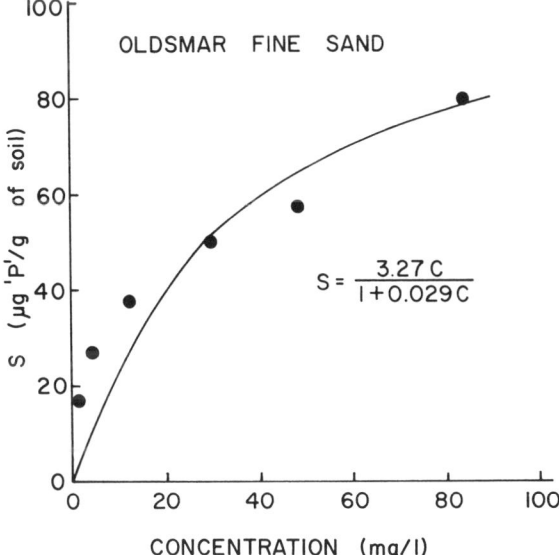

Fig. 4. Sorption isotherm for Oldsmar fine sand after 24 of equilibration adapted from Mansell et al., 1977.

sented by Mansell et al. (1977b) for Oldsmar fine sand (sandy, siliceous, hyperthermic alfic Arenic Haplaquod) Al horizon. Figure 5 shows an adaptation of the kinetic data from the same soil sample. Ignoring the initial rapid sorption, the rate of reaction appears to be approximately $0.02\ h^{-1}$. Figure 6 shows experimental data of Mansell et al. (1977b) for a 10 cm column of soil. The experimental conditions are described in Tables 2 and 3. The projection in Fig. 6 is based on Freundlich sorption which has not come to equilibrium (Eqs. [12] and [9]) after the development of

Table 2. Physical parameters for the soil columns.

Soil	Interstitial velocity cm/h	Water content (Θ) (cc/cc)	Bulk density ϱ (g/cm³)	Application rate cm/h	Time of "slug" h	Dispersion coefficient cm²/h	Initial conc. mg/L	Column length cm
Oldsmar†	9.27	0.34	1.39	3.15	10.4	2.92	100	10
Lakeland††	1.0	0.26	1.5	0.3	408	0.8	9.2	5

† Data adopted from Mansell et al., 1977b.
†† Data adopted from Enfield et al., 1981a.

Table 3. Chemical parameters for the soil columns.

Soil	K_d†	Langmuir Max µg/g	Langmuir b	Rate h^{-1}	Freundlich n	Freundlich K
Oldsmar	0.43	112.9†	0.029‡	0.022‡	0.345	17
Lakeland	8.1	86§	0.41§	0.006§		

† Calibrated value based on correct breakthrough time not related to batch sorption studies.
‡ Calculated from 7-day adsorption data and kinetic data reported by Mansell et al., 1977b.
§ Reported by Enfield et al., 1981a.

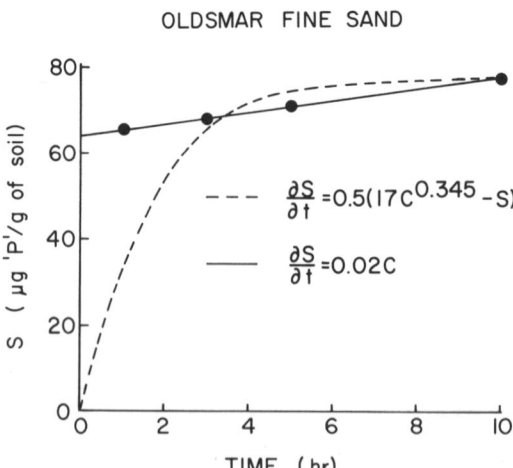

Fig. 5. Kinetics of sorption for Oldsmar fine sand with an initial solution concentration of 100 mg/l adapted from Mansell et al., 1977.

Mansell. The projections in Fig. 7 are for the model projections assuming Langmuir sorption and first-order precipitation where it is assumed the concentration of P at equilibrium will be 0.01 mg L^{-1} and dispersion negligible (Eq. [8] and [9]). Two different precipitation rate coefficients

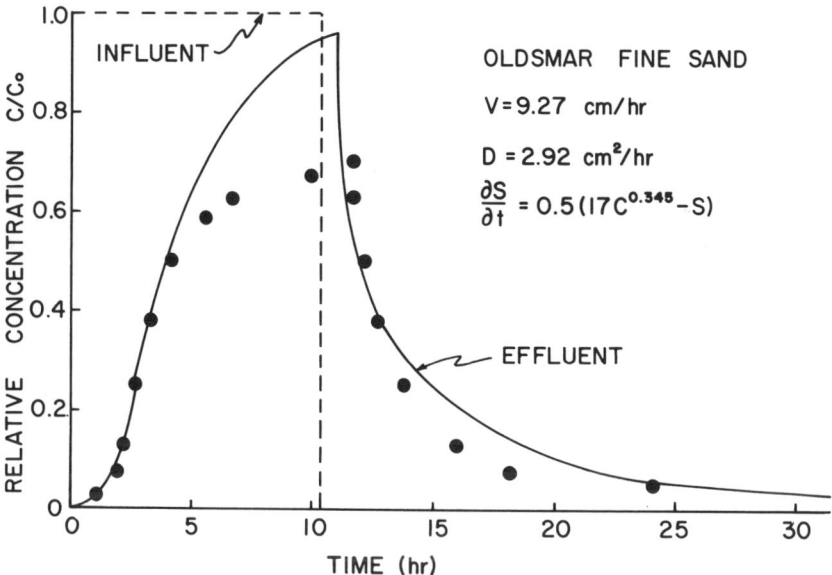

Fig. 6. Measured and projected breakthrough curves assuming Freundlich sorption which had not reached equilibrium adapted from Mansell et al., 1977.

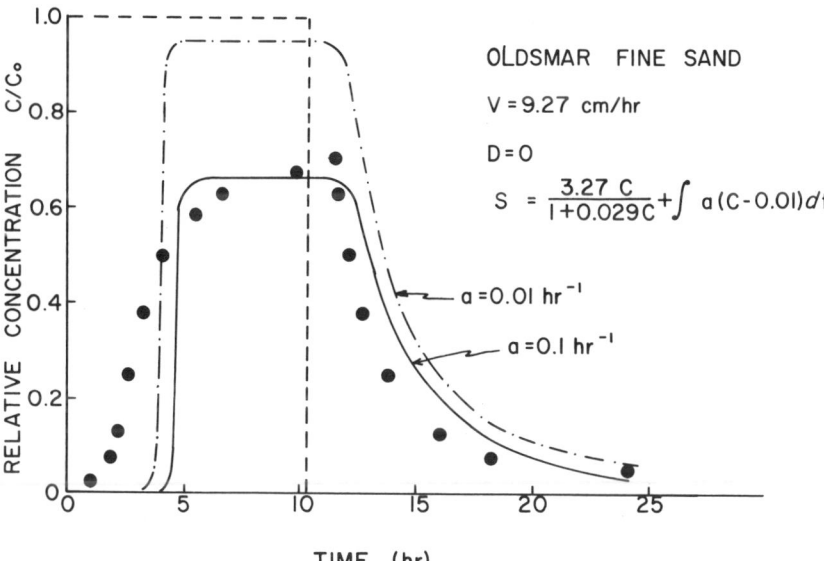

Fig. 7. Measured and projected breakthrough curve assuming instantaneous Landmuir sorption and chemical precipitation and dissolution.

are used. When the projection is based on the rate observed in the batch sorption study, one projects the concentration to be greater than observed. The fast rate (0.1 h^{-1}) is a calibrated value to make the projection look as close as possible to experimental data. Figure 8 shows the same experimental data but assumes linear sorption and first-order dissipation or irreversible precipitation. The linear partition coefficient k_d was calibrated to the column data and not related to any sorption study. Two curves are projected in Fig. 8: (i) the case with no dispersion and (ii) dispersion as reported by Mansell. The rate of irreversible precipitation was 0.1, as used in Fig. 7.

If we define breakthrough time as the time required to reach one-half of the quasi-steady-state effluent concentration, the experimental breakthrough time is approximately 3 h. The model used in Fig. 6 does an excellent job of estimating this point. If the Langmuir isotherm was used with no precipitation, the breakthrough time for the experimental conditions would have been 3.7 h. The overestimation of the breakthrough time is likely due to using batch data after 24 h of equilibration to develop the Langmuir coefficients. Adding precipitation increases the breakthrough time only a small amount in the range being considered in Fig. 7. Adding dispersion, as demonstrated in Fig. 8, greatly improves the shape of the breakthrough curve but does not effect the breakthrough time.

Figure 9 is for a different soil than presented in the three previous figures. Lakeland soil sample (thermic, coated Typic Quartzipsamments) has a greater partition coefficient than the Oldsmar sample. The initial applied solution concentration and interstitial water velocity was about 0.1 times that applied to the Oldsmar sample. Even with these large

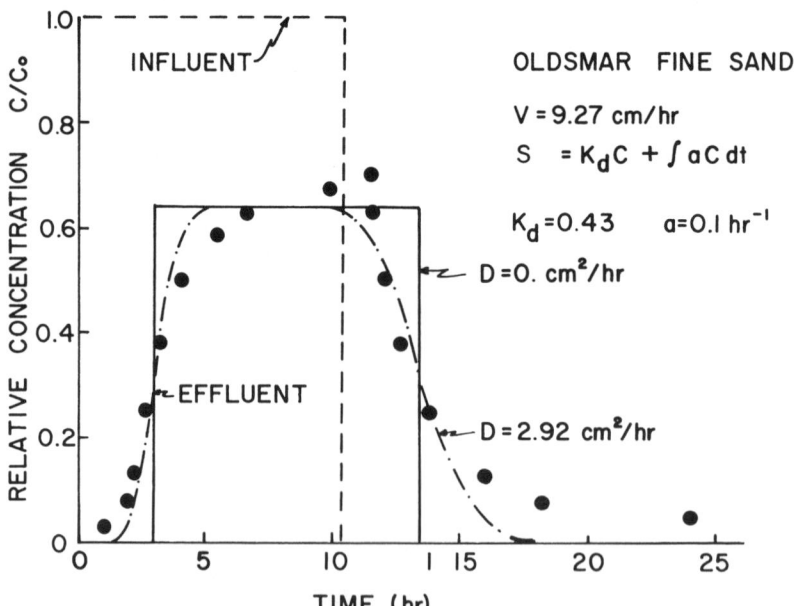

Fig. 8. Measured and projected breakthrough curves assuming instantaneous linear sorption and irreversible precipitation.

differences in input parameters, the modeling results are comparable. At the lower flow rate, the rate of precipitation calculated from the batch sorption studies appears to describe the column breakthrough data much more precisely than was observed at high flow rate. The models appear to give comparable results. There are two significant differences that should be delineated:

(1) When irreversible precipitation is assumed, the solution concentration returns to zero shortly after the pulse of P passes, while with reversible precipitation the concentration is not zero until all the P precipitated is returned to solution, and

(2) The projected time for breakthrough was adequately described using batch studies for the Freundlich or Langmuir equation but the linear sorption coefficient was calibrated to the column data. This simplification allows using a less operationally complex model but considerable experience is needed to understand the consequences of this simplification.

A limitation of the model presented based on Langmuir sorption is its lack of considering the effect of dispersion. This is not a significant problem if one is considering the response of a large quasi-steady-state addition of P as with waste water applications to the land. However, when one is applying an application of P similar to that in normal fertility practice, where the concentration is high for a very short period of time, the influence of dispersion on the concentration of P in the soil solution is dramatic (Enfield and Carsel, 1980). If the activity in the soil solution controls the plant availability rather than the amount in the solid phase, the Langmuir model (Eq. [8] and [9]) should include dispersion to be effective for most agricultural situations.

The term in either model attempting to describe precipitation which appears to be most significant in estimating solution concentration is the rate coefficient. There have been attempts to relate this coefficient to other readily measurable parameters. Figure 10 shows a collection of these results as a function of pH. Enfield et al. (1981b) showed pH vs. rate

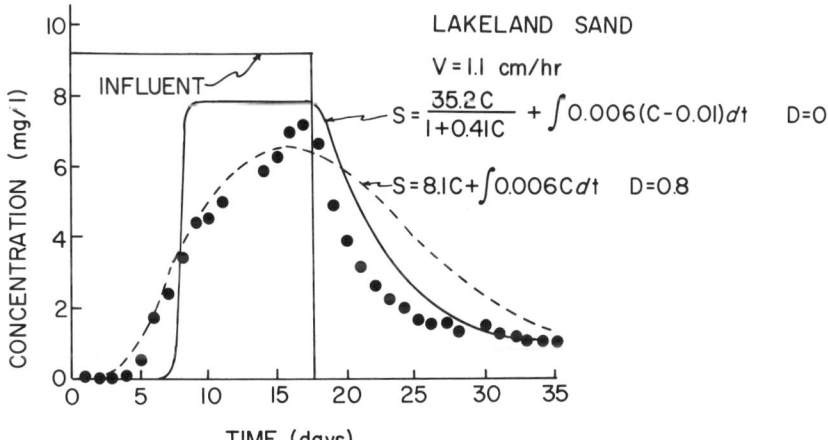

Fig. 9. Measured and projected breakthrough curves assuming either Langmuir or linear sorption plus precipitation.

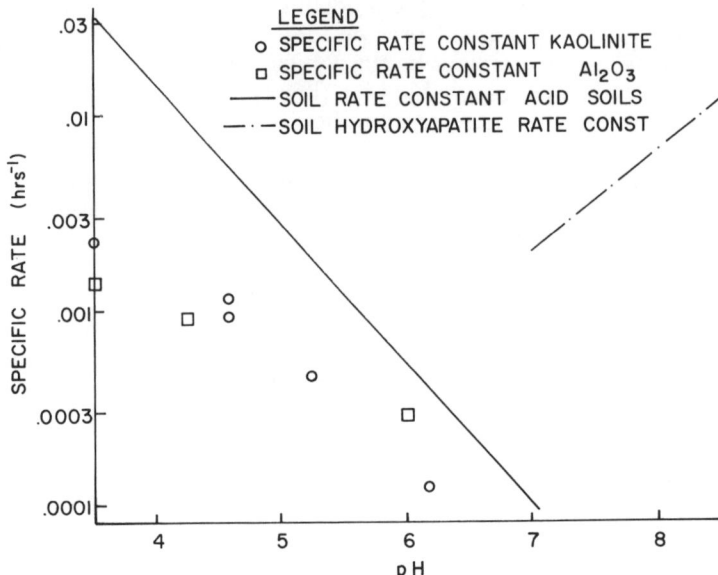

Fig. 10. Specific rate as a function of pH. Data for Kaolinite and Al_2O_3 is referred to the liquid phase; data on acid soils and hydroxyapatite is referred to the solid phase.

coefficient to be statistically significant at the 0.005 level in calcareous soils. Neutral soils appear to have the slowest rate of reaction as well as the greatest solubility for the phosphate chemicals.

CONCLUSIONS

Several different models have been proposed to describe the disappearance of P from soil solutions. None of the models have claimed to fully explain the actual mechanisms for P reaction with soil. Of the kinetic models that have been proposed, two general approaches appear to approximate the majority of reported data: (i) those where sorption sites are assumed not to be at equilibrium having different rates of sorption and desorption; and (ii) those based on kinetics of precipitation and dissolution. Although an understanding of the mechanisms involved is incomplete, approximations can be made for the mobility and activity of P in soil solution.

LITERATURE CITED

Bennett, A. C., and Fred Adams. 1976. Solubility and solubility product of dicalcium phosphate dihydrate in aqueous solutions and soil solutions. Soil Sci. Soc. Am. J. 40:39–42.

Berkheiser, V. E., J. J. Street, P. S. C. Rao, and T. L. Yuan. 1980. Partitioning of inorganic orthophosphate in soil water systems. C.R.C. Critical Reviews in Environmental Control. 10:179–224.

Blanchar, R. W., and L. R. Hossner. 1969. Hydrolysis and sorption of ortho-pryo-tripoly, and trimetaphosphate in thirty-two midwestern soils. Soil Sci. Soc. Am. Proc. 33:622–625.

Brunauer, Stephen, and L. E. Copeland. 1967. Surface tension, adsorption. Part 5. p. 90–115. In E. U. Condon and High Odishaw (ed.) Handbook of physics. McGraw-Hill Book Co., N.Y.

Buckman, H. O., and N. C. Brady. 1970. The nature and properties of soils. 7th Ed. The MacMillan Co., N.Y. 480 p.

de Camargo, O. A., J. W. Biggar, and D. R. Nielsen. 1979. Transport of inorganic phosphorus in an Alfisol. Soil Sci. Soc. Am. J. 43:884–890.

Enfield, C. G., and B. E. Bledsoe. 1975. Fate of wastewater phosphorus in soils. J. of the Irrigation and Drainage Division. ASCE. 101(IR3):145–155.

----, and R. F. Carsel. 1980. Mathematical prediction of toxicant transport through soil. p. 233–250. In Test protocols for environmental fate and movement of toxicants. Association of Official Analytical Chemists, October 21–22, Washington, D.C.

----, C. C. Harlin, and B. E. Bledsoe. 1976. Comparison of five kinetic models for orthophosphate reaction in mineral soils. Soil Sci. Soc. Am. J. 40:243–249.

----, T. Phan, and D. M. Walters. 1981a. Kinetic model for phosphate transport and transformation in calcareous soils, II. Laboratory and field transport. Soil Sci. Soc. Am. J. 45:1064–1070.

----, ----, ----, and R. Ellis, Jr. 1981b. Kinetic model for phosphate transport and transformation in calcareous soils, I. Kinetics of transformation. Soil Sci. Soc. Am. J. 45:1059–1064.

Fiskell, J. G. A., R. S. Mansell, H. M. Selim, and F. G. Martin. 1979. Kinetic behavior of phosphate sorption by acid sandy soil. J. Environ. Qual. 8:579–584.

Freeman, J. S., and D. L. Rowell. 1981. Adsorption and precipitation of phosphate onto calcite. J. Soil Sci. 32:75–84.

Griffin, R. A., and J. J. Jurnak. 1973. Estimation of activity coefficients from electrical conductivity of natural aquatic systems and soil extracts. Soil Sci. 116:26–30.

Holford, I. C. R., R. W. M. Wedderburn, and G. E. G. Mattingly. 1974. A Langmuir two-surface equation as a model for phosphate adsorption by soil. J. Soil Sci. 25:242–256.

Hsu, Pa Ho. 1979. Effect of phosphate and silicate on the crystallization of gibbsite from OH-AL solution. Soil Sci. 127:219–226.

Langmuir, I. 1918. The adsorption of gasses on plane surfaces of glass, mica, and platinum. J. Am. Chem. Society 40:1361–1403.

Lindsay, W. L. 1979. Chemical equilibra in soils. John Wiley and Sons, N.Y. 162–209.

Mansell, R. S., H. M. Selim, and J. G. A. Fiskell. 1977a. Simulated transformations and transport of phosphorus in soil. Soil Sci. 124:102–109.

----, ----, P. Kanchanasut, J. M. Davidson, and J. G. A. Fiskell. 1977b. Experimental and simulated transport of phosphorus through sandy soils. Water Resour. Res. 13:189–194.

Novak, L. T., D. C. Adriano, G. A. Coulman, and P. B. Shah. 1975. Phosphorus movement in soils theoretical aspects. J. Environ. Qual. 4:93–99.

Overman, A. R., Ro-Lan Chu, and Yung Le. 1978. Kinetic coefficients for phosphorus transport in packed bed reactor. Journal WPCF. 50:1905–1910.

Sadiq, M., and W. L. Lindsay. 1979. Selection of standard free energies of formation for use in soil chemistry. Colorado State Univ., Technical Bull. 134. 1069 p.

Sawhney, B. L. 1977. Predicting phosphorus movement through soil columns. J. Environ. Qual. 6:86–89.

Shayan, A., and B. G. Davey. 1978. A universal dimensionless phosphate adsorption isotherm for soil. Soil Sci. Soc. Am. J. 42:878–882.

Sibbesen, E. 1981. Some new equations to describe phosphate sorption by soils. J. Soil Sci. 32:67–74.

Subbarao, Y. V., and Roscoe Ellis, Jr. 1977. Determination of kinetics of phosphorus mineralization in soils under oxidizing conditions. Robert S. Kerr Environmental Res. Lab., Ada, Okl. EPA-600/2-77-180.

van Riemsdijk, W. H., F. A. Westrate, and G. H. Bolt. 1975. Evidence for a new aluminum phosphate phase from reaction rate of phosphate with aluminum hydroxide. Nature 257:473–474.

Chapter 8

The Movement of Micronutrients in Soils[1]

B. G. ELLIS, B. D. KNEZEK, AND L. W. JACOBS[2]

For years agronomists and soil scientists have been interested in describing the movement of nutrients, both macro and micro, to plant roots. In recent years an added interest has been created by our concern for environmental quality and the movement of micronutrients to groundwater or surface waters. Basically two mechanisms—diffusion and convection—may be operational in the movement of micronutrients to either plant roots or to groundwater. Seldom is there a clear division between the two mechanisms. In general, both occur to some degree whenever micronutrients are transported within a soil system. Which mechanism will be dominant will depend on the rate and direction of water movement, upon the micronutrient involved, the plant species, and the environment surrounding the system.

This discussion will consider movement to plant roots and to groundwaters of the micronutrients Zn, Cu, Mn, Fe, B, and Mo. We will, in general, emphasize the practical aspects of micronutrient movements.

---

[1] Contribution of the Crop and Soil Sci. Dep., Michigan Agric. Exp. Stn. J. Article No. 10201.

[2] Professors, and associate professor, respectively, Crop and Soil Sci. Dep., Michigan State Univ., East Lansing, MI 48824.

Copyright © 1983 ASA, SSSA, 677 South Segoe Road, Madison, WI 53711. *Chemical Mobility and Reactivity in Soil Systems.*

DIFFUSION

Since all ions or molecules tend toward randomness, an activity gradient becomes the driving force for net transfer of ions or molecules from a region of higher activity to one of lower activity. This transfer of ions or molecules is known as diffusion and has been described in terms of concentration by Fick's law for steady state diffusion in a pure liquid:

$$(\partial Q/\partial t) = -DA(\partial C/\partial x), \qquad [1]$$

where Q = the quantity of nutrient diffusing across the cross-sectional area, A, normal to the activity or concentration gradient; t = time; D = diffusion coefficient; x = the distance in the direction of net movement; and C = concentration of ions in the bulk solution. As the system $\partial C/\partial x$ approaches zero, the rate of movement of ions will approach equality in all directions and net diffusion approaches zero.

Diffusion coefficients for the various micronutrients in water have been measured or calculated. Values for the micronutrients under discussion are given in Table 1. In general Do (the diffusion coefficient in bulk water) varies from 3×10^{-6} to 10×10^{-6} for the micronutrients. Apparent diffusion coefficients in soils are lower than in water, with the cations generally having much lower diffusion coefficients than anions. Although it is difficult to compare diffusion coefficients reported in the literature because the experimental methods for determining them are not constant and the method of reporting is different, Table 2 has been prepared to give some of the values reported. In Table 2, De is the apparent diffusion in the soil uncorrected; Dp is the diffusion coefficient as determined in soil corrected for the porosity; and Dp/B is the diffusion coefficient corrected for porosity and divided by the capacity factor for the soil (for a discussion of capacity factor see Olsen and Kemper, 1968).

Since more studies have been reported on diffusion of Zn^{2+} in soils than for other micronutrients, we will discuss factors associated with soils that alter the diffusion of Zn as a general case.

Soil moisture directly affected the diffusion of ions in soils. Increasing the soil moisture content will make the path of diffusion less tortuous and hence increase the rate of diffusion (Porter et al., 1960; Warncke and Barber, 1972a). In addition Warncke and Barber (1972a) point out that the Zn^{2+} concentration in solution may decrease or remain the same as soil moisture increases. As will be discussed later, decreasing the Zn concentration will decrease the De for Zn. In four of the six soils they studied, increasing volumetric moisture from 0.13 to 0.45 greatly increased De (by as much as 10^2). However, for two soils there was relatively little change in De with moisture content. Tortuosity appeared to be the primary factor in the four soils whose De increased with moisture content. In two soils, the effect of reduced tortuosity in increasing De was apparently offset by other factors which reduced De. These two soils had a higher Zn content and lower tension at 10% moisture than the other four soils.

Table 1. Diffusion coefficients of micronutrients in water.

Nutrient	Do	Reference
	$cm^2\ s^{-1} \times 10^6$	
Zn^{2+}	8.8	Elgawhary et al., 1970a
Zn^{2+}	7.1*	Elgawhary et al., 1970a
Zn^{2+}	9	Robinson and Stokes, 1959
Zn^{2+}	7.3	Phillips and Ellis, 1970
Cu^{2+}	10.9	Phillips and Ellis, 1970
Mn^{2+}	10.4	Phillips and Ellis, 1970
Fe^{2+}	3.9	International Critical Tables, 1962
Fe^{3+}	5.7	International Critical Tables, 1962
H_3BO_3	10.1	International Critical Tables, 1962

* Calculated.

Table 2. Diffusion coefficients of micronutrients in soils.

Nutrient	Dp	Dp/B	De	Reference
		$cm^2\ s^{-1}$		
Zn^{2+}			$10^{-9}-10^{-12}$	Warncke and Barber, 1972a
Zn^{2+}			$10^{-8}-10^{-11}$	Melton et al., 1973
Cu^{2+}			$1-55 \times 10^{-7}$	Ellis et al., 1970a
Mn^{2+}			$2-60 \times 10^{-7}$	Ellis et al., 1970a
Mn^{2+}			$0.3-2.2 \times 10^{-7}$	Halstead et al., 1968
Fe^{2+}	$7-10 \times 10^{-7}$			Ellis et al., 1970b
Fe^{3+}	$1.6-4.3 \times 10^{-7}$			Ellis et al., 1970b
Fe^{3+}	1.5×10^{-7}			O'Connor et al., 1971
B		$2.7-4.6 \times 10^{-6}$		Scott et al., 1975
B	$2.3-2.7 \times 10^{-6}$	$0.9-1.3 \times 10^{-6}$		Sulaiman and Kay, 1972
Mo		$0.46-8.4 \times 10^{-7}$		Lavy and Barber, 1964

Bulk density of a soil is related to the water-holding capacity; therefore, the effect of bulk density on De may be related to soil moisture. Warncke and Barber (1972b) have shown that De, in general, increased as bulk density increased from 1.1 to 1.5 but decreased sharply as bulk density increased from 1.5 to 1.6. They reported a highly significant interaction between soil moisture levels and soil bulk density.

The complex nature of the relationship between De and bulk density was suggested by Phillips and Brown (1965, 1966) to occur for three reasons. First, increasing bulk density for a soil at or less than field capacity should increase the volumetric moisture content which should increase the continuity of the liquid phase and therefore, reduce the tortuosity of the diffusion path. Secondly, increasing bulk density will increase the solids per unit volume which should increase tortuosity of the diffusion path. A third possibility is that adsorption sites for Zn^{2+} will be brought closer together and, if diffusion through the adsorbed phase is significant relative to diffusion through the liquid phase, increasing bulk density would increase diffusion rate. Generally evidence favors a combination of the first two factors as being most important.

Soil pH has been shown to alter the De for Zn (Melton et al., 1973; Clark and Graham, 1968). In both studies major increases were found in

Zn diffusion when pH was reduced below pH 7.0. These changes are likely the result of pH changing the ratio of soil adsorbed to soil solution Zn or changing the mechanism by which Zn is held by soils. As such, the effect of pH may be ion specific and also influenced by soil type.

The De for Zn^{2+} has been shown to depend upon concentration of Zn^{2+}. There have been attempts to quantify this relationship; for example, Nye (1966) suggested the following equation:

$$De = D_1 \theta f_1 (dC_1/dC) + R, \qquad [2]$$

where
- De = effective Zn diffusion coefficient in soil,
- D_1 = diffusion coefficient of Zn in soil solution,
- f_1 = tortuosity factor,
- θ = volumetric moisture content of the soil,
- C_1 = concentration of Zn in solution (g/ml),
- C = concentration of Zn in soil (g/cc), and
- R = residual term for diffusion through the adsorbed phase.

If θ and f_1 are held constant and R is negligible, De should be directly related to the slope of the adsorption isotherm. This was investigated by Warncke and Barber (1973), who found that De was linearly related to dC_1/dC. But calculated values of De were 5 to 10 times larger than those measured. The reason for this difference was not obvious.

Other ions may also affect the diffusion of an ion. To maintain electrical neutrality during diffusion of a charged species (i.e., as in the case of Zn^{2+}) either co-diffusion, where an associated ion of opposite charge moves with the ion, or counter-diffusion, where ions of the same charge move in different directions, must occur (Low, 1962). Thus, the measurement of De will reflect these processes. The specific effect of P on Zn^{2+} diffusion was reported by Melton et al. (1973). Although the effects were not large, De did decrease with increasing levels of P added to the soil for a constant level of Zn addition. One cannot rule out that the effect of P and Zn diffusion may be indirect.

Any factor which will increase the level of an ion in solution should increase its effective diffusion rate. Thus chelates and organic complexes, either natural or synthetic, could be expected to increase the De for an ion, even though the complex formed is larger and may have a slower diffusion rate in pure solution than the uncomplexed ion. This was shown for Zn by Elgawhary et al. (1970a, 1970b) who found that De for Zn was increased from 0.36×10^{-9} to 3.0×10^{-9} $cm^2 s^{-1}$ with the addition of 1.53×10^{-8} mol of EDTA/g soil and to 9×10^{-9} $cm^2 s^{-1}$ with the addition of 1.53×10^{-8} moles of ZnEDTA/g soil. The effect may well explain how plants obtain certain micronutrients (i.e., Fe) when the level of inorganic micronutrient in solution is far too low for adequate growth. Total Fe in soil solution greatly exceeds the inorganic Fe content because Fe exists in solution as an organic complex. This increased level of Fe in solution will increase the diffusion rate and therefore, increase the plant uptake of Fe.

MISCIBLE DISPLACEMENT

The fundamental principle of micronutrient movement by miscible displacement is very simple—whenever water moves (in response to gravity, hydraulic head, or suction), it will carry any ions, molecules, or colloidal particles in solution with it. Thus, micronutrients may be leached or moved to plant roots by mass flow.

Movement of Micronutrients to Plant Roots

Olsen and Kemper (1968) have reviewed movement of nutrients to plant roots. Although their review is largely developed for ions other than micronutrients, for those interested in a thorough treatment of micronutrient movement it is recommended that their reference be used as a starting point. Wilkinson (1972) reviewed the movement of micronutrients to plant roots and we will draw heavily on his review. The concept of differentiating between ions based on their mobility within soils, that is ions which are relatively mobile as opposed to those that are relatively immobile, was introduced by Bray in 1954. The general concept is that certain ions are adsorbed strongly by the soil and hence are not susceptible to rapid movement. Other ions are adsorbed only weakly and maintain a higher concentration in the soil solution which allows them to be relatively mobile in the soil by diffusion and convection. Therefore, in our discussion Zn, Cu, Mn, and Fe are considered immobile, because they are strongly adsorbed by soils, and the anions B and Mo mobile, because anions and neutral molecules are less strongly adsorbed. But as pointed out later, this is a relative split and considerable variability in mobility exists within the subgroups.

Autoradiographs have been used to qualitatively identify the major mechanism of ion movement to plant roots (Barber et al., 1963; Lavy and Barber, 1964; Wilkinson et al., 1968a, 1968b; Barber and Ozanne, 1970; Walker and Barber, 1961; Vasey and Barber, 1963; Lewis and Quirk, 1967). If the autoradiographs show depletion of a nutrient around the root, absorption by the plant was assumed to be more rapid than transport to the plant by convection; thus, diffusion is a major contributor to movement of the ion. But an accumulation of a particular ion at the root surface indicates that convection is moving the ion to the root surface at a rate greater than absorption by the plant, so convection may be assumed to be the major mechanism for moving this ion to the plant root surface.

The amount of a nutrient reaching a plant via convection has been estimated by Barber (1962) by multiplying the amount of water transpired by the plant times the soil solution concentration of an element. This method depends upon assigning an accurate "average" value for the concentration of a nutrient during the growth period. For a critical discussion of this approach see Wilkinson (1972).

Zinc. It was concluded by Oliver and Barber (1966) that diffusion was the principle means of moving Zn to the root surface. This was confirmed by Wilkinson et al. (1968a) who found that a three-fold increase in transpiration did not change the depletion zone in the soil or the Zn absorption by the plant. They also concluded that soil adsorbed Zn was the major source of Zn to plants even though the adsorbed Zn was not leachable by 0.01 M $CaCl_2$. Halstead et al. (1968) also found that convection (mass flow) and root interception could not account for Zn uptake by lettuce (*Lactuca sativa* L.), wheat (*Triticum aestivum* L.), tomatoes (*Lycopersicon esculentum* Mill.), or soybeans (*Glycine max* (L.) Merr.), which also indicated that diffusion is the dominant process for Zn movement to plant root surfaces.

Copper. Although total Cu levels in soil solution are similar to those of Zn for many soils (Hodgson et al., 1966) and the De for Cu is similar to that for Zn, total Cu uptake by plants is an order of magnitude lower than for Zn. Therefore, movement of Cu to a plant root can be accounted for by convection (aided by root interception). According to studies by Oliver and Barber (1966) diffusion accounted for less than 5% of the total Cu absorbed by the plant.

Iron. Oliver and Barber (1966) reported that Fe mobility to plant roots was a function of diffusion, convection, and root interception and the relative contribution of each varied with experimental conditions. Under low transpiration with a 1 part soil + 4 parts quartz sand mixture, diffusion accounted for as little as 4% of the Fe uptake. But with 100% soil and low transpiration, diffusion accounted for 71% of the Fe uptake.

O'Connor et al. (1971) suggested that convection was not a significant mechanism for moving Fe to a plant root unless pH was less than 4.5 or the Fe in solution was complexed, thus increasing the total Fe in solution. They found that Fe uptake increased linearly with the concentration of Fe in solution. Since convection was assumed to be negligible in their systems, the increased uptake was attributed to increased diffusion caused by chelates increasing the level of Fe in solution.

Manganese. The mechanism of Mn movement to plant roots depends upon the soil properties (Oliver and Barber, 1966; Barber et al., 1966; and Halstead et al., 1968). When the Mn content of the soil solution is low (i.e., saturation extract containing <0.4 μ molar Mn), diffusion accounted for most of the movement of Mn to the plant root surface. When the soil solution content of Mn was high (i.e., saturation extract of soil solution containing >14 μ molar Mn), convection (aided by root interception) accounted for all of the uptake of Mn.

Boron. Although Oliver and Barber (1966) reported that mass flow was the major mechanism moving B to a plant root surface, others emphasized the importance of diffusion (Scott et al., 1975; Sulaiman and Kay, 1972). Although B exists largely as H_3BO_4 or as $B(OH)_4^-$ at pH values commonly observed in soils, it is adsorbed more strongly by soils than many other anions (Ellis and Knezek, 1972). Thus, Sulaiman and Kay (1972) found that the relationship between adsorbed and solution B was important in determining the rate of diffusion. They found that the addition to the soil of H_3BO_4 solutions containing less than 1 ppm B caused the capacity factor and the solution B concentrations to be lower than the corresponding values when water without H_3BO_4 was added. Under these conditions, the quantity of B diffusing to a "sink" was less for the B treated soil than for an untreated soil. A practical conclusion of this observation is that one must add sufficient B to a soil to increase its solution concentration in order to increase the B supply for plants. Low B additions to soils could be detrimental to B uptake by plants.

Molybdenum. If the saturation extract contains more than 4 ppb Mo, convection is the major process moving Mo to plant roots (Lavy and Barber, 1964). When the saturation extract contains less than 4 ppb Mo, autoradiographs indicate that diffusion is important.

Leaching of Micronutrients

Although leaching may be viewed as movement of water through a soil profile carrying with it the micronutrients in solution, this is a gross over-simplification. As the soil solution moves through a soil profile, the micronutrients are constantly encountering a new environment which may potentially add to or remove from solution certain of the micronutrients. In some cases the rate of exchange of micronutrients between solution and solid (either as adsorbed or precipitated forms) can be very rapid. But other reactions may be slower and the rate of leaching may influence the quantity of a micronutrient moved to lower depths in the soil profile. Thus, adequate understanding of movement of micronutrients to groundwater or drainage water by leaching requires knowledge of a combination of factors affecting water movement and the chemistry of each micronutrient in soils including their kinetic reactions.

Attempts have been made to model micronutrient leaching but we will not review these. Rather we will address the practical data for individual ions where measurements have been made to relate the movement of micronutrients within soil profiles.

Zinc. As with all of the cationic micronutrients, Zn levels in most soil solutions are very low (Hodgson et al., 1966). Zinc in native soil profiles is rather uniformly distributed reflecting parent material rather than relocation as soils developed.

The source of Zn that is most likely to increase the surface soil content of Zn to a sufficiently high level to cause leaching is waste byproducts, particularly municipal or industrial sludges. These waste materials are quite variable in Zn content (Blakeslee, 1973; Horvath and Koshut, 1981). And in addition the Zn they contain is not uniformly leachable. When leaching with 0.06 N $CaCl_2$, Lagerwerff et al. (1976) found that 80% of the Zn would leach from a Baltimore sludge but only 0.4% leached from a Washington D.C. sludge. The principle difference in sludges appeared to be the pH (Baltimore pH = 4.0; Washington D.C. pH = 6.5).

Many studies have shown almost no movement of Zn from the plow layer after application of sludge. Loading a forest area (soil pH 5.0) with sludge containing 28 and 49 kg Zn/ha produced no movement below 7.5 cm (Sidle and Kardos, 1977). Although Zn was reported to be more mobile than Cr, Cu, or Pb, little movement of Zn below 20 cm was shown

Table 3. Zinc content of soil after application of sludge.†

Depth	Control		22 t/ha in 1976		99 + 81 + 140 t/ha	
	DTPA	Total	DTPA	Total	DTPA	Total
	mg·kg⁻¹					
0–15	5.2	16	24.4	52	145	1337
15–30	1.6	7	3.0	10	14.2	148
30–45	1.1	9	3.7	10	11.2	250
45–60	0.9	8	1.7	9	4.0	113
60–75	0.8	7	1.1	8	1.3	48
75–90	0.6	8	1.2	10	2.8	79
Total Zn applied	none		150 kg/ha		3490 kg/ha	

† Ellis et al. (1981).

Table 4. Zinc and copper content of ground water under sludge plots.†‡

Treatment	Zn	Cu
	µg·kg⁻¹	
Forage corn (no sludge)	76	35
Corn-Sludge§	711	69

† Ellis et al. (1981).
‡ Each value is a mean of four replications and determinations by two laboratories.
§ 99 + 81 + 140 T/ha applied in a 3-year period.

in a leaching study where sludge was added to two soils (pH 8.5, pH 6.3) (Chang and Broadbent, 1980).

Others have reported movement of Zn (Peterson and Gschwind, 1972; King and Morris, 1972; Boswell, 1975; Kuo, 1981). In general, these soils were acid (pH 2.5 to pH 6.2). Movement of Zn is related to rate of water percolating through the soil, soil pH, soil type, and Zn loading rate. Data from studies of loading rates of metal contaminated sludge are reported by Ellis et al. (1981). In this study, metal contaminated sludge was incorporated into a sandy (sandy, mixed, mesic Entic Haplaquod) soil at the Muskegon Wastewater Treatment System (Michigan). The study site received a total of 289 cm of water over the 3 years of the experiment; thus, increasing the potential for leaching. Table 3 gives data from this study. Little movement of Zn occurred when 22 tons/ha of sludge containing 150 kg Zn was applied in 1976. The surface 15 cm contained significantly higher DTPA extractable and total Zn than for control soils, and there was some evidence of DTPA extractable Zn increasing throughout the profile. But total Zn showed little increase below the 15 cm layer. As the sludge application rate increased (rates of 11 t annually for 3 years; 22 t in 1976; 22 t annually for 3 years; 54 t in 1976; 99 t in 1976 and approximately 100 t annually for 3 years were applied) Zn movement increased. At the very high loading rate movement of Zn occurred as shown by the increase in total and DTPA extractable Zn at all depths in the profile (Table 3). That Zn was in fact reaching the groundwater at 6 to 8 ft. was verified (Table 4). The Zn content of the groundwater under the high sludge plot was increased tenfold from 76 to 711 ppb Zn).

Table 5. Copper content of soil after application of sludge.†

Depth	Control		22 t/ha in 1976		99 + 81 + 140 t/ha	
	DTPA	Total	DTPA	Total	DTPA	Total
cm			mg·kg^{-1}			
0–15	0.3	8	9.0	25	19.6	62
15–30	nd	8	0.8	8	0.6	14
30–45	nd	9	0.6	8	0.9	10
45–60	nd	9	0.3	7	1.3	9
60–75	nd	9	0.2	7	0.3	8
75–90	nd	9	0.2	7	0.4	8
Total Cu applied	none		98 kg/ha		930 kg/ha	

† Ellis et al. (1981).

In summary, Zn is one of the more mobile of the cationic micronutrients. But applications up to 150 kg Zn/ha will not result in leaching of Zn on most soils.

Copper. Soil solutions generally contain low concentrations of Cu. But almost all of the Cu in solution is in a complexed form (Hodgson et al., 1966). Examination of the distribution of Cu in podzol (spodosol) profiles suggests that it moves with organic matter.

Copper is bound to sludge and is not susceptible to direct leaching; thus, little Cu could be leached from either a Baltimore (acidic) or Washington D.C. (neutral) sludge (Lagerwerff et al., 1976). The quantity of Cu that could be extracted was also found to increase several fold after pretreatment with 3% H_2O_2 showing the prevalence of easily oxidizable organocopper complexes in sludge or sludge leachates.

Addition of manures or sludge containing low levels of Cu has resulted in little or no leaching of Cu (Kuo, 1981; Sidle et al., 1979; Amoozegar-Fard et al., 1980; Sidle and Kardos, 1977). Another source of Cu (preservative-treated wooden stakes) was examined by DeGroot et al. (1979) who found relatively little movement of Cu from the wooden stakes even after 35 years.

Copper movement to the groundwater was detected where the highest loading of metal-contaminated sludge was applied at the Muskegon Wastewater Treatment Facility (Table 4) (Ellis et al., 1981). Total Cu content of the soil profiles does not appear to show movement (Table 5). But only 21% of the added Cu can be accounted for in this research plot suggesting that the sludge was oxidized and that the Cu may have leached from the profile. The lack of accumulation of Cu with depth in the profile may have resulted from the failure of the very sandy material in this soil to retain Cu.

One can conclude from these reports that Cu will be retained by the organic fraction of soils and should not leach. But the possibility of leaching organic complexed Cu does exist, particularly in very sandy soils.

Manganese. Total Mn is higher in the surface horizons of soils suggesting that it associates with the organic fraction. However, insoluble inorganic Mn compounds exist in soils. Since Mn exists in several oxidation states, redox as well as pH exerts a great influence on the solubility of Mn

Table 6. DTPA extractable Mn from 0 to 15 cm soil after sludge application.†

Treatment	Date					
	May 76	Oct 76	May 77	Nov 77	May 78	Oct 78
	mg·kg^{-1}					
Control	4.6	1.4	4.3	1.3	2.7	2.1
22 t sludge/ha‡ (1976)	3.2	1.3	4.2	1.4	2.3	1.4
100 t sludge/ha§ (annually)	5.1	1.6	4.8	1.0	1.9	0.9

† Ellis et al. (1981).
‡ 10 kg Mn/ha applied.
§ 44, 14, and 32 kg Mn/ha applied in 1976, 1977, and 1978, respectively.

compounds. From the discussion of Mn^{2+} concentration as a function of pe + pH published by Lindsay (1979), it is evident that soil properties are often more important in determining the level of Mn in solution than the quantity of Mn added to the soil.

Several studies have found Mn to be mobile but its mobility was generally not related to Mn applied in sludge. Chang and Broadbent (1980) reported that Mn was the most mobile of the heavy metals they measured. But movement was greater in the control soil than in a sludge treated soil. Data from Ellis et al. (1981) showed no effect of sludge application on levels of DTPA soluble Mn in soil (Table 6). Their data indicated a decrease in DTPA extractable Mn both in the surface soil and throughout the profile with increasing sludge application. This was related to a higher pH on the sludge treated soils (after 3 years the pH of the control was 6.5 and the pH of the highest sludge treatment was 7.0).

In a column leaching study, Mn has been found to increase in the leachate after addition of fluidized bed combustion waste to an acid silt loam soil (Sidle et al., 1979). In fact Mn was the only heavy metal to leach in this study. The percolate concentration of Mn increased relative to controls after 2 leaching periods and then returned to the level of the control after 11 leaching periods (10 cm of water/leaching).

Movement of Mn may relate to changing soil conditions; thus, short-term experiments may not give a true picture of what could occur in field situations. For example, Amoozegar-Fard et al. (1980) found in a column study that Mn in a small aggregate system kept increasing as leaching progressed. This change may be due to the columns becoming more reducing as time progressed.

In summary, Mn should not be susceptible to leaching if soil pH is near neutrality, and the soil is well aerated. But acid, anaerobic soils will readily leach Mn.

Iron. In most well-aerated natural soils, Fe is so insoluble that levels in solution would support only negligible movement by convection (O'Connor et al., 1971). But leaching does occur in one of three ways. First, organic complexes can increase the levels of Fe in solution. Podzol (spodosols) soils have long been known to accumulate Fe associated with organic matter at the surface of the B horizon (B_{ihr}) known as an ortstein layer. Second, anaerobic soils will reduce Fe^{3+} to Fe^{2+} thus increasing the levels of Fe in solution. A very practical illustration of Fe movement due to this chemical reduction occurs in drainage systems. The occurrence of

Fe and Mn deposits in tile drains has been reported by Grass (1969) and MacKenzie (1962). The deposits may be explained by the fact that anaerobic conditions reduce Fe^{3+} to Fe^{2+} and Mn^{4+} to Mn^{2+} allowing Fe and Mn to move to the tile drain with the percolating water. But the zone surrounding the tile is aerobic, thus oxidizing Fe^{2+} to Fe^{3+} and Mn^{2+} to Mn^{4+} leading to their precipitation. Data by Grass et al. (1973a, 1973b) showed that reducing conditions near the soil surface after irrigation were favorable for dissolving both Fe and Mn, thus leading to the movement and eventual precipitation of these metals as discussed above.

Third, Fe may be mobilized by soils which become very acid. For example, a waterlogged soil may accumulate reduced S (i.e., as FeS or other compounds) which will produce H_2SO_4 if the soil becomes well-aerated after drainage. The result is a great increase in the Fe levels in solution and in leaching waters (Harmsen and Van Breemen, 1975a, 1975b). This effect was observed in Granby soils at the Muskegon Wastewater Treatment Facility during the first few months of operation. Quantities of Fe leaching from these soils were quite high and precipitated in the drainage ditch after solution pH increased due to mixing with other water.

Boron. Levels of B that are toxic to plant growth have occurred in many soils, particularly in arid areas. Consequently, a number of studies have been designed to determine the ease with which B may be leached from soils. Although B exists as H_3BO_3 or $B(OH)_4^-$ in soil solution, it is adsorbed much more strongly than other anions such as Cl^- or NO_3^- (Ellis and Knezek, 1972). In addition, native B is fixed in forms that are released slowly into soil solution, so it is more difficult to leach B than most soluble salts (Eaton and Wilcox, 1939; Reeve et al., 1955).

The relationship between solution and adsorbed B has been described by the Langmuir equation (Hatcher and Bower, 1958; Tanji, 1970) and by the Freundlich equation (Wierenga et al., 1975). By combining the Langmuir equation with an equation from chromatographic displacement theory, Tanji (1970) was able to predict accurately B leaching from soil columns. Combination of B concentrations predicted by the Freundlich equation and movement of water described by a differential equation describing movement through a porous medium under steady state conditions allowed Wierenga et al. (1975) to predict that it would take 160 to 1,800 years for B to reach the groundwater (86 m) at levels equal to one-half of the input rate.

It may be concluded that soils, particularily those that contain Fe and Al oxides-hydroxides, will adsorb B. But B in the solution phase will be subject to leaching.

Molybdenum. There is little information published on leaching of Mo. In acid soils Mo is adsorbed rather strongly and should not be leached. But in calcareous soils Mo is not adsorbed as strongly and thus would be subject to leaching. Sixty to 90% of Mo (0.75 ppm) applied in water was found to leach through 16 cm of soil; thus, Jones and Belling (1967) concluded that it was highly mobile in calcareous soils. Soil profile data presented by Jackson et al. (1975) also indicated that Mo is mobile in calcareous soils.

SUMMARY

Cationic micronutrients (Zn, Mn, Cu, and Fe) are generally considered to be rather immobile in soils because they partition strongly to the solid phase rather than to the soil solution. Thus, their movement in soils is restricted. Even when they are moved to a lower horizon with the water, they are adsorbed and further movement is restricted. Soil chemical reactions which shift the equilibrium to favor the solution phase (i.e., pH changes, complex or chelate formation, or reducing conditions) will greatly increase mobility of some or all of these micronutrients. Recent studies with waste materials have shown that if massive quantities of micronutrients added to soils some leaching will occur.

Boron and Mo exist as anions or uncharged molecules in soils and as such are expected to be mobile. However, soils high in Fe and Al-oxides-hydroxides may strongly adsorb B or Mo and reduce their mobility.

LITERATURE CITED

Amoozegar-Fard, A., W. H. Fuller, and A. W. Warrick. 1980. The movement of salts from soils following heavy application of feedlot wastes. J. Environ. Qual. 9:269–273.

Barber, S. A. 1962. A diffusion and mass-flow concept of soil nutrient availability. Soil Sci. 93:39–49.

----, E. H. Halstead, and R. F. Follett. 1966. Significant mechanisms controlling the movement of manganese and molybdenum to plant roots growing in soil. p. 299–304. *In* Int. Soil Sci. Soc. (Aberdeen, Scotland) Trans. Comm. II and IV.

----, J. M. Walker, and E. H. Vasey. 1963. Mechanisms for the movement of plant nutrients from the soil and fertilizer to the plant root. J. Agric. Food Chem. 11:204–207.

----, and P. G. Ozanne. 1970. Autoradiographic evidence for the differential effect of four plant species in altering the calcium content of the rhizosphere soil. Soil Sci. Soc. Am. Proc. 34:635–637.

Blakeslee, P. A. 1973. Monitoring considerations for municipal wastewater effluent and sludge application to land. p. 183–198. *In* Proc. of the Joint Conference on Recycling Municipal Sludges and Effluents on Land. Champaign, Ill.

Boswell, F. C. 1975. Municipal sewage sludge and selected element applications to soil: Effect on soil and fescue. J. Environ. Qual. 4:267–273.

Bray, R. H. 1954. A nutrient mobility concept of soil plant relationships. Soil Sci. 78:9–22.

Chang, Fu-Hsian, and F. E. Broadbent. 1980. Effect of nitrification on movement of trace metals in soil columns. J. Environ. Qual. 9:587–592.

Clark, A. L., and E. R. Graham. 1968. Zinc diffusion and distribution coefficients in soil as affected by soil texture, zinc concentration and pH. Soil Sci. 105:409–418.

DeGroot, R. C., T. W. Popham, L. R. Gjovik, and T. Forehand. 1979. Distribution gradients of arsenic, copper, and chromium around preservative-treated wooden stakes. J. Environ. Qual. 8:39–41.

Eaton, F. M., and L. V. Wilcox. 1939. The behavior of boron in soils. U.S. Dep. of Agric. Tech. Bull. 696.

Elgawhary, S. M., W. L. Lindsay, and W. D. Kemper. 1970a. Effect of EDTA on the self-diffusion of zinc in aqueous solution and in soil. Soil Sci. Soc. Am. Proc. 34:66–70.

----, ----, and ----. 1970b. Effect of complexing agents and acids on the diffusion of zinc to a simulated root. Soil Sci. Soc. Am. Proc. 34:211–214.

Ellis, B. G., A.E. Erickson, L. W. Jacobs, J. E. Hook, and B. D. Knezek. 1981. Cropping systems for treatment and utilization of municipal wastewater and sludge. NTIS PB81-187254.

----, and B. D. Knezek. 1972. Adsorption reactions of micronutrients in soils. p. 59–78. *In* J. J. Mortvedt, P. M. Giordano, and W. L. Lindsay (ed.) Micronutrients in agriculture. Soil Sci. Soc. Am., Madison, Wis.

Ellis, J. H., R. I. Barnhisel, and R. E. Phillips. 1970a. The diffusion of copper, manganese, and zinc as affected by concentration, clay mineralogy, and associated anions. Soil Sci. Soc. Am. Proc. 34:866–870.

----, ----, and ----. 1970b. The diffusion of iron in montmorillonite as determined by x-ray emission. Soil Sci. Soc. Am. Proc. 34:591–595.

Grass, L. B. 1969. Tile clogging by iron and manganese in Imperial Valley, California. J. Soil Water Conserv. 24:135–138.

----, A. J. MacKenzie, B. D. Meek, and W. F. Spencer. 1973a. Manganese and iron solubility changes as a factor in tile drain clogging: I. Observations during flooding and drying. Soil Sci. Soc. Am. Proc. 37:14–17.

----, ----, ----, and ----. 1973b. Manganese and iron solubility changes as a factor in tile drain clogging: II. Observations during the growth of cotton. Soil Sci. Soc. Am. Proc. 37:17–21.

Halstead, E. H., S. A. Barber, D. D. Warncke, and J. B. Bole. 1968. Supply of Ca, Sr, Mn and Zn to plant roots growing in soil. Soil Sci. Soc. Am. Proc. 32:69–72.

Harmsen, K., and N. van Breemen. 1975a. A model for the simultaneous production and diffusion of ferrous iron in submerged soils. Soil Sci. Soc. Am. Proc. 39:1063–1068.

----, and N. van Breemen. 1975b. Translocation of iron in acid sulfate soils: II. Production and diffusion of dissolved ferrous iron. Soil Sci. Soc. Am. Proc. 39:1148–1153.

Hatcher, J. T., and C. A. Bower. 1958. Equilibria and dynamics of boron adsorption by soils. Soil Sci. 85:319–323.

Hodgson, J. F., W. L. Lindsay, and J. F. Trierweiler. 1966. Micronutrient cation complexing in soil solution: II. Complexing of zinc and copper in displaced solution from calcareous soils. Soil Sci. Soc. Am. Proc. 30:723–726.

Horvath, D. J., and R. A. Koshut. 1981. Proportions of several elements found in sewage effluent and sludge from several municipalities in West Virginia. J. Environ. Qual. 10:491–497.

International Critical Tables. 1962. National Res. Counc. McGraw-Hill Book Co., N.Y.

Jackson, D. R., W. L. Lindsay, and R. D. Heil. 1975. The impact of molybdenum-enriched irrigation water on agricultural soils near Brighton, Colorado. J. Environ. Qual. 4:223–229.

Jones, G. B., and G. B. Belling. 1967. The movement of copper, molybdenum, and selenium in soils as indicated by radioactive isotopes. Aust. J. Agric. Res. 18:733–740.

King, L. D., and H. D. Morris. 1972. Land disposal of liquid sewage sludge: II. The effect on soil pH, manganese, zinc, and growth and chemical composition of rye (*Secale cereale* L.). J. Environ. Qual. 1:425–429.

Kuo, Shiou. 1981. Effects of drainage and long-term manure application on nitrogen, copper, zinc and salt distribution and availability in soils. J. Environ. Qual. 10:305–308.

Lagerwerff, J. V., G. T. Biersdorf, and D. L. Brower. 1976. Retention of metals in sewage sludge I: Constituent heavy metals. J. Environ. Qual. 5:19–23.

Lavy, T. L., and S. A. Barber. 1964. Movement of molybdenum in the soil and its effect on availability to the plant. Soil Sci. Soc. Am. Proc. 28:93–97.

Lewis, D. G., and J. P. Quirk. 1967. Phosphate diffusion in soil and uptake by plants. III. P^{31} movement and uptake by plants as indicated by P^{32} autoradiography. Plant Soil 26:445–453.

Lindsay, W. L. 1979. Manganese, p. 151–161. *In* Chemical equilibria in soils. John Wiley and Sons, N.Y.

Low, P. F. 1962. Effect of quasi-crystalline water on rate processes involved in plant nutrition. Soil Sci. 93:6–15.

MacKenzie, A. J. 1962. Chemical treatment of mineral deposits in drain tile. J. Soil Water Conserv. 17:124–125.

Melton, J. R., S. K. Mahtab, and A. R. Swoboda. 1973. Diffusion of zinc in soils as a function of applied zinc, phosphorus, and soil pH. Soil Sci. Soc. Am. Proc. 37:379–381.

Nye, P. H. 1966. The measurement and mechanism of ion diffusion in soil. I. The relation between self-diffusion and bulk diffusion. J. Soil Sci. 17:16–23.

O'Connor, G. A., W. L. Lindsay, and S. R. Olsen. 1971. Diffusion of iron and iron chelates in soil. Soil Sci. Soc. Am. Proc. 35:407–410.

Oliver, S., and S. A. Barber. 1966. Mechanisms for the movement of Mn, Fe, B, Cu, Zn, Al, and Sr from one soil to the surface of soybean roots. Soil Sci. Soc. Am. Proc. 30:468–470.

Olsen, S. R., and W. D. Kemper. 1968. Movement of nutrients to plant roots. Adv. Agron. 20:91–151.

Peterson, J. R., and John Gschwind. 1972. Leachate quality from acidic mine spoil fertilized with liquid digested sewage sludge. J. Environ. Qual. 1:410–412.

Phillips, R. E., and D. A. Brown. 1965. Ion diffusion III. The effect of soil compaction on self diffusion of Rb^{86} and Sr^{89}. Soil Sci. Soc. Am. Proc. 29:657–661.

----, and ----. 1966. Counter diffusion of Rb^{86} and Sr^{89} in compacted soil. J. Soil Sci. 17:200–211.

----, and J. H. Ellis. 1970. A rapid method of measurement of diffusion coefficients in aqueous solutions. Soil Sci. 110:421–425.

Porter, L. K., W. D. Kemper, R. D. Jackson, and B. A. Stewart. 1960. Chloride diffusion in soils as influenced by moisture content. Soil Sci. Soc. Am. Proc. 24:460–463.

Reeve, R. C., A. F. Pillsbury, and L. V. Wilcox. 1955. Reclamation of a saline and high boron soil in the Coachella Valley of California. Hilgardia 24:69–91.

Robinson, R. A., and R. H. Stokes. 1959. Electrolyte solutions. Butterworth's, London.

Scott, H. D., S. D. Beasley, and L. F. Thompson. 1975. Effect of lime on boron transport to and uptake by cotton. Soil Sci. Soc. Am. Proc. 39:1116–1121.

Sidle, R. C., and L. T. Kardos. 1977. Transport of heavy metals in a sludge-treated forested area. J. Environ. Qual. 6:431–437.

----, W. L. Stout, J. L. Hern, and O. L. Bennett. 1979. Solute movement from fluidized bed combustion waste in acid soil and mine spoil columns. J. Environ. Qual. 8:236–241.

Sulaiman, Wan, and B. D. Kay. 1972. Measurement of the diffusion coefficient of boron in soil using a single cell technique. Soil Sci. Soc. Am. Proc. 36:746–752.

Tanji, K. K. 1970. A computer analysis on the leaching of boron from stratified soil columns. Soil Sci. 110:44–51.

Vasey, E. H., and S. A. Barber. 1963. Effect of placement on the absorption of Rb^{86} and P^{32} from soil by corn roots. Soil Sci. Soc. Am. Proc. 27:193–197.

Walker, J. M., and S. A. Barber. 1961. Ion uptake by living plant roots. Science 133:881–882.

Warncke, D. D., and S. A. Barber. 1972a. Diffusion of zinc in soil: I. The influence of soil moisture. Soil Sci. Soc. Am. Proc. 36:39–42.

----, and ----. 1972b. Diffusion of zinc in soil: II. The influence of soil bulk density and its interaction with soil moisture. Soil Sci. Soc. Am. Proc. 36:42–46.

----, and ----. 1973. Diffusion of zinc in soils: III. Relation to zinc adsorption isotherms. Soil Sci. Soc. Am. Proc. 37:355–358.

Wierenga, P. J., M. Th. van Genuchten, and F. W. Boyle. 1975. Transfer of boron and tritiated water through sandstone. J. Environ. Qual. 4:83–87.

Wilkinson, H. F. 1972. Movement of micronutrients to plant roots. p. 139–169. In J. J. Mortvedt, P. M. Giordano, and W. L. Lindsay (ed.) Micronutrients in agriculture. Soil Sci. Soc. Am., Madison, Wis.

----, J. F. Loneragan, and J. P. Quirk. 1968a. The movement of zinc to plant roots. Soil Sci. Soc. Am. Proc. 32:831–833.

----, ----, and ----. 1968b. Calcium supply to plant roots. Science 161:1245–1246.

Chapter 9

Principles of Salt Movement in Soils[1]

R. J. WAGENET[2]

The movement of dissolved materials through the soil has substantial implications when one considers that almost all organic and inorganic chemicals dissolve to some extent in water. Fertilizers and other soil amendments, organic and inorganic wastes, and soluble salts in water are all subject to displacement once applied to the soil surface. The description of this displacement depends upon consideration of physical, chemical, and biological processes related to water flow and solute properties. Generally, the nature of these processes has been studied in detail in laboratory situations and to a lesser degree in the field. The purposes of this discussion are to review the theoretical approaches presently used to describe solute displacement and to indicate in general terms the ability of these approaches to describe both laboratory and field conditions. The emphasis will be on inorganic salts that are not greatly subject to microbial transformation, with consideration of such topics as N, P, and organic chemicals left to other presentations at this symposium.

The physical and chemical mechanisms that must be considered in describing solute movement are summarized in miscible displacement theory. This mechanistic approach to solute transport considers the fundamental processes operating during displacement to be chemical diffusion, hydrodynamic dispersion, and mass flow of water. The basic equation derived from this theory can be altered to include a wide range

---

[1] Prepared for presentation at joint S1-S2 Symposium, "Chemical Mobility and Reactivity in Soil Systems," annual meeting of the Am. Soc. of Agron., Atlanta, Georgia, 30 Nov.–4 Dec. 1981.
[2] Associate professor, Dep. of Agronomy, Cornell Univ., Ithaca, NY 14853.

Copyright © 1983 ASA, SSSA, 677 South Segoe Road, Madison, WI 57311. *Chemical Mobility and Reactivity in Soil Systems.*

of sources or sinks of solute (i.e., biological and chemical transformations, ion-soil interactions, and plant extraction). Depending on their complexity, the resultant equations are solvable by analytic or numeric methods, with the result that a wide variety of basic equations and solutions now exist in the scientific literature. Whether these solutions are for one-dimensional cases, as most are, or for two- and three-dimensional cases, they most often require that the researcher have knowledge of one or more of the following characterizations: (i) the magnitude of source or sink processes, (ii) the average pore water velocity, and (iii) the apparent diffusion coefficient. In laboratory experiments, these relationships have been measured quite accurately, yet the implications of temporal and spatial variations of these relationships in the field are only beginning to be studied. An accurate prediction of solute fluxes on a field basis using miscible displacement theory presently is limited by this lack of knowledge about these field variations.

Miscible displacement theory describes well the movement of solutes through homogeneous, relatively non-structured soils that have a fairly narrow pore size distribution. It has been observed that in a number of cases solute movement does not behave according to these assumptions. In highly aggregated soils, or in soils with large cracks, the movement of water through "preferred paths" or "macropores" apparently predominates in displacing solute from one point to another within the soil profile. These macropores have been hypothesized to be points of intersection between adjacent soil peds, with strong soil structure therefore being a physical prerequisite to the phenomenon. A subset of miscible displacement theory, here termed "macropore transport" theory, has been developed to describe these cases. In a field situation, the geometry of the porous media dictates that both macropore transport in the larger pores and miscible displacement in the bulk soil between these larger pores occur simultaneously to some degree with interaction often occurring between these two flow realms. The relative importance of each transport process is a function of soil structure, pore size distribution, and soil water content.

PRINCIPLES

Chromatographic Theory

It was reported as early as 1905 (Slichter, 1905) that water and solutes do not move at the same rate in soil. It was not until the early 1940's and the development of chromatographic theory that a foundation was laid for describing this differential movement. Chromatographic separations, based upon the differential rates of migration of solutes through an adsorbing medium (resin column) provided the first studies to experimentally and theoretically consider solute movement through a porous media. Description of this movement first represented the porous media as a series of discontinuous plates (Martin and Synge, 1941; DeVault, 1943). This approach artificially partitioned the flow regime into discrete units; yet, if these units could be made theoretically small enough, solute migration was described. Several investigators (Rible and Davis, 1955; Bower et al., 1957; Thomas and Coleman, 1959) applied these models to the description of ion distribution between exchange and solution phases during flow. These studies failed to consider the mixing that takes place as a result of pore water velocity distributions (resulting from pore size and shape distributions). Additionally, in all of the above cases displacement and ion exchange under partially water-saturated conditions, such as normally prevail in soil, were not considered. Even so, fairly good agreement between measurement and theory was obtained. Therefore, although physically not complete in the description of a soil matrix, the chromatographic models represented an important step forward in the development of more complete representations of solute transport. The chromatographic models as applied to soil are reviewed by Frissel and Poelstra (1967) and Reiniger and Bolt (1972).

Recognition of the deficiencies of the discrete plate chromatographic models soon led to development of a "continuous plate model" (Glueckauf, 1955) that described solute movement as a function of distance, time, flow velocity, theoretical plate height, and solute retention. This theory assumes that at the boundary between leaching water and soil solution, the two are in equilibrium and that the diffuse boundary is caused by ionic or molecular diffusion. Van der Molen (1956) applied this theory to the desalinization of slowly permeable soils, and found agreement between theory and observation. Gardner and Brooks (1957) modified the Glueckauf theory and considered soil water to be composed of both mobile and relatively immobile phases. They concluded that soil leaching processes are dominated by dispersion at the flow velocities that normally occur during leaching of soils.

Miscible Displacement Theory

Chromatographic descriptions of solute movement in soil soon evolved (Lapidus and Amundson, 1952; Nielsen and Biggar, 1961, 1962; Biggar and Nielsen, 1962, 1963) into a comprehensive theoretical approach that considers solute transport to be governed by the combined processes of convection (Movement with the bulk soil solution) and diffusion (thermal motion within the soil solution). A substantial number of experimental studies (reviewed by, e.g., van Genuchten and Cleary, 1979; Nielsen et al., 1980) have given much credibility to this conceptualization, which is most often referred to as miscible displacement theory. The essential aspects of this theory should be recognized by anyone considering the description of solute transport in soil.

According to miscible displacement theory, the flux of solute is the result of the combined effects of diffusion and convection. That is:

[1] $$J_S = J_D + J_C,$$

where J is the mass of solute transported through a cross-sectional area in a unit time, g cm^{-2} h^{-1}, and the subscripts S, D, and C represent total solute, solute transported by diffusion, and solute transported by convection, respectively. Consideration of J_D and J_C separately will contrast the separate effects.

Fick's first law states that

$$J_D = -D_o \frac{dC}{dx}, \qquad [2]$$

where C is the solute concentration (g cm^{-3}), x is distance (cm) and D_o is the ionic diffusion coefficient in a pure water system (cm^2 h^{-1}). The frame of reference for Fick's law in a soil system can be taken as either the entire bulk soil or the solution phase within which diffusion occurs. Considering the entire soil volume, Eq. [2] becomes

$$J_D = -D_p \frac{dC}{dx} \qquad [3]$$

in which the effective diffusion coefficient, D_p (cm^2 h^{-1}), for any given ion is related to D_o by

$$D_p = D_o \theta \left(\frac{L}{L_e}\right)^2 \omega \varkappa, \qquad [4]$$

where θ is the volumetric soil water content (cm^3 cm^{-3}), $(L/L_e)^2$ is a tortuosity factor (Olsen and Kemper, 1968) and $\varkappa$ and ω contain the effects of anion exclusion and the charged soil matrix on water viscosity. Values of D_p are always less than D_o. Using Eq. [4], Eq. [2] can be rewritten as

$$J_D = -D_o \theta \left(\frac{L}{L_e}\right)^2 \omega \varkappa \frac{dC}{dx} = -D_p(\theta) \frac{dC}{dx}. \qquad [5]$$

Estimation of D_p for soils has been the subject of a number of studies (e.g., Porter et al., 1960; Kemper and Van Schaik, 1966). One empirical representation of D_p (Kemper and Van Schaik, 1966) is

$$D_p(\theta) = D_o\, a e^{b\theta}, \qquad [6]$$

where a and b are empirical constants reported (Olsen and Kemper, 1966) to be approximately b = 10 and $0.005 < a < 0.01$.

Macroscopic convective solute transport is usually described by considering the two components of the convective flow to be (i) mean porewater velocity and (ii) deviations from the mean as a result of local variations of the flow velocity in individual pores. The latter creates a dispersion effect which is similar to diffusion in that there is a net movement of solute from zones of high to low concentration. This net movement now results from interaction between large and small pores through the connecting local velocities. This effect is commonly represented in the same general form as Fick's law, where now the molecular diffusion coefficient is replaced by a mechanical dispersion coefficient. Scheidegger (1960) has described this similarity in form between diffusion and dispersion to be more or less accidental.

Assuming steady water movement in one direction through a homogeneous soil of uniform water content, the total amount of solute transported by convection across a unit area in the direction of flow is given by

$$J_C = -\theta D_m(v) \frac{dC}{dx} + v\theta C, \quad [7]$$

where v is the average interstitial flow velocity (cm h^{-1}), D_m is the mechanical dispersion coefficient (depending only on v), and the other terms have been defined. The first term represents flow by mechanical dispersion and the second term considers transport due to the average flow velocity. For one-dimensional flow in a homogeneous non-aggregated porous material, D_m can be considered proportional to the first power of the average velocity,

$$d_m(v) = \lambda |v|, \quad [8]$$

where $|v|$ is the absolute value of v, and λ ranges from 0.2 to 2 cm or more, with a value of 0.4 cm found in uniformly packed soil columns.

Combining Eq. [6] and [7] into [1] allows representation of the joint effects of diffusion and convection:

$$J_S = -[\theta D_m(v) + D_p(\theta)]\frac{dC}{dx} + v\theta C = -\theta D(v,\theta)\frac{dC}{dx} + qC, \quad [9]$$

where D is the apparent diffusion coefficient (cm^2 h^{-1}) and q is the volumetric water flux (cm h^{-1}). At high average pore water velocities, D_m is usually much larger than D_p, and diffusion is completely obscured (Kirda et al., 1973).

The continuity equation (Kirkham and Powers, 1972) states that the rate of change of solute within a finite volume element must equal the difference between the amounts of solute that enter and leave the element. Applying these relationships to Eq. [9], and including consideration of ion-soil interaction (i.e., cation adsorption) and sources or sinks of solute (i.e., chemical precipitation-dissolution reactions) one obtains

$$\frac{\partial}{\partial t}(S + \theta C) = \frac{\partial}{\partial z}\left[\theta D(v,\theta)\frac{\partial C}{\partial z}\right] - \frac{\partial(qC)}{\partial z} + \phi, \quad [10]$$

where S is the concentration of solute in the "adsorbed" phase (mg cm^{-3} soil), ϕ is solute source or sink, t is time (h), and z has replaced x to specifically designate a positive downward space coordinate. If the solute does not interact with the soil (S = 0) and no gains or losses exist (ϕ = 0), Eq. [10] is the historical representation of miscible displacement theory.

Most of the experimental studies on miscible displacement have been performed under laboratory conditions with temporally and spatially constant flow velocities and water contents. Under such conditions, with a non-interacting solute, Eq. [10] reduces to

$$\frac{\partial C}{\partial t} = D \frac{\partial^2 C}{\partial z^2} - v \frac{\partial C}{\partial z}, \qquad [11]$$

where $D = [D_m(v) + D_p(\theta)/\theta]$. The solution to this equation for a range of boundary conditions has shed light on the basic phenomena characterizing solute movement in soil. These processes can be illustrated by a simple graphical method, called an elution curve. When a salt-free soil solution is displaced through a soil column by a continuously applied solution containing an inert (non-interacting) solute of concentration C_o, the fraction of this solute in the effluent at time t can be designated by C/C_o. Plots of C/C_o vs. pore volumes of effluent (ratio of volume of effluent to volume of solution contained in the soil column) are commonly called breakthrough curves (BTC). If piston displacement were operative, no mixing would occur between the displacing and displaced solutions, and a vertical boundary would represent solute breakthrough (Fig. 1). A sigmoidal shaped BTC indicates mixing. Shifting of the curve to the left indicates exclusion of or bypass from a significant portion of the soil solution, while shifting to the right indicates adsorption or solute retention by the soil.

Transport in Macropore Systems

Soils with large cracks and well aggregated soils present physical situations more difficult to describe than that incorporated into basic miscible displacement theory. In the dispersion-convection equation (Eq. [10]), it is assumed that a rather narrow pore size distribution exists which can be described by an average macroscopic pore water velocity and dispersion coefficient. This assumption does not hold for well aggregated and cracking soils. The large pores in these soils, sometimes termed macropores,

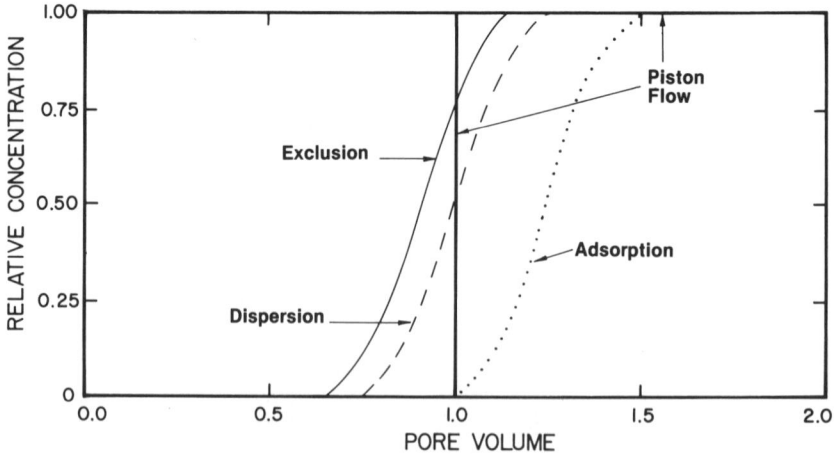

Fig. 1. Solute breakthrough curves illustrating several patterns of solute elution.

may appear in the form of cracks in shrinking-swelling clays, as earthworm channels, old root holes, or as inter-aggregate pores and interpedal voids. A large fraction of the flow in these cases may occur in the macropores (inter-aggregate porosity) while the soil solution in the smaller, intra-aggregate pores will not be as mobile, but can actually act as a distributed source or sink for solute (Passioura, 1971; Green et al., 1972; Addiscott, 1977). Several conceptualizations of this flow system have been presented (Table 1). These approaches generally consider the macropores to contain mobile water through which the bulk of solute transport occurs, while the micropore water is relatively immobile or non-moving. Transfer between these two regions is by diffusion. The solutions of the equations in Table 1 have been limited to steady state cases of water flow, and there has yet been no application of these models to field cases. Their potential use is not limited to strongly aggregated media, as almost any soil can be envisioned to consist of both relatively mobile and immobile pore water phases (e.g., water films on and between solid particles; liquid-filled dead-end pores).

Table 1. Conceptual approaches to description of flow in aggregated soils.

Model	Literature cited
[12] $\theta_A \dfrac{\partial C_A}{\partial t} + \theta_M \dfrac{\partial C_M}{\partial t} = D_M \theta_M \dfrac{\partial^2 C_M}{\partial z^2} - v_M \theta_M \dfrac{\partial C_M}{\partial z}$	Coats and Smith (1964)
where $\theta_A \dfrac{\partial C_A}{\partial t} = \gamma(C_M - C_A)$	van Genuchten and Wierenga (1976)
[13] $\dfrac{\partial}{\partial t}(\theta_A C_A) = k(C_M - C_A)$	Philip (1968)
[14] $\dfrac{\partial C_M}{\partial t} = -v_M \dfrac{\partial C_M}{\partial z} - \dfrac{D_A}{L}\left(\dfrac{\partial C_A}{\partial y}\right)_{y=d}$	Skopp and Warrick (1974)
y is normal to the direction of v_M, and $\partial C_A/\partial t = D_A \dfrac{\partial^2 C_A}{\partial y^2}$ and at the interface between regions A and M $C_A = \gamma C_M$	
[15] $\dfrac{\partial C_M}{\partial t} = D_M \dfrac{\partial^2 C_M}{\partial z^2} - v_M \dfrac{\partial C_M}{\partial z} - \dfrac{W(z,t)}{\theta_M}$	Rao et al. (1980)
where $W(z,t) = \dfrac{N}{\Delta t}[M(z_i, t+\Delta t) - M(z_i, t)]$ and $M = f(C_A, z, t)$	

θ = volumetric water content.
c = solute concentration.
D,v = apparent diffusion coefficient and average pore water velocity, respectively.
M,A = denote macropore (mobile) and micropore (immobile) phases, respectively.
t,z = time, distance, respectively.
γ,k = partition coefficients.
M = mass of solute within one micropore at time t and position z.
N = total number of micropores.

REACTIVE SOLUTES

Most solutes are subject to chemical or biological reactions during transport through the profile. Movement of inorganic salts (Ca, Mg, Na, K, Cl, SO_4, HCO_3) can be affected by cation exchange, adsorption-desorption type reactions, anion exclusion from a liquid volume near the soil particle surfaces, or by chemical precipitation-dissolution reactions. Plant uptake and microbiological transformation of these inorganic solutes will not be considered here.

Adsorption-Desorption

A number of adsorption-desorption representations have been developed (Table 2). They describe an equilibrium (not time dependent; Eq. [17], [18], and [20] in Table 2) between the concentration of solution and adsorbed, or the approach towards an equilibrium condition (time dependent; Eq. [16], [19], [21], [22], and [23]).

In general, the linear type equilibrium models were first used in attempts to describe reactive solute transport, and were found inadequate

Table 2. Models describing cation adsorption-desorption during transport.

	Model	Literature cited
[16]	$\frac{\partial S_\alpha}{\partial t} = k \left(C_\alpha S_\beta - \frac{1}{k_1} S_\alpha C_\beta \right)$	Bower et al. (1957)
[17]	$S = k_2 C + k_3$	Lapidus and Amundson (1962)
[18]	$S = S_{max} \left(\frac{C}{C_o}\right) \left[\frac{C}{C_o} + \left(1 - \frac{C}{C_o}\right)\right]^{-1}$	Lai and Jurinak (1971)
[19]	$\frac{\partial S'}{\partial t} = k_d' \left[\left(\frac{k_a^m}{k_d'}\right) S_{max}^{1-m} C^n - S' \right]$	Hornsby and Davidson (1973)
[20]	$S = RC^n;\ \frac{\partial S}{\partial t} = RnC^{n-1} \frac{\partial C}{\partial t}$	van Genuchten et al. (1974)
[21]	$\frac{\partial S}{\partial t} = (k_a \theta/\varrho)C^n - k_d S$	van Genuchten et al. (1974)
[22]	$S = S_1 + S_2;\ \frac{\partial S}{\partial t} = k_4 \frac{\theta}{\varrho} C - k_5 S_1 + k_6 \frac{\theta}{\varrho} \frac{\partial C}{\partial t}$	Cameron and Klute (1977)
[23]	$\frac{\partial S}{\partial t} = f \frac{\partial S_M}{\partial t} + (1-f) \frac{\partial S_A}{\partial t}$	van Genuchten and Wierenga (1976)

S = concentration of adsorbed solute or solute located in immobile zone.
C = solute concentration in solution, with subscript zero indicating applied concentration.
α,b = subscripts denoting two different cations.
k = rate coefficient, with different numerical subscripts to indicate different reactions.
R = partition coefficient.
m,n = superscripts denoting reaction orders.
a,d = subscripts denoting adsorption and desorption processes, respectively.
θ,ϱ = volumetric water content and bulk density, respectively.
′ = primed symbols indicate a desorption process.

in a number of studies (e.g., Kay and Elrick, 1967; Davidson et al., 1968). The introduction of hysteretic adsorption-desorption isotherms and nonlinear isotherms improved predictive capabilities (Swanson and Dutt, 1973; van Genuchten et al., 1974). The use of kinetic non-equilibrium models (Lindstrom and Boersma, 1970; Hornsby and Davidson, 1973) improved predictive capabilities in experiments conducted at low pore velocities. At higher velocities, solute adsorption has best been described by a combination of the mobile-immobile water theory with diffusion controlled transfer between the two liquid phases (Skopp and Warrick, 1974; van Genuchten and Wierenga, 1976). A third approach, successful in several cases, considers adsorption on one fraction of the adsorption sites to be instantaneous, and the balance of adsorption on the remaining sites time-dependent (Selim et al., 1976; Cameron and Klute, 1977).

In most of these models, the rate constants have physical significance, but seldom can be studied independent of the flowing system. Curve fitting procedures (van Genuchten, 1981) are often used to evaluate the parameters in a specific situation. This is necessary considering the substantial differences between flowing and batch-type systems (Ardakani and McLaren, 1977; Wagenet et al., 1977), and the sensitivity of measurement required for some of the coefficients, a sensitivity that in fact prevents their measurement and requires they be inferred from curve fitting procedures.

Cation Exchange

Adsorption-desorption models that consider only one or two cations interacting with the solid phase are simplifications of most soil systems. Probably the more realistic case is the situation where at least three (Ca, Mg, Na) or four (Ca, Mg, Na, K) cations compete at once for adsorption onto a negatively charged clay mineral surface. Cation exchange models describe this simultaneous competition. Traditionally, these models have considered only Ca-Mg-Na exchange reactions (Paul et al., 1966; Dutt et al., 1972; Jury et al., 1978) and with Ca and Mg sometimes being combined and considered as a single species (USSL, 1954). As numerical modeling techniques have become more sophisticated, approaches to description of multication exchange have been expanded to consider several other cations.

Robbins et al. (1980a) present a cation exchange model that can be used instead of the $\partial S/\partial t$ term in Eq. [10]. They propose that the model not be included as part of the numerical differencing of Eq. [10], but that their cation exchange model be used as a subroutine that adjusts cation solution concentrations according to the proportions of cations in both the solution and adsorbed phases. The basic assumption is

$$\text{CEC} = X_{1/2\,\text{Ca}} + X_{1/2\,\text{Mg}} + X_{\text{Na}} + X_{\text{K}}, \qquad [24]$$

where X denotes an exchangeable cation and CEC is the cation exchange capacity, both in units of meq/100 g. Using the Gapon convention, selectivity coefficients are written for each of the six possible combinations of the four cations. For example, for Ca-Na exchange the equation would be

$$\frac{(Na)X_{1/2\,Ca}}{(Ca)^{1/2}X_{Na}} = K_{Na\text{-}Ca}, \quad [25]$$

where the parentheses denote soil solution activities (meq/L) corrected for ion pairing. The equation for calculating exchangeable calcium then becomes

$$X_{1/2\,Ca} = CEC \div \left[\frac{(Mg)^{1/2} K_{Ca\text{-}Mg}}{(Ca)^{1/2}} + \frac{(Na)}{(Ca)^{1/2} K_{Na\text{-}Ca}} + \frac{(K)}{(Ca)^{1/2} K_{K\text{-}Ca}} + 1\right]. \quad [26]$$

Values of K for each combination of two cations were calculated by Robbins et al. (1980a) using measured exchangeable cation values and calculated cation activities in the soil solution. This approach produced improved simulations (Robbins et al., 1980b) of transient soil chemistry regimes in lysimeter experiments (Fig. 2).

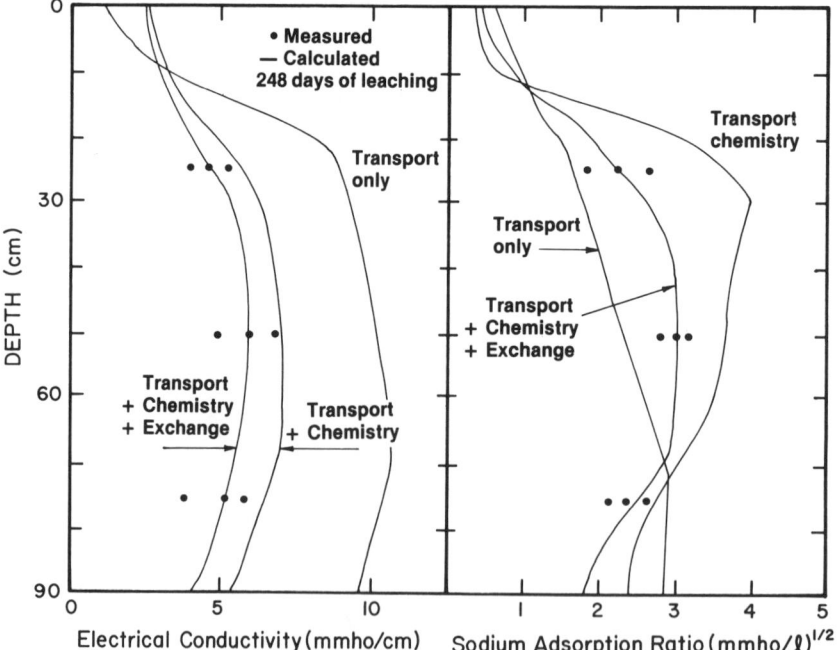

Fig. 2. Measured and simulated values of electrical conductivity and sodium adsorption ratio calculated by a numerical model of water and solute movement that includes description of cation exchange and precipitation-dissolution reactions (Robbins et al., 1980b).

Anion Exclusion

Anions are generally assumed to be non-reactive during their displacement through clay mineral dominated soils. This is in fact not the case, as the negatively charged ions are effectively excluded from a region near the negatively charged clay mineral surface. The magnitude of this anion exclusion varies from no effect in sandy soils or in the presence of sesquioxides to a larger effect in highly negatively charged soils. Anion exclusion from a discrete volume of the pore (known as the exclusion volume) has been considered in anion transport (Krupp et al., 1972; Bresler, 1973a). The model by Krupp et al. is very similar in structure to the mobile-immobile water theory, but now is modified to include an anion exclusion volume. As presented by van Genuchten (1981), this model is

$$\theta_m \frac{\partial C_m}{\partial t} + \theta_A \frac{\partial C_A}{\partial t} = \theta_m D_m \frac{\partial^2 C_m}{\partial z^2} - \theta_m v_m \frac{\partial C_m}{\partial t} \qquad [27]$$

and

$$\frac{\partial C_A}{\partial t} = k_r (C_m - hC_A), \qquad [28]$$

where $k_r \cong \gamma$ (Eq. [12]) and $h = \theta_A(\theta_A - \theta_{ex})^{-1}$, with $\theta_{ex} = \varrho \, d_{ex} A_o$ (d_{ex} = equivalent exclusion distance (cm); ϱ = bulk density (g/cm^3), A_o = specific surface area (cm^2/g)).

The value of d_{ex} is a simplification of the diffuse double layer description of an exponentially increasing anion distribution away from the charged surface into solution. Here, the exponential curve is replaced by a step function with range d_{ex} (Krupp et al., 1972; Bolt and deHaan, 1979). The exclusion volume furthermore is considered to be limited to the immobile phase. The value of θ_{ex} is assumed to be always less than θ_{im}.

Anion exclusion is treated somewhat differently by Bresler (1973a). His model assumes the S term in Eq. [10] to be of the form

$$S(\theta,C) = \Gamma[b(\theta), C] \cdot A_o \cdot \varrho_b = \theta_{ex}(\theta,C) \cdot C, \qquad [29]$$

where $\Gamma[b(\theta),C]$ is the calculated negative adsorption (me/cm^2), a function of $b(\theta)$ and C. Methods for calculating Γ are given by Bresler (1970), whereas the product $(\theta_{ex} \cdot C)$ can be calculated as was done in the model of Krupp et al. Bresler (1973a) further included a correction for the effect of osmotic gradients on the hydraulic conductivity. Using numerical methods, it was concluded that anion adsorption at times was of importance in predicting solute movement, but that salt concentration gradients played a minor role in determining the gradient for water flow.

Solution Chemistry

Soil chemical regimes are usually transient. If irrigation water quality is variable, or if rainfall and irrigation follow in alternating events, the soil chemical regime will not reach steady state. Similarly, plant extraction of water will concentrate ions in the soil solution, thus producing changes in ion solubility and in the composition of the exchange phase. Description of solute transport must consider these processes, particularly in those cases where gypsum or lime are already present as components of the soil profile.

An empirically based representation of soil chemistry has been used in place of the ϕ term in Eq. [10] by Melamed et al. (1977), such that

$$\phi = \lambda K(R - C), \qquad [30]$$

where K is a transfer coefficient related to salt composition and soil properties, R is the maximum soil solution concentration at which there is no precipitation or dissolution, and $\lambda = 0$ for $S = 0$; $\lambda = 1$ when $S > 0$ or $R > C$. This assumes that R is independent of both the solution and exchange phases. Values of K were determined by curve-fitting so that calculated solute distributions matched those that were observed. The adjustment of K thereby corrected for many processes, including cation exchange and precipitation-dissolution reactions. Each application of this model to a new case would require adjustment of K to describe the new conditions.

Mechanistic models of soil chemistry have been developed and coupled to descriptions of solute transport (e.g., Tanji et al., 1972; Dutt et al., 1972; Oster and Rhoades, 1975; Jury and Pratt, 1980; Robbins et al., 1980b). These models are constructed upon the general principles of chemical equilibrium, and usually consider ionic strength to calculate ion activities, represent the relationships between $CaCO_3$-pH-CO_2, and precipitate and dissolve various solid phases depending upon solubility relationships. These models are intended to be used as subroutines in numerical solutions of Eq. [10], so that solution ion concentrations predicted by transport alone can be corrected for chemical equilibrium. In most cases, these models are transferrable to description of other situations, as it is assumed that the fundamental relationships of equilibrium chemistry are not site specific. All these models deal with relatively dilute solutions, as are found in most soils up to an electrical conductivity of about 20 mmho/cm. The models of Oster and Rhoades (1975) and Robbins et al. (1980b) are probably the most comprehensive in their consideration of fundamental processes. The former has a more complete description of pH-CO_2 equilibria, but without this component, the latter was able to describe transient Ca, Mg, SO_4, and EC transport in a mixed calcareous-gypsiferous system. The details of these models are too complex to be elaborated here.

VALIDATION OF THEORETICAL MODELS

Substantial validation of both miscible displacement theory and macropore transport theory has been accomplished for controlled laboratory conditions (van Genuchten and Cleary, 1979; Nielsen et al., 1980).

These experiments, usually conducted under constant flux and constant water content conditions, have provided soil scientists with some confidence in the formulations of fundamental processes represented by the theories. A much larger and presently still unresolved question focuses on the description of transient solute fluxes in the field using these models.

In the past few years, several models describing solute transport in the field based on the relationships of Eq. [10] have been proposed (e.g., Bresler, 1973b, 1975; Childs and Hanks, 1975; Melamed et al., 1977; Robbins et al., 1980b). Accurate calculation of the apparent diffusion coefficient and the average pore water velocity is crucial to the accurate description of solute movement. These calculations in the above models use predictions of water contents and water fluxes that are derived from assumed or measured soil water retention $h(\theta)$ and unsaturated hydraulic conductivity $K(\theta)$ functions. In fact, field measurements illustrate that both these water flow associated functions as well as the resulting solute transport parameters are spatially quite variable (Nielsen et al., 1973; Biggar and Nielsen, 1976; Van de Pol et al., 1977; Carvello et al., 1976). Whether these variations will preclude field estimation of fluxes using deterministic models built upon either the miscible displacement or macropore transport theories is not clear.

Two solute displacement studies deserve attention in light of the above discussion. The original field study that demonstrated the spatial variability of solute transport properties (Nielsen et al., 1973; Biggar and Nielsen, 1976) was conducted using appropriate initial and boundary conditions such that the analytical solution of Eq. [11] could be employed in data analysis. Twenty 0.01 ha leaching plots were established in a 150-ha field. Steady state water flow was established in each plot, followed by the application of a pulse of chloride, after which water was again continuously applied for the duration of the experiment. Triplicate samples were collected at regular time intervals at six depths in each plot during the leaching of the chloride. Analysis of this data using the appropriate solution of Eq. [10] provided a population of D and v values for the entire field (as represented by the 120 locations sampled). These populations both were found to be skewed and best characterized by a logarithmic normal distribution. The mode value of D was two orders of magnitude less than the mean, and the mode value of v was only 1/10 the mean. These results indicated that modeling solute displacement over the entire field using only one relationship between D, θ, and v (the required information according to the structure of the deterministic models listed above) was an improper procedure (Fig. 3). It was soon appreciated from these studies that some consideration of spatial variation in solute transport processes need be considered in describing transient solute movement in the field.

Recognizing these issues, Bresler et al. (1979) attempted the description of field measured solute movement using a model built upon Eq. [11] but incorporating a description of spatially variable $h(\theta)$ and $K(\theta)$ relationships. This was accomplished using a dimensionless scaling factor α_r (Peck et al., 1977):

$$\alpha_r = \lambda_r/\overline{\lambda}, \qquad [31]$$

where λ_r is the microscopic dimension of each of the parameters $h(\theta)$ and $K(\theta)$ and $\overline{\lambda}$ is some reference value of λ_r. The pressure head h_r and hydraulic conductivity K_r at a given water content θ and location r in the field can then be related to their respective "average" values $\overline{h}$ and $\overline{K}$ by

$$h_r = \overline{h}/\alpha_r \qquad [32]$$

and

$$K_r = \overline{K}\alpha_r^2. \qquad [33]$$

Values of α_r determined from several soils (Warrick et al., 1977; Bresler et al., 1979) fall within the 95% confidence range of ($0.2 \leq \alpha_r$

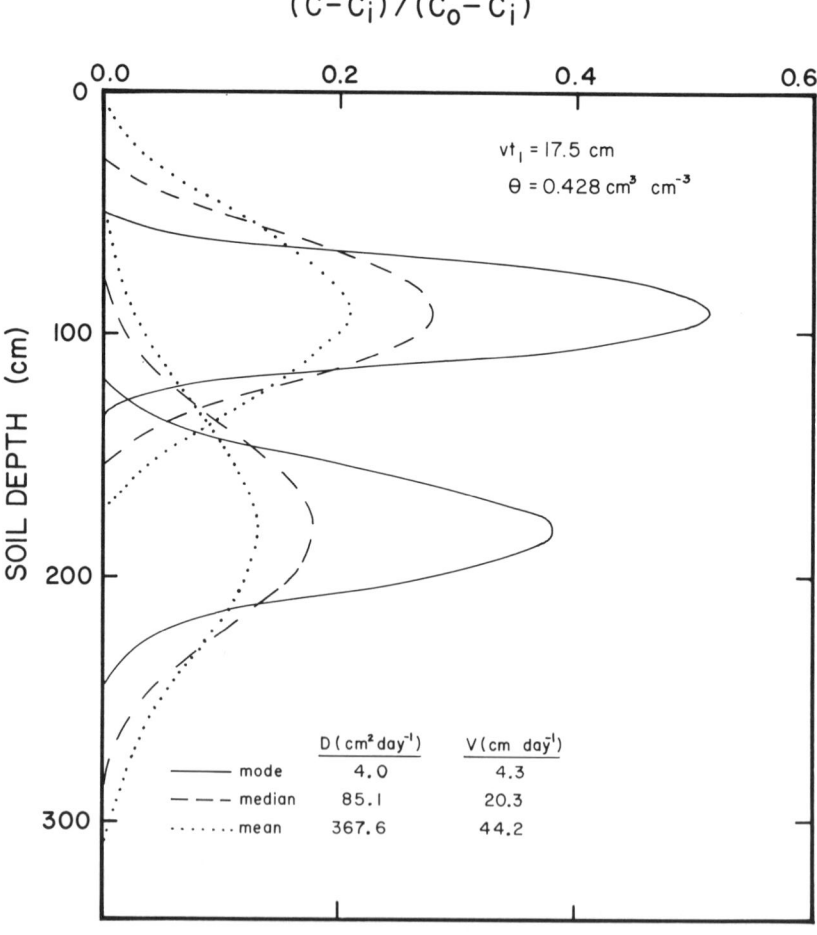

Fig. 3. Solute (chloride) concentration profiles calculated using field-measured values of D and ν (Biggar and Nielsen, 1976).

≤ 2.5). When these limits were used for α_r to generate $K(\theta)$ and $h(\theta)$ relationships, field solute concentrations were predicted better in some situations (Fig. 4). The conclusion was presented that approximate estimates of average field values of water and salt contents could be made using numerical models. Unfortunately, there were no measurements or estimates of the reliability of these models in predicting solute fluxes over the field.

SUMMARY

The development of both miscible displacement theory and macropore transport theory has greatly contributed to an understanding of solute displacement in isotropic, homogeneous porous media. Such media are closely approximated by carefully packed soil columns. Functional relationships have been developed and studied for adsorption-desorption, cation exchange, and soil solution chemical reactions that influence salt movement. Although the same fundamental principles are operative also under field conditions, present model formulations are not very effective in describing solute movement in the field. One reason is that models which use single-form functional relationships of water and solute transport parameters do not appropriately represent the large variation in these processes under field conditions.

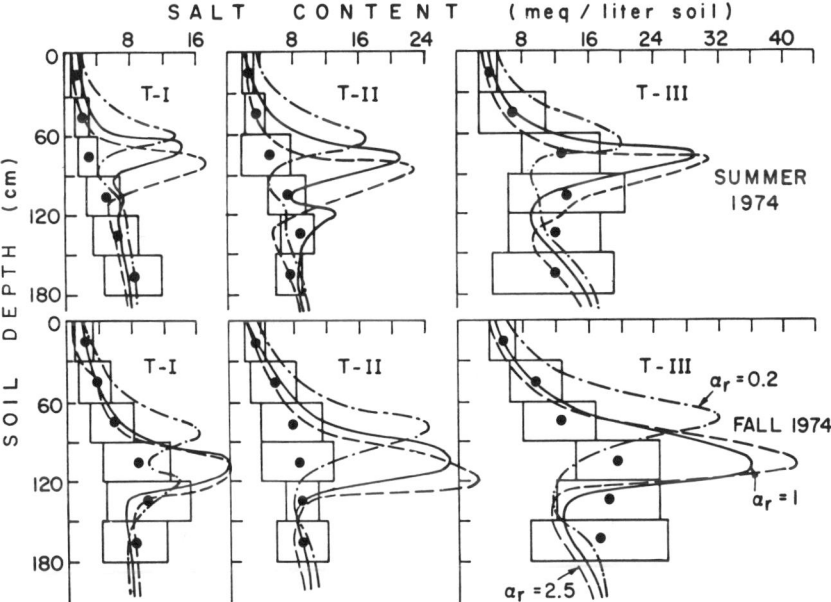

Fig. 4. Predicted and measured solute (chloride) concentration profiles for a transient field case. Three values of α_r were used to calculate solute distribution at three times (T-I, T-II, T-III) during each of two seasons (from Bresler et al., 1979).

Describing salt movement on a field basis is presently one of the most challenging problems facing soil scientists. One hundred years after the first development of the Darcy equation for water flow and 75 years after the first recognition that water and solutes do not always move at the same rate (Slichter, 1905), it is still questionable whether or not soil scientists can predict with a reasonable degree of certainty on a mechanistic basis what the rate of water and salt movement will be in a field. Only within the last decade has this problem been approached directly by measuring, on a large scale, solute distributions in field soil profiles. The first results indicate that while we understand basic principles, we do not yet have a good grasp of how to summarize and integrate their effect on a field basis.

LITERATURE CITED

Addiscott, T. M. 1977. A simple computer model for leaching in structured soils. J. Soil Sci. 28:554-563.

Ardakani, M. S., and A. D. McLaren. 1977. Absence of local equilibrium during ammonium transport in a soil column. Soil Sci. Soc. Am. J. 41:877-879.

Biggar, J. W., and D. R Nielsen. 1962. Miscible displacement: 2. Behavior of tracers. Soil Sci. Soc. Am. Proc. 26:125-128.

----, and ----. 1963. Miscible displacement: 5. Exchange processes. Soil Sci. Soc. Am. Proc. 27:623-627.

----, and ----. 1967. Miscible displacement and leaching phenomenon. In R. M. Hagan, H. R. Haise, and T. W. Edminster (ed.) Irrigation of agricultural lands. Agronomy 11:254-274. Am. Soc. of Agron., Madison, Wis.

----, and ----. 1976. Spatial variability of the leaching characteristics of a field soil. Water Resour. Res. 12:78-84.

Bolt, G. H., and F. A. M. deHaan. 1979. Anion exclusion in soil. p. 233-257. In G. H. Bolt (ed.) Soil chemistry: B. Physico-chemical models. Elsevier Scientific Publishing Co., Amsterdam.

Boast, C. W. 1973. Modeling the movement of chemicals in soils by water. Soil Sci. 115:224-230.

Bower, C. A., W. R. Gardner, and J. O. Goertzen. 1957. Dynamics of cation exchange in soil columns. Soil Sci. Soc. Am. Proc. 21:20-24.

Bresler, E. 1970. Numerical solution of the equation for interacting diffuse layers in mixed ionic systems with non-symmetrical electrolytes. J. Colloid Interface Sci. 33:278-283.

----. 1973a. Anion exclusion and coupling effects in nonsteady transport through unsaturated soils: I. Theory. Soil Sci. Am. Proc. 37:663-669.

----. 1973b. Simultaneous transport of solute and water under transient unsaturated flow conditions. Water Resour. Res. 9:975-986.

----. 1975. Two-dimensional transport of solutes during nonsteady infiltration from a trickle source. Soil Sci. Soc. Am. Proc. 39:604-613.

----, H. Bielorai, and A. Laufer. 1979. Field test of solution flow models in a heterogeneous irrigated cropped soil. Water Resour. Res. 15:645-652.

Cameron, D. A., and A. Klute. 1977. Convective-dispersive solute transport with a combined equilibrium and kinetic adsorption model. Water Resour. Res. 13:183-188.

Carvello, H. O., D. K. Cassel, J. Hammond, and A. Bauer. 1976. Spatial variability of in-situ unsaturated hydraulic conductivity of Maddock sandy loam. Soil Sci. 121:1-8.

Childs, S. W., and R. J. Hanks. 1975. Model for soil salinity effects on crop growth. Soil Sci. Soc. Am. Proc. 39:617-622.

Coats, K. H., and B. D. Smith. 1964. Dead-end pore volume and dispersion in porous media. Soc. Pet. Eng. J. 4:73-84.

Davidson, J. M., P. S. C. Rao, R. E. Green, and H. M. Selem. 1980. Evaluation of conceptual models for solute behavior in soil-water systems. p. 241-251. In A. Banin and

U. Kafkafi (ed.) Agrochemicals in soils. Proc. Int. Congress of ISSS, Jerusalem. 1976. Pergammon Press, London.

----, C. M. Rieck, and P. W. Santleman. 1968. Influence of water flux and porous material on the movement of selected herbicides. Soil Sci. Soc. Am. Proc. 32:629–633.

DeVault, D. 1943. The theory of chromatography. J. Am. Chem. Soc. 65:534–540.

Dutt, G. R., M. J. Schaffer, and W. J. Moore. 1972. Computer simulation model of dynamic bio-physicochemical processes in soils. Univ. Ariz. Agric. Exp. Stn. Tech. Bull. No. 196.

Frissel, M. J., and P. Poelstra. 1967. Chromatographic transport through soils. I. Theoretical evaluations. Plant Soil 26:285–301.

Gardner, W. R., and R. H. Brooks. 1957. A descriptive theory of leaching. Soil Sci. 83:295–304.

Glueckauf, E. 1955. Theory of chromatography, part 9. The "theoretical plate" concept in column separations. Trans. Faraday Soc. 51:34–44.

Green, R. E., P. S. C. Rao, and J. C. Corey. 1972. Solute transport in aggregated soils: tracer zone shape in relation to pore-velocity distribution and adsorption. p. 732–752. In Vol. 2. Proc. 2nd Symp. Fundamentals of Transport Phenomena in Porous Media. IAHR-ISSS. Guelph, Canada. 1972.

Hornsby, A. G., and J. M. Davidson. 1973. Solution and adsorbed fluometuron concentration distribution in a water-saturated soil: experimental and predicted evaluation. Soil Sci. Soc. Am. Proc. 37:823–828.

Jury, W. A. 1978. Transient changes in the soil-water system from irrigation with saline water: I. Theory. Soil Sci. Soc. Am. J. 42:579–585.

----, and P. F. Pratt. 1980. Estimation of the salt burden of irrigation drainage water. J. Environ. Qual. 9:141–146.

Kay, B. D., and D. E. Elrick. 1967. Adsorption and movement of lindane in soils. Soil Sci. 104:314–322.

Kemper, W. D., and J. C. Van Schaik. 1966. Diffusion of salts in clay-water systems. Soil Sci. Soc. Am. Proc. 30:534–540.

Kirda, C., D. R. Nielsen, and J. W. Biggar. 1973. Simultaneous transport of chloride and water during infiltration. Soil Sci. Soc. Am. Proc. 37:339–345.

Kirkham, D., and W. F. Powers. 1972. Advanced soil physics. Wiley-Interscience, N.Y.

Krupp, H. K., J. W. Biggar, and D. R. Nielsen. 1972. Relative flow rates of salt and water in soil. Soil Sci. Soc. Am. Proc. 36:412–417.

Lai, S. H., and J. J. Jurinak. 1971. Numerical approximation of cation exchange in miscible displacement through soil columns. Soil Sci. Soc. Am. Proc. 35:894–899.

Lapidus, L., and N. R. Amundson. 1952. Mathematics of adsorption in beds. VI. The effect of longitudinal diffusion in ion exchange and chromatographic columns. J. Phys. Chem. 56:984–988.

Lindstrom, F. T., and L. Boersma. 1970. Theory of chemical transport with simultaneous sorption in a water saturated porous medium. Soil Sci. 110:1–9.

Martin, A. J. P., and R. L. M. Synge. 1941. A new form of chromatogram employing two liquid phases. I. A theory of chromatography. Biochem. J. 55:1358–1364.

Melamed, D. R., R. J. Hanks, and L. S. Willardson. 1977. Model of salt flow in soil with a source-sink term. Soil Sci. Soc. Am. J. 41:29–33.

Nielsen, D. R., and J. W. Biggar. 1961. Miscible displacement in soils: 1. Experimental information. Soil Sci. Soc. Am. Proc. 25:1–5.

----, and ----. 1962. Misciple displacement: 3. Theoretical considerations. Soil Sci. Soc. Am. Proc. 26:216–221.

----, ----, and K. T. Erh. 1973. Spatial variability of field measured soil-water properties. Hilgardia 42:215–260.

----, ----, and C. S. Simmons. 1981. Mechanisms of solute transport in soils. p. 115–135. In I. K. Iskander (ed.) Modeling waste water rennovation: Land treatment. Wiley-Interscience, N.Y.

Olsen, S. R., and W. D. Kemper. 1968. Movement of nutrients to plant roots. Adv. Agron. 20:91–151.

Oster, J. D., and J. D. Rhoades. 1975. Calculated drainage water compositions and salt burdens resulting from irrigation with river waters in Western United States. J. Environ. Qual. 4:73–79.

Passioura, J. B. 1971. Hydrodynamic dispersion in aggregated media. I. Theory. Soil Sci. 111:339–344.

Paul, J. L., K. K. Tanji, and W. D. Anderson. 1966. Estimating soil and saturation extract composition by a computer method. Soil Sci. Soc. Am. Proc. 30:15–17.

Peck, A. J., R. J. Luxmoore, and J. L. Stolzy. 1977. Effects of spatial variability of soil hydrologic properties on water budget modeling. Water Resour. Res. 13:348–354.

Porter, L. K., W. D. Kemper, R. J. Jackson, and B. A. Stewart. 1960. Chloride diffusion in soils as influenced by moisture content. Soil Sci. Soc. Am. Proc. 24:460–463.

Rao, P. S. C., D. E. Rolston, R. E. Jessup, and J. M. Davidson. 1980. Solute transport in aggregated porous media: Theoretical and experimental evaluation. Soil Sci. Soc. Am. J. 44:1139–1146.

Reiniger, P., and G. H. Bolt. 1972. Theory of chromatography and its application to cation exchange in soils. Neth. J. Agric. Sci. 20:301–313.

Rible, J. M., and L. E. Davis. 1955. Ion exchange in soil columns. Soil Sci. 79:41–47.

Robbins, C. W., J. J. Jurinak, and R. J. Wagenet. 1980a. Calculating cation exchange in a salt transport model. Soil Sci. Soc. Am. J. 44:1195–1200.

————, R. J. Wagenet, and J. J. Jurinak. 1980b. A combined salt-transport chemical equilibrium model for calcareous and gypsiferous soils. Soil Sci. Soc. Am. J. 44:1191–1194.

Scheidegger, A. E. 1960. The physics of flow through porous media. Univ. of Toronto Press.

Selim, H. M., J. M. Davidson, and R. S. Mansell. 1976. Evaluation of a two-site adsorption-desorption model for describing solute transport in soils. p. 444–448. *In* Proc. Summer Computer Simulation Conf. Washington, D.C.

Skopp, J., and A. W. Warrick. 1974. A two-phase model for the miscible displacement of reactive solutes through soils. Soil Sci. Soc. Am. Proc. 38:545–550.

Slichter, C. S. 1905. Field measurement of the rate of movement of underground waters. U.S.G.S. Water Supply and Irrigation Paper No. 140. Washington, D.C.

Swanson, R. A., and G. R. Dutt. 1973. Chemical and physical processes that affect atrazine movement and distribution in soil systems. Soil Sci. Soc. Am. Proc. 37:872–876.

Tanji, K. K., L. D. Doneen, G. V. Ferry, and R. S. Ayers. 1972. Computer simulation analysis on reclamation of salt-affected soils in Jan Joaquin Valley, California. Soil Sci. Soc. Am. Proc. 36:127–133.

Thomas, G. W., and N. T. Coleman. 1959. A chromatographic approach to the leaching of fertilizer salts in soils. Soil Sci. Soc. Am. Proc. 23:113–116.

U.S. Salinity Laboratory Staff. 1954. USDA Handb. No. 60. U.S. Government Printing Office, Washington, D.C.

Van de Pol, R. M., P. J. Wierenga, and D. R. Nielsen. 1977. Solute movement in a field soil. Soil Sci. Soc. Am. J. 41:10–13.

Van der Molen, W. H. 1956. Desalinization of saline soils as a column process. Soil Sci. 81:19–27.

van Genuchten, M. Th. 1981. Non-equilibrium transport parameters from miscible displacement experiments. Research Rep. No. 119. U.S. Salinity Lab., Riverside, Calif.

————, and R. W. Cleary. 1979. Movement of solutes in soil: Computer simulated and laboratory results. p. 349–386. *In* G. H. Bolt (ed.) Soil chemistry: B. Physico-chemical models. Elsevier Scientific Publishing Co., Amsterdam.

————, and P. J. Wierenga. 1976. Mass transfer studies in sorbing porous media. I. Analytical solutions. Soil Sci. Soc. Am. J. 40:473–480.

————, J. M. Davidson, and P. J. Wierenga. 1974. An evaluation of kinetic and equilibrium equations for prediction of pesticide movement through porous media. Soil Sci. Soc. Am. Proc. 38:29–35.

Wagenet, R. J., J. W. Biggar, and D. R. Nielsen. 1977. Tracing the transformations of urea fertilizer during leaching. Soil Sci. Soc. Am. J. 41:896–902.

Warrick, A. W., G. J. Mullen, and D. R. Nielsen. 1977. Scaling field-measured soil hydraulic properties using a similar media concept. Water Resour. Res. 13:355–361.

Chapter 10

Irrigation Management and Crop Production as Related to Nitrate Mobility[1]

R. J. HANKS, D. W. JAMES, AND D. W. WATTS[2]

Much of past irrigation practice has been with little consideration to NO_3-N mobility. Nitrate mobility has been ignored because soil N losses were inexpensive to replace as well as the fact that some storage of NO_3-N occurs in the lower part of the root zone. Also, costs of irrigation equipment and water often were low enough that careful management was not required.

Traditional practices still persist in much of irrigated agriculture in spite of increased costs for energy and water as well as fertilizer. The NO_3-N pollution of groundwater has been a growing concern and certainly is an issue. Until now there has been considerable uncertainty regarding the relationship between irrigated agriculture and NO_3-N pollution of the environment. Certainly future irrigation management will need to be more concerned for both irrigation water and N fertilizer costs as well as environmental issues.

---

[1] Contribution from the Dep. of Soil Sci. and Biomet., Utah State Univ., and the Agricultural Eng. Dep., Univ. of Nebraska.

[2] Professors, Dep. of Soil Sci. and Biomet., Utah State Univ., Logan, UT 84322; and Agric. Eng. Dep., Univ. of Nebraska, Lincoln, NE 68583.

Copyright © 1983 ASA, SSSA, 677 South Segoe Road, Madison, WI 53711. *Chemical Mobility and Reactivity in Soil Systems.*

IRRIGATION, FERTILIZATION, AND YIELD

There are important interrelations between irrigation and fertilizer application on crop yield. This pronounced influence is, of course, the main reason for irrigation and fertilization in the first place. A classic example of this interrelation is shown in Fig. 1 from the data of Stutler et al. (1981) from an experiment done in El Salvador. Under this rather extreme condition, corn (*Zea mays* L.) yield was very low with low irrigation and no N fertilizer. With low irrigation, increasing N rates had a small influence on yield with high rates of N causing slight yield decreases. As irrigation was increased, yields generally increased except at the zero level. In this experiment the yield increased continuously as the irrigation level increased. These data represent an extreme in that the soil had very little storage capacity for either water or N so that both were needed before any appreciable yield resulted. Yield was independent of irrigation method for similar water use although less water was applied with trickler irrigation as compared to the furrow method.

A similar experiment was conducted by Bauder et al. (1975) using a continuous variable approach both with N fertilizer (22 levels) and irrigation water (9 levels). Figure 2 shows a plot of the production function for corn derived from their data.

The data in Fig. 2 illustrate many features of the interaction among N fertilization, irrigation, and corn yield. First the yield with no fertilization increased greatly as the water level (irrigation level) increased. At this location there was considerable stored soil moisture at the time of planting and there was some limited rainfall during the season. The field

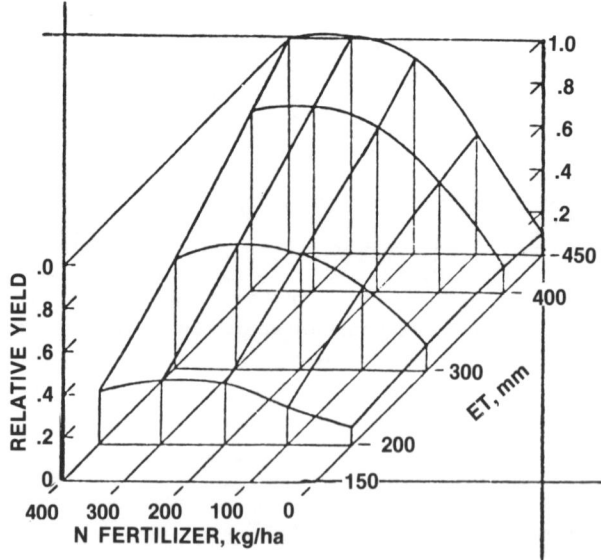

Fig. 1. Relation of evapotranspiration (ET) and N-fertilizer application to relative yield. Different irrigation regimes were used to cause different ET (Stutler et al., 1981).

area had been cropped with no fertilizer the previous year but there was still some residual N available to crops.

The second principle illustrated by Fig. 2 is that there is a broad range of fertilizer levels that gave approximately the same yield. For example, the data upon which Fig. 2 is based would indicate a yield of 92% of the maximum was attained with a relative fertilizer level range of 0.3 to 0.95 (165 to 520 kg/ha N) with a maximum at a fertilizer level of 0.6 (330 kg/ha N) for a water, or irrigation level of 1.0. Thus, if yield were the only criteria, best management may be to fertilize at about 0.3 of the maximum used in this study. While the maximum yield at any one irrigation treatment was not attained at the same fertility level, the differences in N levels that provide the maximum yield were small.

The third principle illustrated by Fig. 2 when considered together with Fig. 1 is that the data are site and time specific. The data probably would not apply to another location because soil water and nitrate storage would be different. Moreover, the rainfall before planting and during the season usually is different from year to year so even at the same location the results would not be the same from one year to the next. Also NO_3-N carry-over from year-to-year would be different depending on the fertilization and cropping practice history.

Thus, production function equations are of limited use because of site specificity. What is needed is a more general method that would allow one to predict the consequences of a particular fertilizer and irrigation management regime, taking into account all of site specific inputs. Recently, much research has been directed towards this need to develop a model that can be used for management decisions. This subject will be discussed later in more detail.

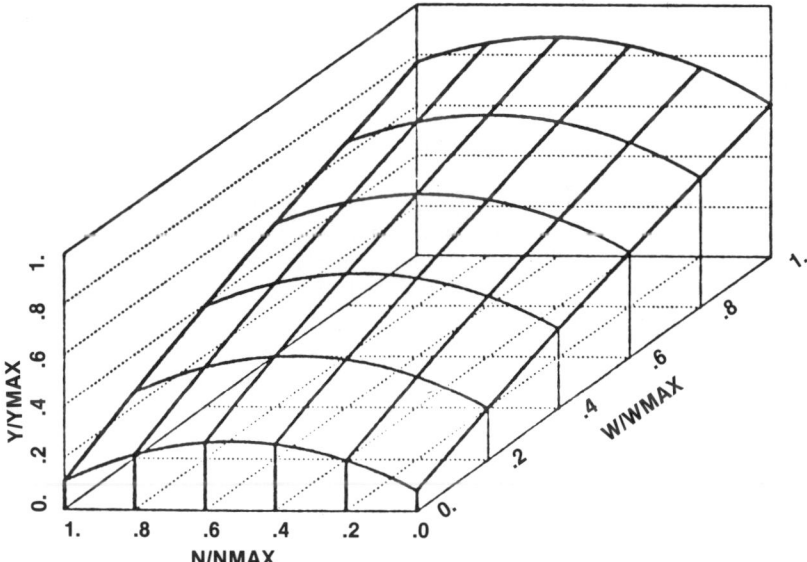

Fig. 2. Relation of relative water or irrigation application level, W/WMAX and relative N-fertilizer application level, N/NMAX, to relative corn yield Y/YMAX (Bauder et al., 1975).

IRRIGATION AND NITRATE LEACHING

As discussed earlier, the problem of NO_3-N leaching due to over-irrigation is a major concern with the extent of the problem still uncertain. It is common irrigation experience to see lighter colored or N-deficient crops where irrigation has been excessive. Watts and Hanks (1978) have discussed this problem relative to irrigation of the Nebraska Sandhills region. Where leaching occurred, corrective action could be taken, if crop yields were the only consideration, by adding N (for example, urea-ammonia nitrate) to the irrigation water with the resulting increased costs. However, this practice may increase pollution of the groundwater beneath the irrigated area.

Since traditional irrigation practice in many irrigated areas involves intentional leaching for salinity control, it would appear that much more sophisticated irrigation management schemes will need to be developed to minimize NO_3-N leaching out of the root zone without salt buildup in the soil.

Middleton et al. (1975) found that the furrow method of irrigation caused inefficient use of N fertilizer in Washington on sandy soils. Yields were less than maximum with fertilizer rates as high as 446 kg/ha N because of leaching of nitrate from the root zone. Where sprinkler irrigation was used, with frequency and amount tied to ET demand, yield of potatoes was as high as that obtained on loam soils with the use of 180 to 360 kg/ha N. Furrow irrigation required about twice as much water as sprinkler irrigation. They concluded furrow irrigation was impractical on very sandy soils.

The problem is aggravated because of irrigation application nonuniformity, especially with gravity (furrow) systems. Because of the nonuniformity of water application, even though no leaching would be occurring on parts of a field where the average amount of irrigation has been applied, there are other areas of the same field that will be leached as well as some areas that will receive inadequate irrigation. This problem was illustrated by Childs and Hanks (1975) comparing two different irrigation distributions as shown in Table 1. The gravity system is a very poor system in terms of uniformity of application approximating the wild flooding practice still being used in some areas. The solid set system has one of the best uniformity patterns currently available. The data in Table 1 indicate that even with good uniformity with no leaching on the average segment (53.1 cm of irrigation and rain), there was some leaching in some parts of the field. There were also some field areas with insufficient moisture so that overall yield was decreased. This illustrates that for an irrigation management scheme where water is applied to keep yields near maximum, there will almost always be some degree of leaching in part of the field. For the very poor uniformity situation (Table 1), both over and under-irrigation are emphasized to an even greater extent.

It should be emphasized that the amount of NO_3-N leaching would not necessarily be a direct function of the amount of water leached because many factors affect the N balance in soils such as a time and method

Table 1. An example of estimated leaching and relative crop yield for two different irrigation systems (Childs and Hanks, 1975).

Irrigation plus rain (cm)	Percent of area	Relative yield	Leaching water (cm)
Solid set sprinkler irrigation − coeff. of uniformity = 0.88 (parabola)			
40.4	10.4	0.85	0
46.7	24.8	0.92	0
53.1	29.6	0.96	0
59.5	24.8	0.99	3.6
65.8	10.4	1.00	8.8
	Weighted average	0.95	1.8
Gravity irrigation − coefficient of uniformity = 0.42 (rectangular)			
10.6	20.0	0.50	0
31.9	20.0	0.75	0
53.1	20.0	0.96	0
74.3	20.0	1.00	17.3
95.6	20.0	1.00	38.0
	Weighted average	0.84	11.1

of fertilization, decomposition of organic matter, and plant uptake. The timing of fertilizer application to timing of irrigation is particularly important.

IRRIGATION AND RESIDUAL NITROGEN CARRYOVER

Another problem relating to irrigation and N management is seasonal N carryover. James (1978) found that in Utah the type of crop, irrigation management, time, method, and form of N application all had an effect on carryover from fall to spring. Also, winter precipitation had a significant influence on N carryover—especially on coarse-textured soils. It was found that whenever sufficient N fertilizer was applied to obtain maximum crop yields in one season, there was significant N carryover to the next season. Many farmers routinely apply about 180 kg/ha N on irrigated land which results in considerable carryover. This practice is often used to offset poor irrigation management. It was found that with reasonable soil moisture control, maximum yields for corn could be obtained by applying 80 to 110 kg/ha N once in the season by banding at planting time. A soil test for N was found to successfully predict supplemental N needs. This test requires sampling below the top foot of soil to 4 ft.; N found below 4 ft. had little influence on yield.

MODELING THE COMBINED EFFECTS OF IRRIGATION NITROGEN FERTILIZATION ON YIELD AND NO_3-N LEACHING

Models have been under development for many years to evaluate the influence of the many complicated interactions of irrigation timing and amount of nitrate behavior in soil-plant systems. A review of some of these models is given by Tanji and Gupta (1978). Bresler and Laufer (1974) simulated the movement of nitrate in a uniform soil and considered nitrification but did not consider plant uptake of water or N. Saxton et al. (1977) have reported a research-type model where one-dimensional transport of water and water soluble N-species as a result of irrigation or rain events were considered. Since this model required a large number of input parameters, most of which were frequently unavailable, a simple management-type model was also developed. This simple model requires a minimum of input data but provides a gross description of the fate of N species in the plant root zone. Tanji et al. (1977) reported a steady-state N model which considered water and N flows on an annual or cropped cycle time scale. Davidson et al. (1978) have developed both research and simple models to describe the fate of N in the plant root zone. The authors indicated good agreement with limited field data but emphasized the value of the models to design future experiments to study basic N cycling processes.

While many models are available, the complexity of the problem as well as the input data requirements lead to difficulties for practical application. Much needs to be done to develop practical methods of evaluating and verifying the complex interactions of irrigation and nitrate mobility. As indicated by Tanji and Gupta (1978), although the theoretical aspects of N transformation and movement are fairly well developed, field verification of the models is very limited.

For the purposes of illustrating or simulating the many consequences of irrigation management and nitrate mobility, the model developed by Watts and Hanks (1978) is used. This model was designed to simulate field conditions using available data. Like other models, this model is limited in application, particularly in the need to include "active plant uptake" to simulate realistic conditions. Also, denitrification is not considered. A more comprehensive model building on the same foundation of water flow and plant uptake (Childs and Hanks, 1975) has been reported by Tillotson and Wagenet (1982).

The factors considered by the Watts and Hanks (1978) model are as follows in the sequence of computation:
1. Soil water flow and root uptake (predicts soil water content vs. depth and time).
2. Soil heat flow (predicts soil temperature vs. depth and time).
3. Mineralization rate of organic N forms to NH_4-N, conversion rate of NH_4-N to NO_3-N and the hydrolysis rate of urea to NH_4-N, if urea is applied.
4. Balance of organic N and the amount of NH_4-N on the exchange complex at different times and depths.
5. Transformation sink terms for urea, NH_4-N and NO_3-N.
6. Change in amount of NO_3 as a result of mass flow, dispersion diffusion, plant uptake, and biological transformations.
7. Increment time and repeat sequence adjusting addition of fertilizer, water, and diurnal variation of both evapotranspiration (ET) and N uptake demand.

One of the main difficulties encountered in developing this model so that it simulated field results was the realization that N uptake could involve an active uptake mechanism under certain conditions. When most of the mineral N had been moved to the lower half of the root zone due to over-irrigation, predicted root N uptake, based on concentration of NO_3-N in the soil water withdrawn for ET, underpredicted actual N uptake by about 50%. An additional circulation routine was therefore added to estimate active uptake.

Figure 3 shows a comparison of measured and computed NO_3-N concentration at the bottom of the root zone as function of time for a preplant N application. Generally, early season agreement was poor and late season agreement was excellent. Part of the poor agreement may be due to difficulty in assessing the beginning value of various N species.

Figure 4 illustrates the comparison of N uptake as a function of different fertilizer and water applications. The computed results agree well with the measured ones. Also shown is the potential uptake curve which was developed from field measurements under controlled irrigation and high soil N levels. This potential uptake function is one of the most critical components of this model.

Figure 5 shows good agreement between measured seasonal NO_3-N leaching loss and that estimated from water balance-concentration considerations.

Watts and Martin (1981) studied the effects of water and N management on leaching loss in sandy soils. They simulated three situations: (i) a preplant broadcast [disked in application of ammonium nitrate (NH_4NO_3)], (ii) a preplant application of anhydrous ammonia (NH_3), (iii) urea-ammonium nitrate (UAN) solution injected in the irrigation water in

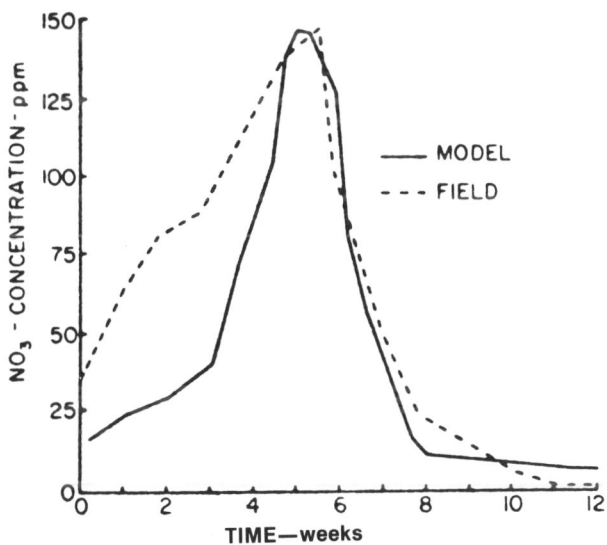

Fig. 3. Computed and measured nitrate concentration in the soil solution at a depth of 150 cm (Watts and Hanks, 1978).

five equal injections at 10-day intervals beginning at the eight-leaf stage. Control of irrigation amounts to minimize percolation and proper selection of N amount, timing, and source all significantly reduced leaching. However, simulation shown in Fig. 6 indicates that it is impossible to reduce leaching losses to zero and still maintain present production levels. It

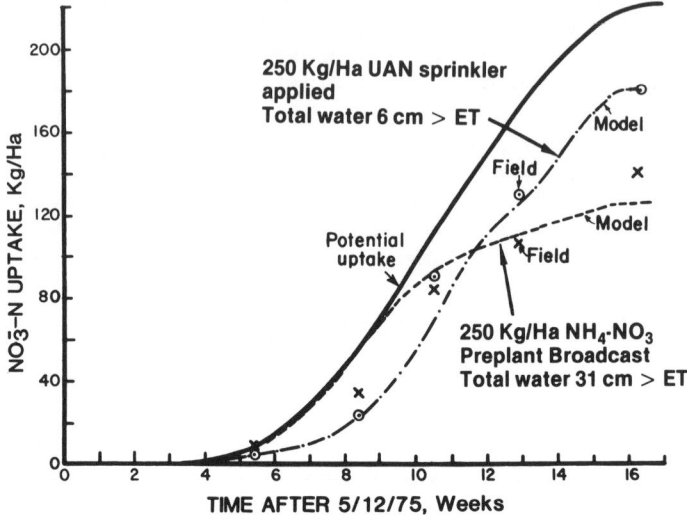

Fig. 4. Computed and measured N uptake by a growing corn crop compared with potential uptake (Watts and Hanks, 1978).

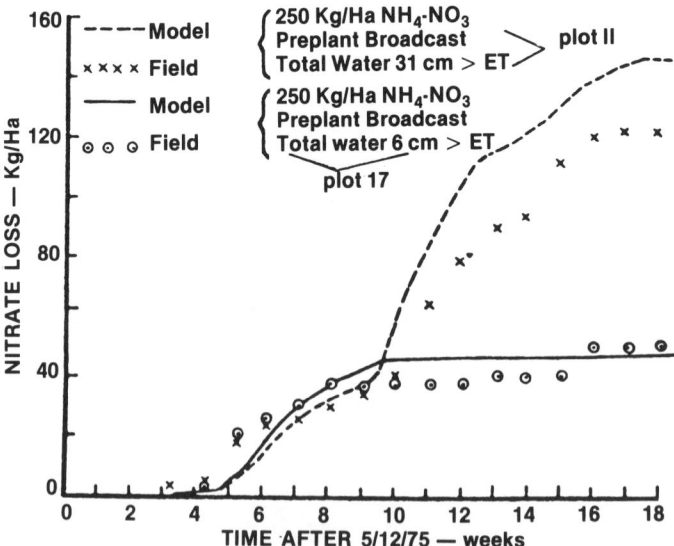

Fig. 5. Nitrate leaching losses with time as estimated by the model and water balance concentration calculations (Watts and Hanks, 1978).

should be noted that this simulation does not consider denitrification. Under conditions of high water table and low aeration denitrification may occur and minimize nitrate leaching into the groundwater.

Martin and Watts (1980) have used a modification of the Watts and Hanks (1978) model to investigate the possibility of reducing NO_3-N of the groundwater through irrigation management. Simulations were made for the sandy soils of the central Platte Valley of Nebraska where the groundwater had 10 to 40 ppm of NO_3-N. Water was pumped from the shallow groundwater and used to irrigate corn. The simulations indicated that both irrigation and NO_3-N management had a strong influence on corn production and on the increase or decrease of the groundwater NO_3-N content. The predictions indicate that under Platte Valley conditions, about 20 ppm NO_3-N in the groundwater will accompany reasonable corn yields. The potential for reducing NO_3-N (the authors use the term purification) increases as the groundwater NO_3-N concentration increases, provided that proper management is used. Above 20 ppm of NO_3-N, the crop can extract more NO_3-N from the irrigation water than is returned to the groundwater through leaching. Excessive irrigation amplifies the NO_3-N purification potential, over a wide range of water applications, as long as fertilizer rates are low and irrigation water NO_3-N concentrations are high. It is clear that if excessive fertilization is applied, the opportunity for purification is forfeited (Fig. 7).

The Nebraska experience is probably generally representative of problems encountered with irrigated crop production on coarse-textured soils. However, with finer soil textures and water-holding capacity increases, the required irrigation frequency decreases, and the leaching potential decreases. As leaching potential decreases, the plant water and

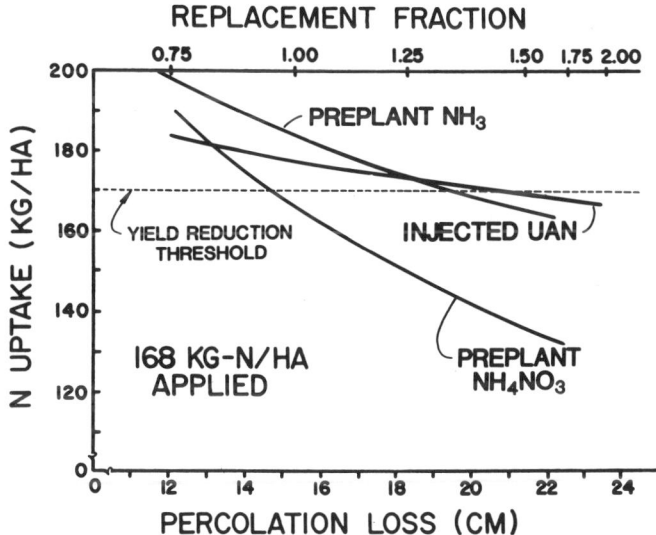

Fig. 6. Influence of fertilizer application method and irrigation amount on computed N uptake and percolation (leaching) (Watts and Martin, 1981).

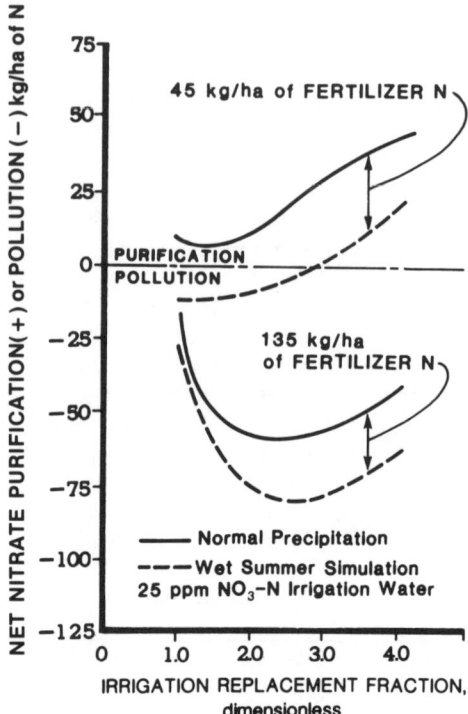

Fig. 7. Simulation of the influence of fertilization and irrigation amount on net nitrate decrease (+) (purification) or increase (−) (pollution) for different precipitation situations (Martin and Watts, 1980).

N use efficiency increase. The effect of increased plant N and water use efficiency is both decreased production costs and decreased hazard of NO_3-N loss to the environment.

LITERATURE CITED

Bauder, J. W., R. J. Hanks, and D. W. James. 1975. Crop production function determination as influenced by irrigation and nitrogen fertilization using a continuous variable design. Soil Sci. Soc. Am. Proc. 39:1187–1192.

Bresler, E., and A. Laufer. 1974. Simulation of nitrate movement in soils under transient unsaturated flow conditions. Israel J. Agric. Res. 23:141–153.

Childs, S. W., and R. J. Hanks. 1975. Model of soil salinity effects on crop growth. Soil Sci. Soc. Am. Proc. 39:617–622.

Davidson, J. M., D. A. Graetz, P. Suresh, C. Rao, and H. M. Selim. 1978. Simulation of nitrogen movement, transformation, and uptake in plant root zone. EPA-600/3-78-029.

James, D. W. 1978. Diagnostic soil testing for nitrogen availability: The effects of nitrogen fertilizer rate, time, method of application, and cropping pattern on residual soil nitrogen. Utah Agric. Exp. Stn. Bull. 497. 28 p.

Martin, D. L., and D. G. Watts. 1980. Potential nitrogen purification of groundwater through irrigation management. Preprint Paper No. 80-2028. Presented at 1980 meetings of Am. Soc. Agric. Eng., San Antonio, Tex., 15–18 June 1980.

Middleton, J. E., S. Roberts, D. W. James, T. A. Cline, B. L. McNeal, and B. L. Carlile. 1975. Irrigation and fertilizer management for efficient crop production on a sandy soil. Wash. State Univ. College of Agric. Res. Ctr. Bull. 811.

Saxton, K. E., G. E. Schuman, and R. E. Burwell. 1977. Modeling nitrate movement and dissipation in fertilized soils. Soil Sci. Soc. Am. J. 41:265-271.

Stutler, R. K., D. W. James, T. M. Fullerton, R. F. Wells, and E. R. Shipe. 1981. Corn yield functions of irrigation and nitrogen in Central America. Irrig. Sci. 2:79-81.

Tanji, K. K., M. Fried, and R. M. Van de Pol. 1977. A steady state conceptual nitrogen model for estimating nitrogen emissions from cropped lands. J. Environ. Qual. 6:155-162.

----, and S. K. Gupta. 1978. Computer simulation modeling for nitrogen in irrigated croplands. p. 79-130. *In* D. R. Nielson and J. G. MacDonald (ed.) Nitrogen in the environment. Academic Press, N.Y.

Tillotson, W. R., and R. J. Wagenet. 1982. Simulation of fertilizer nitrogen under cropped conditions. Soil Sci. 133:133-143.

Watts, D. G., and R. J. Hanks. 1978. A soil-water-nitrogen model for irrigated corn on sandy soils. Soil Sci. Soc. Am. J. 42:492-499.

----, and D. L. Martin. 1981. Effects of water and nitrogen management on nitrate leaching loss from sands. Trans. ASAE, p. 911-916.

Chapter 11

Principles of Microbial Processes of Chemical Degradation, Assimilation, and Accumulation[1]

DENNIS R. KEENEY[2]

There are many biological reactions in soils that can affect mobility of inorganic and organic compounds. The effects can be direct through reactions such as mineralization or indirect such as changes in redox or pH. This brief review will concentrate on the principles of various reactions rather than attempt to enumerate all possible reactions which may come into play.

MICROBIAL BIOMASS AND ACTIVITY

The soil microbial population is a fascinating array of organisms with diverse enzyme systems capable of deriving energy from metabolism of organic or inorganic compounds (Alexander, 1977). Most consider that microbial life in soils exists largely under starvation conditions (Gray and Williams, 1971; Alexander, 1977) except in the rhizosphere (Martin and Focht, 1977). Microbial life is adapted to the available energy and nutrient supply under widely differing environmental conditions.

---

[1] Contribution of the College of Agricultural and Life Sciences, Univ. of Wisconsin-Madison.
[2] Professor and chairman, Dep. of Soil Science, 1525 Observatory Drive, Univ. of Wisconsin, Madison, WI 53706.

Copyright © 1983 ASA, SSSA, 677 South Segoe Road, Madison, WI 53711. *Chemical Mobility and Reactivity in Soil Systems.*

Table 1. Microbial biomass in 0 to 23 cm of an unmanured calcareous plot under continuous wheat (*Triticum aestivum* L.).†

Weight of soil	2600 t ha^{-1}
Organic C in soil	26 t ha^{-1}
Annual input of organic matter	1.2 t C ha^{-1} year^{-1}
Gross turnover time of soil organic C	22 years
Volume of spherical organisms	0.71 mm^3 g^{-1}
Volume of hyphae	0.97 mm^3 g^{-1}
Number of bacteria and actinomycetes (direct count)	1600 × 10^6 g^{-1}
Fraction of soil pore space occupied by biovolume	0.35%
Microbial biomass from biovolume	190 µg C g^{-1}
Microbial biomass from CHCl$_3$ flush	530 kg C ha^{-1}
Microbial biomass, % of soil organic C	2.0%
Gross turnover time of biomass C‡	0.74 years

† Jenkinson (1977); values are for Broadbalk plot 03, Rothamsted.
‡ Minimal figure, assuming that at most some 60% of the annual input of C is converted to microbial C at some stage in the conversion process, and that steady-state conditions prevail.

Most microbial activity centers around organic compounds as the energy source (heterotrophy), but important autotrophic and photosynthetic reactions occur. Many attempts have been made to develop meaningful relationships between numbers of specific microbial groups in soil and a related activity, but most of these relationships have not proved useful due to the extreme difficulty in obtaining valid population estimates in a dynamic system. Recent research has emphasized biomass estimates. Jenkinson and Powlson (1976) estimated that the biomass in nine Rothamsted soils ranged from 1.7 to 3.7% of the soil organic C or from 440 to 2020 kg biomass C ha^{-1}. Anderson and Domsch (1980) obtained similar biomass values for 29 cultivated soils (range of 0.27 to 4.8% of total soil C). They estimated that, on average, about 25% of the biomass was associated with bacteria (range, 10 to 40%) and 75% with fungi (range, 60 to 90%), even though the former are of much higher total numbers in soils. Anderson and Domsch (1980) cite several references giving similar distribution of bacteria and fungal biomass. They further found that about 5% (range of 0.5 to 15.3%) of the soil total N was found in microbial biomass. On average, for the widely divergent agricultural soils they examined (0.08 to 38.0% organic C), the biomass bound mineral elements were: P, 83 kg ha^{-1}; K, 70 kg ha^{-1}; C, 11 kg ha^{-1}. The proportion of total C and N in the biomass of forest soils was much lower.

Jenkinson (1977) summarized numerous related values for a well-studied Broadbalk plot (Table 1). The annual input of organic C is only about twice the standing crop of biomass. He calculated the gross turnover time of biomass C as 0.74 year, or less than two cell divisions per year. This suggests a very low maintenance energy for the soil population.

It is commonly estimated that 1 to 4% of the total soil N is mineralized each year (e.g., Alexander, 1977). This is remarkably similar to the proportion of N as biomass N (Anderson and Domsch, 1980). Jenkinson's (Table 1) estimate of a long turnover time lends credance to the N-miner-

Table 2. Some enzymatic activities observed in soils.[†]

Enzyme grouping	Reaction catalyzed
Dehydrogenases	$XH_2 + A \rightarrow X + AH_2$
Oxidases	Reduced compound + $O_2 \rightarrow$ more oxidized compound + CO_2
Phosphatases	mono or diesters of orthophosphate + $H_2O \rightarrow$ R–OH + $PO_4^=$
Sulfatases	Sulfate ester + $H_2O \rightarrow$ R–OH + $SO_4^=$
Cellulases	Hydrolysis of β-1,4 glucan links in cellulose
Lipases	Lipid $\rightarrow$ glycerol + fatty acids
Proteinases	Protein $\rightarrow$ peptides + amino acids
Urease	Urea $\rightarrow CO_2 + NH_3$

[†] Adapted from Skujins (1967), Alexander (1977), and Ladd (1978).

alization values and indicates that the greater part of the N and likely other nutrients in microflora are continuously recycling, albeit at a slow rate.

Jenkinson (1977) states the situation well: "The soil biomass is the eye of the needle through which all natural organic materials that enter the soil must pass, often more than once, as they are degraded to inorganic compounds. . ." Soil biomass is a large, mostly dormant population with great diversity and the ability to survive hard times. To more fully visualize this "eye of the needle," we can assume that ca. 2.5% of the soil C is biomass. At steady-state (no fresh C residues), the microbial biomass of a mineral soil with 2% C, 2% as biomass, 25% of the biomass as bacteria, and 10% C fresh weight of bacteria, a standing crop bacterial biomass of about 1400 kg ha^{-1}, and fungal biomass of about 4200 kg ha^{-1} can be calculated.

When the starving microbial population is offered an energy source, specific populations rapidly increase in numbers and activity to utilize this energy. Microbes possess enormous genetic potential for enzyme synthesis (Gray and Williams, 1971; Martin and Focht, 1977). While all enzymes are produced within cells, some are excreted or released on cell lysis (the extracellular or soil enzymes) (Skujins, 1967; Ladd, 1978) and others are intracellular. Some extracellular enzymes, e.g., proteinases and polysaccharidases, hydrolyze large, insoluble polymers to utilizable monomers while other soil enzymes such as urease and the phosphatases catalyze the hydrolytic cleavage of low molecular weight compounds. Reviews such as Skujins (1967) and Burns (1978a) discuss the soil enzymes in considerable detail. Some exoenzymes such as urease accumulate in soil due to as-yet undefined stability mechanisms (Bremner and Mulvaney, 1978), while others are rapidly induced or re-repressed when presented with an energy-yielding or growth-inhibiting substrate. Most synthetic compounds such as hydrocarbons and organic pesticides are also degraded, albeit sometimes slowly (Miller, 1973; Kaufman, 1974; Cervelli et al., 1978). Table 2 (Skujins, 1967; Alexander, 1977; Ladd, 1978) lists some of the important enzyme activities observed in soils. Intact enzymes are difficult to extract from soils in active form due to the tight bonding of these protein molecules to soil materials (Burns, 1978b). Many of these enzymes involve conversion of relatively high molecular weight immobile compounds to lower molecular weight and more mobile compounds.

Table 3. Classification of microbial reactions in relation to energy sources.†

Class	Electron Donor	Electron Acceptor	Products
Photoautotrophic	H_2O, H_2S, H_2R	CO_2	$(HCHO)_n$ and other reduced compounds
Respiration			
Aerobic	Organic compounds	O_2	$CO_2 + H_2O$
Anaerobic	Organic compounds	Many, ranging from the same molecule, another molecule, CO_2 or inorganic compound	Many, ranging from reduced compounds, oxidized compounds, CO_2 and H_2O
Chemoautotrophic	Inorganic compound	O_2 or another inorganic compound	Oxidized inorganic compound

† Delwiche (1967).

It would appear that microbial activity in soils would be easily measured from such parameters as CO_2 evolution, biomass, or enzyme activity. But the complex soil system has never been easy to describe, and this appears to hold true for microbial activity parameters. For example, Nannipieri et al. (1978) concluded that no single measurement of microbial activity (CO_2 evolution, phosphatase or urease activity, or ATP content) was sufficient to interpret microbial growth in their glucose or glucose plus N- or P-amended systems, while Sparling et al. (1981) found no correlation between ATP, biomass, and enzymatic activities when soils were amended with glucose.

ENERGY SOURCES FOR MICROBIAL REACTIONS

Energy sources and environmental conditions interact to determine the microbial ecology during any microbial process which affects mobility of a given compound or element. Delwiche (1967) presents a concise review of energy relationships in soil microbiological reactions (Table 3). The fixation of CO_2 in photoautotrophic reactions, which are the same for higher plants and algae, is the primary energy input into soil micro-ecosystems through root exudates and animal and plant remains.

AEROBIC RESPIRATION

The fixed C is released back to the atmosphere as CO_2 by plant, animal and microbial respiratory activity, with soil heterotrophic microorganisms dominating (Gray and Williams, 1971; Wagner, 1975; Alexander, 1977). A varying amount of organic C is retained in the biomass; commonly quoted aerobic respiratory efficiencies (organic C metabolized/biomass C formed) range from 20 to 60% for bacteria and fungi, depending on substrate, while anaerobic C mineralization is much less efficient (Miller, 1973). Most fresh plant residues metabolize rapidly in soil, and within a few weeks to a few months, much of the C is released as CO_2.

Carbon turnover in soil through heterotrophic soil microorganisms also involves the essential nutrients, particularly N. At high rates of addition of low-N containing residues to aerobic soils, N usually is initially limiting the rate of breakdown (Martin and Focht, 1977). As turnover continues and CO_2 is released, the C:N ratio narrows and eventually some of the N will remain in the inorganic form (Alexander, 1977). These same principles hold for mobilization of S and P (Stevenson, 1982). The rate of decomposition also varies with the compound; lignin degrades slowly, followed by the cellulose and hemicelluloses, while water soluble materials degrade rapidly. The large polymeric molecules which compose the bulk of biological residues are converted to smaller units by exoenzymes so that they can be utilized for energy and cell synthesis. Polysaccharides are cleaved to sugars, proteins are converted to peptides and amino acids, fatty acids are oxidized to acetic acid units, and lignin is converted to simple phenolic compounds which can be further converted to aliphatic compounds (Martin and Focht, 1977). Ultimately, soil humus is formed (Martin and Focht, 1977; Stevenson, 1982). Soil humus has many properties which affect mobility of elements and compounds, including slow release of nutrients, sorption, cation exchange, pH buffering, release of nutrients from insoluble soil minerals, and chelation/complexation (Stevenson, 1982).

In addition to the degradation of organic residues, microbial processes are important in the degradation (i.e., change in molecular structure) of organic pesticides that affect their efficacy, mobility and persistence (Hill, 1978; Khan, 1980). Hill (1978) provides an especially complete discussion on microbial transformation of pesticides. Most (but not all) pesticides are degraded fairly rapidly in the environment, and the more persistent pesticides such as the organochlorine insecticides are no longer in use. Numerous factors interrelate to affect the rate of pesticide degradation. Metabolism involves either constitutive or induced enzymes, and while most reactions occur most readily under aerobic conditions (due either to the oxidative nature of the reaction or the activity of the microbial systems, anaerobic degradation can be more rapid for some pesticides such as DDT [1,1,1-trichloro-2,2-bis-(chlorophenyl)ethene] and γ-BHC (γ isomer of 1,2,3,4,5,6-hexachlorocyclohexane) (Yoshida, 1975; Matsumura and Benezet, 1978). Relatedly, it appears that petroleum hydrocarbons are more rapidly degraded in aerobic than in reduced sediments (Hambrick et al., 1980).

Autotrophic (or chemoautotrophic) reactions occupy a special ecological niche. The most important is nitrification, the oxidation of NH_4^+ to NO_2^- and NO_3^-. The energy obtained from oxidation of NH_4^+ or NO_2^- is coupled with CO_2 or HCO_3^- reduction using O_2 mainly as a terminal electron acceptor. This, of course, dramatically increases the mobility of N and adds H^+ to the ecosystem. Nitrification has been the subject of numerous reviews (Verstraete, 1981; Schmidt, 1982), and needs little further elaboration here. Heterotrophic oxidation of reduced N compounds has been amply demonstrated in the laboratory, but there is no evidence of these reactions in nature (Schmidt, 1982).

Sulfur also undergoes energy-yielding autotrophic transformations carried out by the bacterium, *Thiobacillus* (Starkey, 1966; Martin and Focht, 1977). Acid is produced in the reaction, and the pH optimum varies with the species ranging from ca. 7 for *T. thioparus* to 2 to 3 for *T. thiooxidans*. There is remarkable diversity in this group. They all use elemental S or incompletely oxidized S compounds, and SO_4^{-2} is the ultimate end product. *Thiobacillus denitrificans* can also grow anaerobically using NO_3^- as the H acceptor; *T. thiooxidans* and *T. ferrooxidans* are extremely acid tolerant. *Thiobacillus ferrooxidans* is involved in the formation of acid mine drainage by catalyzing iron oxidation during the degradation of pyrites (Walsh and Mitchell, 1972). Pepper and Miller (1978) isolated two heterotrophic bacteria capable of oxidizing S^0 and $S_2O_3^{-2}$ in soils and proposed that heterotrophic sulfur oxidation may be of significance in neutral to alkaline soils. Phototrophic purple and green sulfur bacteria can oxidize H_2S to S^0 or SO_4^{-2} but are important only in aquatic systems (Blackburn et al., 1975; Pfennig, 1975).

Iron (Fe^{+2}) can be oxidized by the chemoautotrophs, *T. ferrooxidans* and *Ferrobacillus ferrooxidans* with the production of metabolic energy (Aristovskaya and Zavarzin, 1971; Martin and Focht, 1977), and by heterotrophic bacteria which form Fe deposits. The heterotrophs form structures which enable attachment to surfaces, and are a major problem in the clogging of pipes and tile drains (Grass et al., 1973). Numerous heterotrophs also degrade mobile organic Fe compounds forming Fe precipitates (Aristovskaya and Zavarzin, 1971).

A number of other inorganic elements and compounds can be oxidized microbially, although it is questionable if any of these reactions yield metabolic energy. Arsenite, which is highly mobile and toxic, is oxidized microbially to the less mobile and relatively nontoxic arsenate (Turner, 1949; Quastel and Scholefield, 1953); insoluble minerals can also be oxidized microbially to more mobile forms (Ehrlich, 1971).

Selenium has many chemical and biological reactions in common with S, and microbial oxidation of H_2Se to SeO_4^{-2} has been reported (Ehrlich, 1971). The microbial oxidation of Mn^{+2} to insoluble MnO_2 is widely known, but it is not clear if this oxidation furnishes any useful energy, especially since no autotrophic Mn-oxidizing organisms have been isolated and organic C is always required (Ehrlich, 1971; Alexander, 1977). There are also reports of microbial oxidation of Mo and I (Alexander, 1977). Bartlett and James (1979) found that the immobile, nontoxic Cr(III) is oxidized to toxic Cr(IV) in soils by Mn^{+4}, giving yet another example of indirect effects of microbial activity on mobility.

There are numerous indirect effects on chemical mobility of compounds and elements arising from aerobic respiration in soils. Organic matter decomposition utilizes O_2, and releases CO_2. This can affect the redox potential, carbonate-bicarbonate equilibria, and pH. Many reactions such as nitrification produce H^+ and may markedly affect pH in N-fertilized fields (Pierre et al., 1971). Soil pH and redox markedly affect the mobility of most metals (Lindsay, 1972) and of metal-chelate equilibria (Sommers and Lindsay, 1979). Low molecular weight complexing agents are continually produced and decomposed (Stevenson, 1982).

ANAEROBIC METABOLISM

Rapidly respiring microbial populations require copious amounts of O_2. Under any condition where the O_2 diffusion rate is insufficient to meet the demand, anaerobic conditions will develop, and alternate electron acceptors are used. While flooded soils and aquatic sediments are the most consistently anaerobic, pockets of O_2 depleted areas can exist in arable field soils especially those amended with organic residues and wastes (Miller, 1973; Knowles, 1981). Anaerobic microbial reactions greatly affect the redox potential (Eh) (Ponnamperuma, 1972). Providing that organic C is present for energy, the various electron acceptors are sequentially exhausted, and electron availability increases (Eh numerically declines). Table 4 gives the sequence of reactions commonly occurring in waterlogged soils and sediments after O_2 disappears.

Denitrification is the most significant anaerobic reaction occurring in soils from an agronomic standpoint. However, since nitrification and NO_3^- reduction to NH_4^+ also produce gaseous N oxides, Firestone (1982) suggests it be defined as "the process in which N oxides serve as terminal electron acceptors for respiratory electron transport leading from a reduced electron donating substrate through numerous electron carriers to a more oxidized N oxide." Denitrification requires the presence of the bacteria, suitable electron donors (usually organic C compounds, but reduced S compounds and H_2 are also utilized by specific bacteria), anaerobic conditions (at least at the microsite) and the N oxide (NO_3^-, NO_2^-, NO, or N_2O).

Reduction of immobile Mn^{+4} to soluble Mn^{+2} occurs at a redox potential of $+300$ mv, close to that for denitrification (Grass et al., 1973). The manganic ion is also solubilized by *Thiobacillus* as it can utilize Mn^{+4} as a terminal electron acceptor during S oxidation (Martin and Focht, 1977). Iron reduction is coupled with organic matter oxidation by a number of ubiquitous bacteria (Aristovskaya and Zavarzin, 1971). The grayish coloration of gley (poorly drained) soils is commonly ascribed to ferrous sulfide deposits.

Table 4. Sequence of reactions occurring as Eh declines in waterlogged soils and sediments.†

Approximate Eh range	Reaction
mv	
$> +350$	O_2 disappears
$+300$ to $+100$	$NO_3^- \rightarrow NO_2^- \rightarrow NO \rightarrow N_2O \rightarrow N_2$
$+100$ to $+200$	Fe^{+2}, Mn^{+2} formed
0 to -200	$SO_4^= \rightarrow S^=$
< -200	CH_4 formed

† Adapted from Keeney (1973).

Sulfate reduction has been widely studied (Bremner and Steele, 1978). Reduction of SO_4^{-2} to $S^=$ is carried out by the ubiquitous *Desulfovibrio* and related bacteria that proliferate under anaerobic conditions using SO_4^{-2} as the terminal electron acceptor. Reduction occurs at a redox potential < 150 to 200 mv, is optimum at pH 7 (Connell and Patrick, 1968) and is delayed by compounds such as NO_3^-, Mn^{+4}, and Fe^{+3} that poise the redox potential at higher levels (Engler and Patrick, 1973). Since Fe^{+3} is reduced before SO_4^{-2}, Fe^{+2} is always present and FeS rather than H_2S normally accumulates (Connell and Patrick, 1969). However, in old rice (*Oryza sativa* L.) paddy soils low in Fe^{+2} due to accumulation of insoluble iron sulfides, H_2S may accumulate to toxic levels (Yoshida, 1975). The formation of CdS in flooded soils also greatly reduces Cd availability compared to aerobic soils (Bingham et al., 1976).

It has been commonly accepted that H_2S is the major volatile S compound released from natural reduced environments such as swamps, but recent studies (reviewed by Bremner and Steele, 1978) suggest that alkyl sulfides and carbonyl sulfide (COS_2) are much more important. These compounds may be produced under aerobic or waterlogged conditions.

Selenium transformations in reduced environments are similar to those of S. Selenite (SeO_3) is readily reduced biologically to elemental Se (Ehrlich, 1971; Doran and Alexander, 1977), and dimethyl selenide formation in anaerobic soils has been demonstrated (Francis et al., 1974).

The biochemistry of H_2 and CH_4 formation in highly reduced environments is complex. Methane bacteria are strict anaerobes and are abundant in anaerobic environments (Stadtman, 1967). While CH_4 can be formed directly from organic substrates such as acetate, it is now well accepted that much of the CH_4 is produced from the reduction of CO_2 by H_2 (Stadtman, 1967). The CH_4-producing bacteria use H_2 for energy. While H_2 is an end product of anaerobic metabolism, its utilization in CH_4 formation minimizes its loss from flooded soils (Yoshida, 1975).

Similar to SO_4^{-2} reduction, CH_4 formation is inhibited by various oxidants (Bollag and Czlonkowski, 1973). In systems such as lake sediments, SO_4^{-2} reduction and CH_4 production are related (Cappenberg, 1975). This is likely due to a substrate relationship involving competition for electrons, H_2 and acetate (Zaiss, 1981); sulfide toxicity may also be involved (Jakobsen et al., 1981).

Fermentation reactions (biological energy-yielding redox reactions in which organic compounds serve as final electron acceptors) are important reactions in flooded soils and sediments (Yoshida, 1975). The primary end products are ethanol, formate, lactate, propionate, butyrate, H_2, and CO_2.

As mentioned previously, some insecticides, most notably γ-BHC and DDT, are degraded more rapidly in anaerobic than in aerobic environments. Mechanisms are discussed by Yoshida (1975).

Biological phosphate reduction to H_3PO_3, H_3PO_2, P_2H_2, and PH_3 has been reported, but probably does not occur in natural systems (Ehrlich, 1971; Burford and Bremner, 1972). However, redox reactions involving Fe- and Mn-oxides can affect P mobility (Keeney and Wildung, 1977). Arsenic, as arsenate, readily undergoes microbial reduction to arsenite

Table 5. Microbial reactions which influence mobility of elements and compounds in soils.†

Reaction
1. Mineralization of inorganic ions during organic matter decomposition when amount of element present is in excess of microbial demand.
2. Immobilization of inorganic ions to organic forms to satisfy needs of microbial growth.
3. Oxidation of inorganic ions, particularly as energy sources by autotrophs although numerous heterotrophs also oxidize compounds but do not obtain energy from the process.
4. Reduction of oxidized elements, particularly when an alternate electron acceptor to O_2 is required. However, some reduction reactions occur in which the oxidized species is not needed as an electron acceptor.
5. Indirect transformations leading to changes in the microenvironment (pH, depletion of O_2 to lower Eh, addition of CO_2).
6. Production or degradation of organic ligands.
7. Methylation to produce more mobile and/or volatile compounds.

† Adapted from Alexander (1977).

(Walsh and Keeney, 1975) and arsenite can easily reach phytotoxic levels in the soil solution of waterlogged, As-contaminated soils (Hess and Blanchar, 1977). Recently Cheng and Focht (1979) demonstrated that under reduced conditions, As is released from soils only as volatile arsine (AsH_3) refuting earlier work (Woolson, 1977) that alkyl arsines are volatilized from As-contaminated soil.

Numerous other elements can be reduced microbially, including Te, V, Hg, and Co (vitamin B_{12} complex) (Ehrlich, 1971; Wood, 1973).

METHYLATION

Microbial methylation reactions which increase mobility of Hg, Sn, Pt, Pd, Au, Te, Se, and Th (Ehrlich, 1971; Wood, 1973) and Pb (Wong et al., 1975) have been recognized. Methylation of Hg is of particular interest in view of the toxicity of Hg and the increased anthropogenic input of Hg in the atmosphere and waters (National Academy of Sciences, 1978), the findings that fish and shellfish concentrate methylmercury, and that methylmercury can be passed to fish-eating birds and mammals. Mercury methylation occurs under aerobic and anaerobic conditions with methyl-B_{12} as a precursor. Dimethylation also occurs, leading to input of Hg into the atmosphere (National Academy of Sciences, 1978).

SUMMARY

Microbial reactions may affect the mobility and toxicity of elements and compounds in several ways. The soil biomass is relatively small and largely static until presented with an energy source, but has tremendous diversity. The effects of microbial processes must be considered in any definitive study of mobility of compounds in soils. Table 5 lists some of the major effects which have been briefly covered in this review.

LITERATURE CITED

Alexander, M. 1977. Introduction to soil microbiology. 2nd ed. John Wiley, N.Y. 467 p.

Anderson, J. P. E., and K. H. Domsch. 1980. Quantities of plant nutrients in the microbial biomass of selected soils. Soil Sci. 130:211–216.

Aristovskaya, T. V., and G. A. Zavarzin. 1971. Biochemistry of iron in soil. p. 385–408. *In* A. D. McLaren and J. J. Skujins (ed.) Soil biochemistry, Vol. 2. Marcel Dekker, N.Y.

Bartlett, R., and B. James. 1979. Behavior of chromium in soils: III. Oxidation. J. Environ. Qual. 8:31–35.

Bingham, F. T., A. L. Page, R. J. Mahler, and T. J. Ganje. 1976. Cadmium availability to rice in sludge-amended soil under "flood" and "non-flood" culture. Soil Sci. Soc. Am. J. 40:715–719.

Blackburn, T. H., P. Kleiber, and T. Fenchel. 1975. Photosynthetic sulfide oxidation in marine sediments. Oikos 26:103–108.

Bollag, J.-M., and S. T. Czlonkowski. 1973. Inhibition of methane formation in soil by various nitrogen-containing compounds. Soil Biol. Biochem. 5:673–678.

Bremner, J. M., and R. L. Mulvaney. 1978. Urease activity in soils. p. 149–197. *In* R. G. Burns (ed.) Soil enzymes. Academic Press, London.

----, and C. G. Steele. 1978. Role of microorganisms in the atmospheric sulfur cycle. Adv. Microb. Ecol. 2:155–201.

Burford, J. R., and J. M. Bremner. 1972. Is phosphate reduced to phosphine in waterlogged soils? Soil Biol. Biochem. 4:489–495.

Burns, R. G. (ed.) 1978a. Soil enzymes. Academic Press, London. 380 p.

----. 1978b. Enzyme activity in soil: some theoretical and practical considerations. p. 295–340. *In* R. G. Burns (ed.) Soil enzymes. Academic Press, London.

Cappenberg, T. E. 1975. Relationships between sulfate-reducing and methane-producing bacteria. Plant Soil 43:125–139.

Cervelli, S., P. Nannipieri, and P. Sequi. 1978. Interactions between agrochemicals and soil enzymes. p. 251–294. *In* R. G. Burns (ed.) Soil enzymes. Academic Press, London.

Cheng, C. N., and D. D. Focht. 1979. Production of arsine and methylarsine in soil and in culture. Appl. Environ. Microbiol. 38:494–498.

Connell, W. E., and W. H. Patrick, Jr. 1968. Sulfate reduction in soil: effects of redox potential and pH. Science 159:86–87.

----, and ----. 1969. Reduction of sulfate to sulfide in waterlogged soil. Soil Sci. Soc. Am. Proc. 33:711–715.

Delwiche, C. C. 1967. Energy relationships in soil biochemistry. p. 173–193. *In* A. D. McLaren and G. H. Peterson (ed.) Soil biochemistry. Marcel Dekker, N.Y.

Doran, J. W., and M. Alexander. 1977. Microbial formation of volatile selenium compounds in soil. Soil Sci. Soc. Am. J. 41:70–73.

Ehrlich, H. L. 1971. Biogeochemistry of the minor elements in soil. p. 361–384. *In* A. D. McLaren and J. J. Skujins (ed.) Soil biochemistry, Vol. 2. Marcel Dekker, N.Y.

Engler, R. M., and W. H. Patrick, Jr. 1972. Sulfate reduction and sulfide oxidation in flooded soil as affected by chemical oxidants. Soil Sci. Soc. Am. Proc. 37:685–688.

Firestone, M. K. 1982. Biological denitrification. *In* F. J. Stevenson (ed.) Nitrogen in agricultural soils. Agronomy 22:289–326. Am. Soc. of Agron., Soil Sci. Soc. of Am., Crop Sci. Soc. of Am., Madison, Wis.

Francis, A. J., J. M. Duxbury, and M. Alexander. 1974. Evolution of dimethylselenide from soils. Appl. Microbiol. 28:248–250.

Grass, L. B., A. J. MacKenzie, B. D. Meek, and W. F. Spencer. 1973. Manganese and iron solubility changes as a factor in tile drain clogging: II. Observations during the growth of cotton. Soil Sci. Soc. Am. Proc. 37:17–21.

Gray, T. R. G., and S. T. Williams. 1971. Soil micro-organisms. Hafner Publishing Co., N.Y. 240 p.

Hambrick, G. A., III, R. D. DeLaune, and W. H. Patrick, Jr. 1980. Effect of estuarine sediment pH and oxidation-reduction potential of microbial hydrocarbon degradation. Appl. Environ. Microbiol. 40:365–369.

Hess, R. E., and R. W. Blanchar. 1977. Dissolution of arsenic from waterlogged and aerated soil. Soil Sci. Soc. Am. J. 41:861–865.

Hill, I. R. 1978. Microbial transformation of pesticides. p. 137–202. *In* I. R. Hill and S. L. J. Wright (ed.) Pesticide microbiology. Academic Press, N.Y.

Jakobsen, P., W. H. Patrick, Jr., and B. G. Williams. 1981. Sulfide and methane formation in soils and sediments. Soil Sci. 132:279–286.

Jenkinson, D. S. 1977. The soil biomass. New Zealand Soil News 25(6):213–218.

----, and D. S. Powlson. 1976. The effects of biocidal treatments on metabolism in soil. V. A method for measuring soil biomass. Soil Biol. Biochem. 8:209–213.

Kaufman, D. D. 1974. Degradation of pesticides by soil microorganisms. p. 133–202. *In* W. D. Guenzi (ed.) Pesticides in soil and water. Soil Sci. Soc. Am., Madison, Wis.

Keeney, D. R. 1973. The nitrogen cycle in sediment-water systems. J. Environ. Qual. 2:15–29.

----, and R. E. Wildung. 1977. Chemical properties of soils. p. 75–97. *In* F. J. Stevenson and L. F. Elliot (ed.) Soils for management of organic wastes and wastewaters. Am. Soc. of Agron., Crop Sci. Soc. of Am., Soil Sci. Soc. of Am., Madison, Wis.

Khan, S. U. 1980. Pesticides in the soil environment. Elsevier, The Netherlands. 240 p.

Knowles, R. 1981. Denitrification. p. 315–330. *In* F. E. Clark and T. Rosswall (ed.) Terrestrial nitrogen cycles: Processes, ecosystem strategies and management impacts. Ecol. Bull. (Stockholm) 33, Swedish Natural Science Research Council.

Ladd, J. N. 1978. Origin and ranges of enzymes in soil. p. 51–96. *In* R. G. Burns (ed.) Soil enzymes. Academic Press, London.

Lindsay, W. L. 1972. Inorganic phase equilibria of micronutrients in soils. p. 41–58. *In* J. J. Mortvedt, P. M. Giordano, and W. L. Lindsay (ed.) Micronutrients in agriculture. Soil Sci. Soc. Am., Madison, Wis.

Martin, J. P., and D. D. Focht. 1977. Biological properties of soils. p. 115–169. *In* F. J. Stevenson and L. F. Elliot (ed.) Soils for management of organic wastes and wastewaters. Am. Soc. Agron., Crop Sci. Soc. Am., Soil Sci. Soc. Am., Madison, Wis.

Matsumura, F., and H. J. Benezet. 1978. Microbial degradation of pesticides. p. 623–668. *In* I. R. Hill and S. L. J. Wright (ed.) Pesticide microbiology. Academic Press, N.Y.

Miller, R. H. 1973. The soil as a biological filter. p. 71–94. *In* W. E. Sopper and L. T. Kardos (ed.) Recycling treated municipal wastewater and sludge through forest and cropland. The Pennsylvania State Univ. Press, University Park, Penn.

Nannipieri, P., R. L. Johnson, and E. A. Paul. 1978. Criteria for measurement of microbial growth and activity in soil. Soil Biol. Biochem. 10:223–229.

National Academy of Sciences. 1978. An assessment of mercury in the environment. Natl. Acad. Sci., Washington, D.C. 185 p.

Pepper, I. L., and R. H. Miller. 1978. Comparison of the oxidation of thiosulfate and elemental sulfur by two heterotrophic bacteria and *Thiobacillus thiooxidans*. Soil Sci. 126:9–14.

Pfennig, N. 1975. The phototrophic bacteria and their role in the sulfur cycle. Plant Soil 43:1–16.

Pierre, W. H., J. R. Webb, and W. D. Shrader. 1971. Quantitative effects of nitrogen fertilizer on the development and downward movement of soil acidity in relation to level of fertilization and crop removal in a continuous corn cropping system. Agron. J. 63:291–297.

Ponnamperuma, F. N. 1972. The chemistry of submerged soils. Adv. Agron. 24:29–96.

Quastel, J. H., and P. G. Scholefield. 1953. Arsenite oxidation in soil. Soil Sci. 75:279–285.

Schmidt, E. L. 1982. Nitrification in soil. *In* F. J. Stevenson (ed.) Nitrogen in agricultural soils. Agronomy 22:253–288. Am. Soc. of Agron., Soil Sci. Soc. of Am., Crop Sci. Soc. of Am., Madison, Wis.

Skujins, J. J. 1967. Enzymes in soil. p. 371–416. *In* A. D. McLaren and G. H. Peterson (ed.) Soil biochemistry. Marcel Dekker, N.Y.

Sommers, L. E., and W. L. Lindsay. 1979. Effect of pH and redox on predicted heavy metal-chelate equilibria in soils. Soil Sci. Soc. Am. J. 43:39–47.

Sparling, G. P., B. E. Ord, and D. Vaughan. 1981. Microbial biomass and activity in soils amended with glucose. Soil Biol. Biochem. 13:99–104.

Stadtman, T. C. 1967. Methane fermentation. Ann. Rev. Microbiol. 21:121–142.

Starkey, R. L. 1966. Oxidation and reduction of sulfur compounds in soils. Soil Sci. 101:297–307.

Stevenson, F. J. 1982. Humus chemistry, genesis, composition, reactions. John Wiley and Sons, N.Y. 443 p.

Turner, A. W. 1949. Bacterial oxidation of arsenite. Nature 164:76–77.

Verstraete, W. 1981. Nitrification. p. 303–314. *In* F. E. Clark and T. Rosswall (ed.) Terrestrial nitrogen cycles: Processes, ecosystem strategies and management inputs. Ecol. Bull. (Stockholm) 33, Swedish Natural Science Research Council.

Wagner, G. H. 1975. Microbial growth and carbon turnover. p. 269–305. *In* E. A. Paul and A. D. McLaren (ed.) Soil biochemistry, Vol. 3. Marcel Dekker, N.Y.

Walsh, F., and R. Mitchell. 1972. A pH-dependent succession of iron bacteria. Environ. Sci. Technol. 6:809–812.

Walsh, L. M., and D. R. Keeney. 1975. Behavior and phytotoxicity of inorganic arsenicals in soils. p. 35–52. *In* ACS Symposium Series, No. 57, Arsenical pesticides. Am. Chem. Soc., N.Y.

Wong, P. T. S., Y. K. Chau, and P. L. Luxon. 1975. Methylation of lead in the environment. Nature (London) 253:263–264.

Wood, J. M. 1973. Metabolic cycles for toxic elements in aqueous systems. Rev. Int. Oceanogr. Med. Tomes. XXXI–XXXII:7–16.

Woolson, E. A. 1977. Generation of alkylarsines from soils. Weed Sci. 25:412–416.

Yoshida, T. 1975. Microbial metabolism of flooded soils. p. 83–122. *In* E. A. Paul and A. D. McLaren (ed.) Soil biochemistry, Vol. 3. Marcel Dekker, N.Y.

Zaiss, U. 1981. Seasonal studies of methanogenesis and desulfurication in sediments of the River Saar. Zbl. Bakt. Hyg., I. Abstr. Orig. C2:76–89.

Chapter 12

Incorporation of the Rhizosphere into Plant Root Models[1]

JOHN H. CUSHMAN[2]

Root surfaces have the amazing capability of being able to both absorb and desorb substances with widely varying characteristics in molecular size and biological and chemical properties. Green vegetation can convert light energy into chemical energy to supply roots with reduced carbon compounds. These compounds are used by the roots as an energy supply for obtaining nutrients from the soil solution phase. The ions in the soil solution phase are in turn transported to the root surface by pore diffusion and transportation induced convection.

It is the purpose of this chapter to briefly summarize i) the structure of the root and its rhizosphere, ii) the root and root surface as an absorption and desorption mechanism, and iii) transport of nutrients to the root surface. This will lead us to introduce a system of coupled equations that simultaneously present a description of flow to the root in both the rhizosphere and bulk soil system.

---

[1] Contribution from the Purdue Agric. Exp. Stn., West Lafayette, IN 47907. Journal Paper No. 8791.

[2] Associate professor of soil physics, Agronomy Dep., Purdue University.

Copyright © 1983 ASA, SSSA, 677 South Segoe Road, Madison, WI 53711. *Chemical Mobility and Reactivity in Soil Systems.*

ROOT AND RHIZOSPHERE STRUCTURE

Root Structure and Ionic Transport Pathways

Figure 1 illustrates the basic cross-sectional view of a typical root. The outer surface of the root is formed by a single layer of thin-walled epidermal cells. Some epidermal cells may expand outward (up to 10 mm) forming what are called root hairs. The number and size of the root hairs depends on many factors: K, H_2O, Ca, P, plant species, microorganisms, pH, and mechanical resistance (Nye and Tinker, 1977). Immediately inside the epidermal layer is the cortex consisting of elongated thin-walled cells and large intercellular spaces (in some plants these cell walls are thick in which case this portion of the cortex is called the hypodermis). The cortex is surrounded on the inside by the endodermis and its associated Casparian band. The endodermis consists of a layer of cells of single cell thickness, with initially thin walls. Shortly after the endodermis forms however, the Casparian band, consisting of a suberin-lignin structure, forms on the transverse and radial walls of the endodermis preventing inward apoplastic movement. Movement through the tangential surface of the endodermal cells can occur until the cells develop into the secondary stage where suberin lamellae is deposited around the entire internal surface of the cell wall.

The remaining interior of the root tissue is vascular and called the stele. The stele is composed of the pericycle, and the phloem and xylem vessels. The phloem transports carbohydrates (as an energy source) and several mineral nutrients to the root from the shoot. Convection moves water and nutrients to the shoot through the xylem vessels.

There are two basic pathways through which water and nutrients are transported (Fig. 2a) to the root vascular system (Russell and Clarkson, 1976). The first pathway is entitled the symplastic route (for a review see Lauchli, 1976b). In the symplastic route plasmodesmata between adjacent cells form a continuous cytoplasmic pathway through which most nutrients other than Ca^{++} pass. The other pathway is called the apoplastic

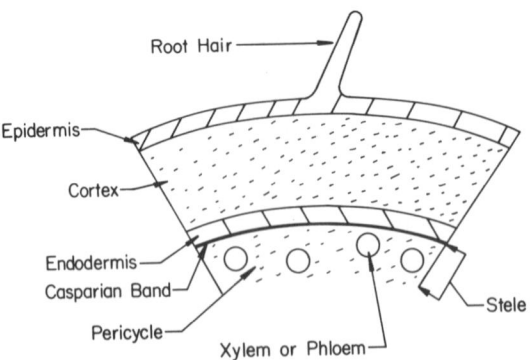

Fig. 1. The basic components of a root cross-section.

route and in this case the nutrients and water move along the free space and in the cell walls to the vascular system, except at the Casparian band (for a review see Lauchli, 1976a). When nutrients and water reach the Casparian strip, if it is possible, they are transported into the endodermal cytoplasm. Once beyond the Casparian band the nutrients may either move back into the apoplastic pathway or remain in the symplastic route. However, with increasing age, as already mentioned, the endodermis cell walls become suberized (Fig. 2b) in which case transport in the apoplastic pathway of water and Ca^{2+} is effectively halted (the evidence for this is that in older roots Ca^{++} absorption is halted and H_2O absorption reduced while the other nutrient ions are still readily absorbed).

Root Rhizosphere

The root rhizosphere is the region immediately next to the root surface extending out to a distance of about 2 mm (Hiltner, 1904) from the root surface. Including the natural soil complex, the root rhizosphere con-

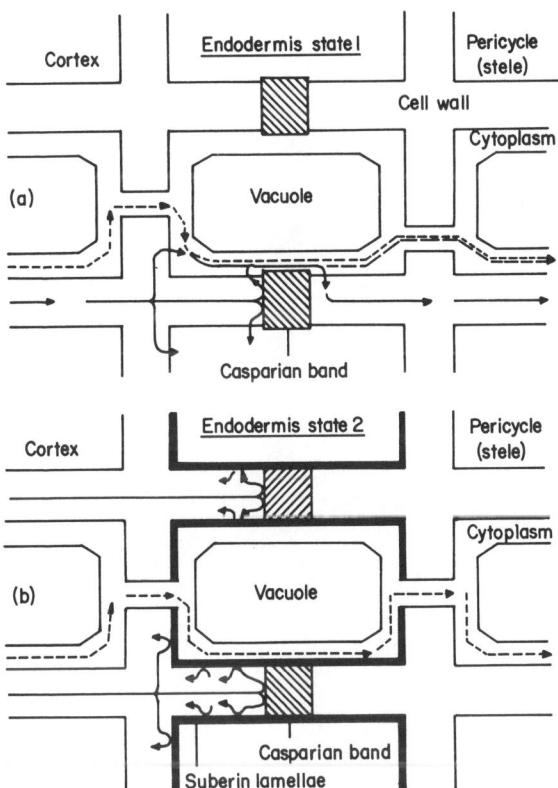

Fig. 2. Pathways of ions across roots. a) before and b) after deposition of suberin lamellae in endodermis. Solid line—apoplastic route. Dotted line—symplastic route (after Russell and Clarkson, 1976).

tains five (see Fig. 3) basic overlapping categories of compounds (Rovira et al., 1979).

i) Exudates: Nonmetabolically released compounds of low molecular weight which leak from cells to the soil by either intercellular spaces or through epidermal walls.

ii) Secretions: Compounds of low molecular weight and high molecular weight mucilages which are released as a result of metabolic processes.

iii) Plant Mucilages: There are basically four types a) root cap mucilage, b) hydrolysates of the polysaccharide of the primary cell wall between epidermal cells and sloughed root cap cells, c) mucilage secreted by the epidermal cells which still only have primary walls (including mucilage from root hairs), and d) mucilage produced by bacterial degradation of the outer multi-lamellate primary cell walls of old, dead epidermal cells.

iv) Mucigel: The gelatinous material at the root surface grown in normal nonsterile soils. It includes natural and modified plant mucilages, bacterial cells and their metabolic products (slimes) as well as colloidal, mineral, and organic matter from the soil.

v) Lysates: Compounds released from autolysis of older epidermal cells when the plasmalemma fails. Microorganisms colonize the cells and release the products of microbial activity into the rhizosphere.

There are currently three concepts of the ultra-structure of the root surface and the region in its immediate vicinity (Foster, 1981); (i) the surface is covered with a granular gel which is so tenuous at its surface

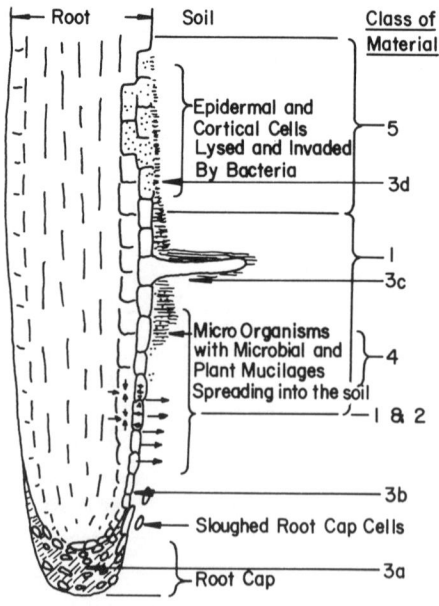

Fig. 3. Diagram of a root rhizosphere (after Rovira et al., 1979).

Table 1. Functional classification of mineral nutrients (after Clarkson and Hanson, 1980).

Covalent		Salts and complexes		
Reduced	Oxidized	Conformational	Redox	Uncertain
N	P	K	Fe	B
S	S	Na	Cu	Cl
		Mg	Mo	
		Ca	Mn	
		Mn		
		Zn		

that its external limits must be demonstrated by use of colloidal, electron dense markers (Jenny and Grossenbacher, 1963); (ii) the root is enclosed by a definite, thin, morphologically-distinct electron dense cuticle (Scott, 1978); (iii) the root surface is fibrous, with fibrils passing at right angles to the root surfaces into the soil (Leppard, 1974). Foster (1981) has given evidence supporting each of the above three concepts for different ages of specific roots.

THE ROOT AND ROOT SURFACE AS AN ABSORPTION AND DESORPTION MECHANISM

Uptake Mechanisms

Before proceeding with our discussion of root absorption it is appropriate to classify the nutrients from a functional standpoint. Various classifications are possible (cf. Mengel and Kirby, 1978); we choose to present the classification of Clarkson and Hanson (1980) Table 1).

The covalently bonded mineral nutrients combine with C, H, and O of organic structures and metabolites of the plant to give them the chemical and physical properties essential to life.

The complexes and salts category covers the alkali, alkaline earth, and transition metals. The interaction of these minerals with biological molecules extends from bonds which are largely ionic to those with a strong covalent nature. Hughes (1972) presents four ways in which metal ions may act: a) trigger and control mechanisms—Na^+, K^+, Mg^+, and Ca^{++} center on membrane properties, e.g., permeability, conductance, transport, etc.; b) structural influences—K^+, Ca^{++}, Mg^{++}, Mn^{++} contribute to structural stability and conformation of macromolecules; c) Lewis acids—Mg^{++}, Ca^{++}, Mn^{++}, Fe^{++}, Cu^{++}, Zn^{++} act as Lewis acid catalysts; d) redox behavior—Cu^{++}, Fe^{++}, Co^{++}, Mo^{++} catalyze valence changes in substrates.

We now turn to a brief description of the physiological and chemical mechanisms which result in the uptake of the above mentioned nutrients. For a detailed discussion see the reviews of Hodges (1974) and Pitman (1977), or the book of Clarkson (1974).

Again, let us delay for a moment and recall the basic structure of a cell (Fig. 4 after Nye and Tinker, 1977). There are seven basic components: nucleus, tonoplast, plasmalemma, vacuole, endoplasmic reticulum, cytoplasm, and cell wall. We are mainly interested in the mechanisms of transport across the plasmalemma. The cell wall itself is composed of chain-like polymers of cellulose and pectin with large intersticial spaces (relative to the dissolved solute molecules) forming a large loose net. Ions and water molecules in general have no difficulty in being transported through the cell wall. On the other hand the membranes do not have the large holes encountered in the cell walls and special mechanisms may be necessary to transport nutrients across them. Simplistically, when an energy source is needed for this transport it comes from the conversion, by photosynthesis, of light energy to chemical energy (NADPH and ATP) which provides the root with an energy source in the form of reduced carbon compounds.

As was mentioned there are basically two pathways (apoplastic and symplastic) that water and nutrients may travel on to reach the stele. We now consider these pathways in more detail.

Let us consider the apoplastic pathway first. As we already know, water and solutes move rather easily through the free space (cell walls and intercellular space) up to the Casparian band where Russell and Clarkson (1976) propose that in young root tissue they pass through the plasmalemma into the cytoplasm. A salient point that we have not found in the literature and we wish to make here is that this hypothesis requires that the plamalemma of the endodermal cells be more selective for Ca^{++} than other ions. Moreover, it requires the plasmalemma of the endodermis to be more permeable to water than the plasmalemma of the cortical cells. If this were not the case, when endodermal cells become suberized we would see little change in water uptake for the water would simply move into the cytoplasm in the cortical cells. Thus to begin to verify the Russell and Clarkson hypothesis one needs to run an experiment to compare the Ca^{++} selectivity and H_2O permeability of the plasmalemmas of cortical and endodermal cells.

We next consider the symplastic pathway. Recall that the symplastic pathway is through the epidermal or cortical cells via the plasmalemma

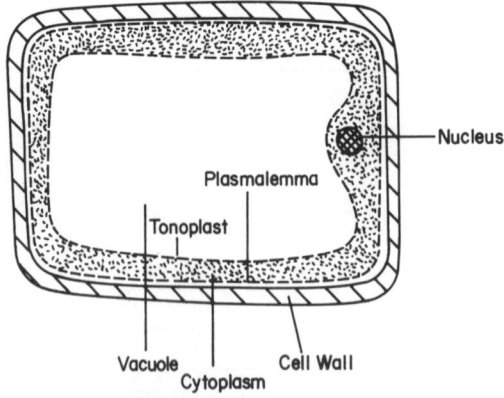

Fig. 4. Cell diagram (after Nye and Tinker, 1977).

into the cytoplasm, then between the various cells radially through the root via plasmodesmata. The major resistance in the symplastic route is encountered at the plasmalemma. Several models of the membrane have been proposed; there is still no definitive proof to define any specific model as the correct model and in fact the membrane structures may vary between different plants. Clarkson (1974) has suggested the following model: the matrix of the membrane is a bilayer with polar groups of phospholipids directed toward the outside. Embedded in the matrix are particulate substructures which are represented as protein and as stabilized lipid micelles. Covering the whole surface is protein and mucopolysaccharide which is hydrogen bonded to the polar groups of the phospholipid. The surface of the membrane is charged and this charge tends to determine the relative permeability of anions and cations by modifying the cation permeability. The membrane is also dynamically selective for various cations.

When metabolism is impaired, the hydrophobic character of the lipid interior limits the ability of highly charged electrolytes from entering the membrane (cf. Cram and Laties, 1971). This is because nonelectrolytes have membrane permeability coefficients on the order of 10^{-1} to 10^{-4} cm s^{-1}, while 10^{-8} cm s^{-1} is not uncommon for an electrolyte (Cram and Laties, 1971). This lends credence to the theory that ion transport across the membrane is metabolically induced. It can also be shown that when metabolism is in process the interior of the cell may have a higher concentration of cations and anions than the exterior. The preceding two points imply energy may be expended to transport the ion across the membrane. This expenditure of energy may either result in passive or active transport, active transport being defined as the movement of an ion against an electrochemical potential gradient (Ussing, 1949).

There are basically two mechanisms in use to determine whether or not an ion is being actively transported across a membrane. If the cell cytoplasm is in equilibrium with the soil solution, then the Nernst equation is valid (Dainty, 1962). The Nernst equation thus provides a check for active transport. It is well known that an actively metabolizing cell usually is in a nonequilibrium state. In this case the Ussing-Teorell (UT) equation should be used (Ussing, 1949; Teorell, 1949). The UT-equation may be written as

$$\frac{I^j_{in}}{I^j_{out}} = \frac{C^j_{out}}{C^j_{in}} \exp\left(z_j \frac{FE}{RT}\right), \qquad [1]$$

where I is the flux, C is the concentration, z an integer with sign corresponding to the valence, F is the Faraday constant, E is the observed electric potential difference across the membrane, R is the gas constant, T is the absolute temperature, and the sub and superscripts in, out, and j represent inside, outside, and the jth species, respectively. If the measured flux ratio is not equal to that predicted by Eq. [1], then there is active transport of the jth species.

Using the Nernst and UT equations researchers have shown that anions are actively absorbed while cations generally are passively absorbed and actively exuded (with the possible exception of K^+, cf. Pierce and Higinbotham, 1970). K^+ may be either actively transported in or out of the cytoplasm.

The flux of ions through the plasmalemma is controlled by three major factors (Clarkson, 1974); a) $\nabla \bar{\mu}_j$ where $\bar{\mu}_j$ is the electrochemical potential of the jth species, b) (a_j^*) activity of the solute in the membrane, c) (U_j^*) the mobility of the solute in the membrane. The activity and mobility help determine the membrane permeability defined as

$$P_j = \frac{U_j^* K_j^* RT}{L}, \qquad [2]$$

where K_j^* is the partition coefficient of the ion between water and the membrane phases (it defines C_j^*), and L is the length of the pathway followed by the ion.

To maintain the root permeability and prevent leaking and deterioration of the plasmalemma a source of Ca^{++} ions is required (Viets, 1944). The actual cause for the membrane deterioration with lack of Ca^{++} can only be speculated.

The electrochemical potential difference across a cell membrane appears to be due to two mechanisms a) a diffusion potential, and b) a metabolically controlled potential (cf. Slayman, 1970). The diffusion potential manifests itself via the different mobilities of the cations and anions in the lipid structure of the membrane (which results in a slight charge separation). On the other hand, the metabolic component is believed to result from ion pumps (Slayman, 1970). These pumps may be either neutral or electrogenic (Smith, 1970). For a more general discussion see Luttge and Pitman (1976).

It is well known that the plasmalemma is a selective membrane. It has been speculated that ion selectivity may be the result of the ion pumps (cf. Papahadjopoulos, 1971). Other theories exist in the literature (cf. Hodges, 1974). It is also known that selectivity changes as the external ion concentration increases (Epstein, 1973), that the selective ion sites may not be stationary, that the sites may be pH dependent (Lucas, 1979), that the sites may be in bands (for example, Lucas has speculated that alkaline bands are involved in the uptake of cations), and that the selectivity may be affected by intersite spacing and charge strength (Clarkson, 1974).

Kinetics

In general, on the root scale and on a temporal scale measured in minutes, the ion uptake rate increases as the external ion concentration increases (cf. Fried and Shapiro, 1961). This increase however, may be in a step-wise fashion as shown in Fig. 5 (from Hodges, 1974. Note the change of scale at about 0.2 mM K^+ concentration). Figure 5 may be explained with a multiphasic model (Nissen, 1971). More often however, the velocity of absorption vs. concentration plot is just a two-step curve (Nye and

Tinker, 1977). It should be noted however, that in a soil system, the higher concentrations (>0.2 mM) are seldom seen (with the exception of when a fertilizer is applied).

Hodges (1974) has proposed an intriguing single carrier model, based on the cooperative enzyme kinetics of Koshland (1970), to account for the following observations: a) anions are actively pumped inward across the plasmalemma and cations are actively exuded, b) the influx of cations and anions exhibits kinetics that are of the pseudo-saturation type discussed above, and c) absorption is selective and this selectivity changes with increasing external ion concentration. Unfortunately this model is fairly complex and it does not admit a simple mathematical interpretation.

Alternatively many scientists employ the simpler Michaelis-Menten reaction kinetics to describe each individual section in Fig. 5 (e.g., the region 0–0.2 mM). This method has the advantage of being represented by a simple mathematical expression

$$I_{in}^j = \frac{V_{max}^j C^j}{Km^j + C^j}, \qquad [3]$$

where V_{max} is the maximum influx (not necessarily constant), Km is the value of the concentration at which $I_{in} = V_{max}/2$, and I_{in}^j is the influx of the jth species (Epstein and Hagen, 1952; Fried and Broeshart, 1967). Although we haven't found any mention of it in the literature, it seems very reasonable to predict from existing evidence that I_{out}^j should also follow Michaelis-Menten kinetics. In either case it is the net flux that is important.

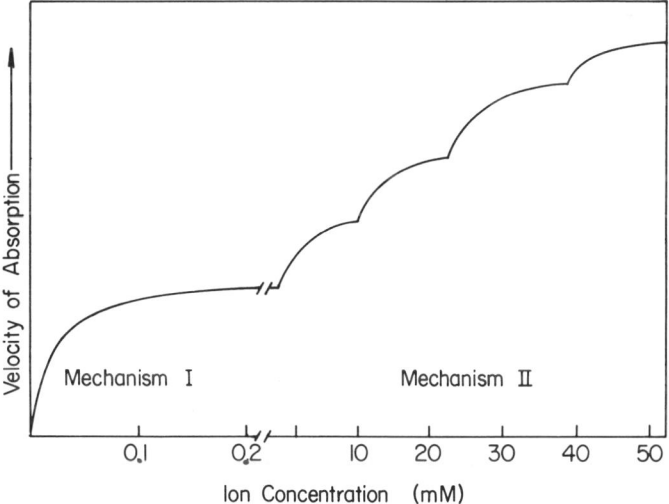

Fig. 5. Velocity of ion influx into roots as a function of external ion concentration (after Hodges, 1974).

Influence of Plant Growth

The various cations and anions react differently to plant growth (Loneragan, 1979 or Drew, 1978). For example K^+ flux into roots has been said to be determined by the growth rate of the plant (Pitman, 1972). Alternatively, Jungk and Barber (1974), and Anghinoni and Barber (1980) have argued that P uptake rate may be governed by the amount of root exposed to P and the P concentration of the root. In general for anions there is no definitive trend, e.g., in some cases (e.g., high P-soil) P-status of the plant dominates while in low P-soil the soil environment may dominate (Edwards, 1970). For an elementary mathematical description considering the growth state of the plant use Nye and Tinker (1977, p. 211).

Nutrient status of the plant tissue may play a major role in nutrient influx and efflux as well as growth rate. In particular the possibility of allosteric control (Levitski, 1978) by cytoplasmic concentrations of an ion must be considered.

TRANSPORT OF NUTRIENTS TO THE ROOT SURFACE

Flow and Transformation in the Rhizosphere

We first consider the contact of the root with the surrounding soil medium and the associated hydraulics. In general the outer epidermal surface of a young root is covered with a mucilage and hence not in direct contact with the soil (Oades, 1978). This mucilage is thought to consist of polysaccharides, lipids, and peptides (Greenland, 1979). The mucilage combines with the soil colloids and organic matter to form a mucigel. Both water and ions appear to be able to diffuse through this mucigel in a manner very similar to free water (Greenland, 1979). It has been suggested in the literature (with no concrete evidence presented) that the mucigel effectively stops convective flow. However, this author and Philip Low (personal communication) doubt this is true, since in clay pastes one still finds convective flow (cf. Mokady and Low, 1968; Banin and Low, 1971).

With increasing elapsed time it is believed that the mucigel forms soil aggregates around the roots (Turcheneck and Oades, 1978). It has been suggested that when the soil dries the mucigel remains attached to the root and forms a buffer (water remains in some of the mucigel pores), keeping the root partially moist (Drew, 1978). During the drying process the mucigel separates from the soil matrix forming a root-soil gap (Russell, 1977) which limits further uptake of water and nutrients from the drying soil. It should also be noted that the root undergoes size fluctuations between day and night that may cause a root-soil gap during the day (cf. Huck, 1979; or Huck et al., 1970). In more moist soils the mucigel may help maintain hydraulic contact between the root and soil matrix.

The pectin component of the mucigel readily binds cations and hence, before the Casparian band is formed, may retard some heavy metal cations from entering young roots (Barlow, 1975).

As an example of the rhizosphere's effects on a specific nutrient, Kepert et al. (1979) suggest there are three main mechanisms by which availability of phosphates is affected in the rhizosphere: a) release of phosphate from insoluble phosphates (Ca, Al, Fe) may be increased by the formation of soluble complexes between metal ions and metabolites (Barber, 1968); b) exuded polycarboxylic acids compete for phosphate adsorbed on kaolinite and aluminum oxides (Nagarajah et al., 1970) and thus retain more phosphate in solution; and c) when cation absorption into the root exceeds anion absorption the rhizosphere pH falls. Similarly hydration of CO_2 to carbonic acid (Brady, 1974) and release of cellular organic acids (Miller, 1974) may alter the pH.

Root hairs in the rhizosphere can increase the "effective radius" of the root, which in turn makes nutrients with very small diffusion coefficients more accessible to the plant. An even larger extension of the effective root surface may be formed by mycorrhizal fungi (cf. Owusu-Bennoah and Wild, 1979). Endotrophic and ectotrophic mycorrhiza hyphae extend for distances much larger than those of root hairs. For example, ectotrophic mycorrhizas of Monterey pine (*Pinus radiata*) translocated phosphate over 12 cm (Skinner and Bowen, 1974). It appears that there is a highly efficient translocation system within the fungal hyphae and an efficient transfer of phosphate from the vacuoles of the fungal hyphae into the cells of the host plant (Clarkson and Hanson, 1980). However, it is not true that microorganisms always have a positive effect on ion uptake (Hale and Moore, 1979). For example, *Trichoderma viride* limits phosphate and sulfate uptake in pea (*Pisum sativum* L.) plants (Brannstrom, 1977). For a recent review of the subject see Tinker (1980).

The Coupled Flow Equations

There are five basic dynamic mechanisms which control the flow of nutrients outside the root rhizosphere: a) solution phase diffusion, b) surface phase diffusion, c) convection, d) mechanical dispersion, and e) solid-liquid exchange phenomena. These mechanisms also work in the rhizosphere but one should also add to this list a sixth mechanism in the rhizosphere; f) microbial (or other) induced solute changes and movement (fungi).

In this article we will take a simplistic yet standard approach and assume the electrochemical potential gradient can be approximated by a concentration gradient. We are well aware that the correct transport forces result in coupled diffusion and counter diffusion, but the corresponding equations are beyond the scope of this article.

Solution phase diffusion is a coupled phenomenon in that it is the result of two basic mechanisms. First, on the molecular scale there is molecular diffusion induced by the thermal motion of the ions in solution. And second, there is a contribution induced by the geometry of the porous media (a form of mechanical dispersion). This second contribution is accounted for by averaging the first over the liquid phase in a representative elementary volume (REV) (Bear, 1972). The solution phase diffusion is governed by a Fick's law type equation

$$\bar{J}_l^d = D_l \nabla C_l, \qquad [4]$$

where D_l is an averaged diffusion tensor in the liquid phase (l), C_l is the solution phase concentration, and $\bar{J}^d$ is the diffusive flux.

Surface phase diffusion is also the result of two factors. On the one hand we again have thermal diffusion on the charged surface of the clay particles (a molecular scale phenomena). But on the other hand there is a contribution due to the geometry of the particle surfaces. The second factor (again a form of mechanical dispersion) may be accounted for by averaging over the surfaces in a REV. This may also give rise to a Fick's type equation

$$\bar{J}_s^d = D_s \nabla C_s = D_s \frac{\partial C_s}{\partial C_l} \nabla C_l, \qquad [5]$$

where s stands for the surface phase.

The convective component of nutrient transport is induced by the bulk movement of water (e.g., in the transpiration stream) which causes a mechanical dispersive effect produced by the geometric tortuosity. For the velocities of interest in this article, the mechanical dispersion effect on the convective flux is negligible and we thus may write this flux as

$$\bar{J}^v = \bar{v} C_l, \qquad [6]$$

where $\bar{v}$ is the Darcy velocity and $\bar{J}^v$ is the convective flux.

The total flux is given by

$$\bar{J} = D_l \nabla C_l + D_s \frac{\partial C_s}{\partial C_l} \nabla C_l + \bar{v} C_l. \qquad [7]$$

If we model the exchange capacity of the soil as a buffering operation, then we may call $\partial C_s / \partial C_l$ the buffer power of the solid phase for the liquid phase and denote it by b. Moreover, we may combine D_l, D_s, and b to form

$$D = D_l/b + D_s \qquad [8]$$

so that

$$\bar{J} = bD\nabla C_l + \bar{v} C_l, \qquad [9]$$

where D is the effective diffusion coefficient. For more detailed and alternative definitions of the effective diffusion coefficient see Olsen and Kemper (1967).

What we would like to do now is write down a tractable mathematic formulation of flow to the root both in and outside the rhizosphere. There are many routes we can take (for reviews see Olsen and Kemper, 1967; Nye and Tinker, 1977) depending on the assumptions we are willing to make. In particular we will make the assumptions of Cushman (1980a) for flow in the rhizosphere and those of Cushman (1979) for flow outside the rhizosphere.

If the root is assumed to be a cylinder and the soil isotropic and homogeneous then $\bar{J}$ reduces to

$$J_r = bD_r \frac{\partial C_l}{\partial r} + v\, C_l, \qquad [10]$$

where J_r is the radial component of $\bar{J}$, and D_r is the constant macroscopic diffusion coefficient in the radial direction. Assuming the buffer power is constant and assuming steady state moisture conditions, one may derive the following equation for nutrient flow in the rhizosphere (Cushman, 1980a):

$$\frac{\partial C_l^{(r)}}{\partial t} = \frac{1}{r}\frac{\partial}{\partial r}\left(rD_r^{(r)}\frac{\partial C_l^{(r)}}{\partial r} + \frac{v_o r_o}{b^{(r)}} C_l^{(r)}\right) + \frac{\alpha^{(r)}}{b^{(r)}} \qquad [11]$$

$$r_0 < r < r_1,$$

where α is the mass of solute in soil solution produced (depleted) per unit time per unit volume via microbial activity, V_o and r_o are respectively the velocity of water at the root surface and the root radius, r_1 is the radius of the rhizosphere, and the superscript r denotes rhizosphere.

If we assume that outside the rhizosphere there is negligible microbial activity, then we have (Cushman, 1979)

$$\frac{\partial C_l^{(o)}}{\partial t} = \frac{1}{r}\frac{\partial}{\partial r}\left(rD_r^{(o)}\frac{\partial C_l^{(o)}}{\partial r} + \frac{v_o r_o}{b^{(o)}} C_l^{(o)}\right), \; r_1 < r < r_2, \qquad [12]$$

where r_2 is some outer radius of influence of the root.

The appropriate boundary bonditions for Eq. [11] and [12] are

$$D^{(r)}b^{(r)} \frac{\partial C_l^{(r)}}{\partial r} + v_o\, C_l^{(r)} = \frac{J_{max} C_l^{(r)}}{K_m + C_l^{(r)}}, r = r_o, t > 0; \qquad [13]$$

$$C_l^{(r)} = C_l^{(o)}, r = r_1, t > 0; \qquad [14]$$

$$D^{(r)}b^{(r)} \frac{\partial C_l^{(r)}}{\partial r} = D^{(o)}b^{(o)} \frac{\partial C_l^{(o)}}{\partial r} \quad r = r_1, t > 0; \qquad [15]$$

and

$$D^{(r)}b^{(r)} \frac{\partial C_l^{(r)}}{\partial r} + \frac{r_o}{r_2} v_o\, C_l^{(r)} = 0, t > 0, r = r_2. \qquad [16]$$

Equation [13] represents continuity of flux at the root surface (here we have used Eq. [3]), Eqs. [14] and [15] represent, respectively, continuity of concentration and flux at the outer surface of the rhizosphere, and Eq. [16] expresses the interroot competition for nutrient (Cushman, 1979). We thus have two coupled equations ([11], [12]) subject to boundary conditions ([13], [14], [15], [16]) and an appropriate initial condition. The solution to this system is beyond the scope of this article and will be discussed further in a later paper.

The reader should note the extreme simplification used in deriving Eq. [11] and [12] (cf. Cushman, 1980a). The importance and significance of the various parameters in Eqs. [11] and [12] may be found in Cushman (1980b).

SUMMARY

If we move from the root axis in an outward radial direction we find the following components of the root soil system: i) stele (xylem, phloem, and pericycle); ii) endodermis and its associated Casparian band; iii) cortex; iv) epidermis; v) rhizosphere (exudates, secretions, plant mucilages, mucigel, lysates); and vi) bulk soil matrix. It is important to consider transport through all of these components for a proper understanding of nutrient movement in the soil root system. The nutrient uptake paths in the root are classified as the apoplastic and symplastic routes. Transport in the apoplastic route is mainly passive and highly dependent on root age, while transport in the symplastic route is more energy dependent and to a lesser extent a function of age. Three major points must be made when considering transport across the plasmalemma of the symplastic route: a) anions are actively pumped inward across the plasmalemma and cations are actively exuded, b) the influx of ions exhibits kinetics of the pseudo-saturation type, and c) absorption is selective and the selectivity changes with external ion concentration. Although the Michaelis-Menten kinetics don't adequately describe the above observations over the entire range of concentrations, the Michaelis-Menten kinetic model is generally a good approximation in soil systems.

Transport of nutrient to the root from the bulk soil matrix is generally a function of five mechanisms: i) solution phase diffusion, ii) surface phase diffusion, iii) convection, iv) mechanical dispersion, and v) solid-liquid exchange phenomena. To this list a sixth mechanism in the rhizosphere must be added: vi) microbial or other induced solute changes. Taking into account the above six points, two coupled equations may be derived representing flow in the rhizosphere and the bulk soil system. The boundary condition at the root surface follows from the Michaelis-Menten kinetics, the boundary conditions at the rhizosphere-bulk soil interface follow by continuity, and the boundary condition at the outer radius of the bulk soil system is a function of interroot competition. Analytical solutions to this system may be derived and will be discussed in a later paper.

ACKNOWLEDGMENTS

I would like to thank S. A. Barber and T. K. Hodges for their interesting discussions on the topic of this paper, and would also like to thank the reviewers for their valuable comments.

Support for this project has been drawn from the "Purdue Onsight Waste Disposal Project."

LITERATURE CITED

Anghinoni, I., and S. A. Barber. 1980. Phosphorus influx and growth characteristics of corn roots as influenced by phosphorus supply. Agron. J. 72:685-688.

Banin, A., and P. F. Low. 1971. Simultaneous transport of water and salt through clays: 2. Steady-state distribution of pressure and applicability of irreversible thermodynamics. Soil Sci. 112(2):69-88.

Barber, D. A. 1968. Microorganisms and the inorganic nutrition of higher plants. Annu. Rev. Plant Physiol. 19:71-88.

Barlow, P. W. 1975. The root cap. p. 21. In J. G. Torrey and D. T. Clarkson (ed.) The development and function of roots. Academic, London.

Bear, J. 1972. Dynamics of fluids in porous media. Elsevier, N.Y.

Brady, N. C. 1974. The nature and properties of soils. MacMillan, N.Y.

Brannstrom, G. 1977. The effect of exudate from Trichoderma viride on ion uptake and translocation in axenic pea plants. Z. Pflanzenphysiol. 83:341-346.

Clarkson, D. T. 1974. Ion transport and cell structure in plants. John Wiley and Sons, N.Y.

----, and J. B. Hanson. 1980. The mineral nutrition of higher plants. Annu. Rev. Plant Physiol. 31:239-298.

Cram, W. J., and G.G. Laties. 1971. The use of short-term and quasi-steady influx in estimating plasmalemma and tonoplast influx in barley root cells at various external and internal chloride concentrations. Aust. J. Biol. Sci. 24:633-646.

Cushman, J. H. 1979. An analytical solution for solute transport near root surfaces for low initial concentration: I. Equations development. Soil Sci. Soc. Am. J. 43(6):1087-1090.

----. 1980a. Analytical study of the effect of ion depletion (replenishment) caused by microbial activity near a root. Soil Sci. 129(2):69-87.

----. 1980b. Completion of the list of analytical solutions for nutrient transport to roots 1. Exact linear models. Water Resour. Res. 16(5):891-906.

Dainty, J. 1962. Ion transport and electrochemical potentials in plant cells. Annu. Rev. Plant Physiol. 13:379-402.

Drew, M. C. 1979. Root development and activities. p. 573-606. In R. A. Perry and D. W. Goodall (ed.) Arid land ecosystems, Vol. I. Cambridge.

----. 1979. Properties of roots which influence rates of absorption. In J. L. Harley and R. S. Russell (ed.) The soil root interface. Academic Press, N.Y.

Edwards, D. G. 1970. Phosphate absorption and long-distance transport in wheat seedlings. Aust. J. Biol. Sci. 23:255-264.

Epstein, E. 1973. Mechanisms of ion transport through plant cell membranes. Int. Rev. Cytol. 34:123-168.

----, and C. E. Hagen. 1952. A kinetic study of the absorption of alkali cations by barley roots. Plant Physiol. 27:457-474.

Foster, R. C. 1981. The ultrastructure and histochemistry of the rhizosphere. New Phytol. 89:263-273.

Fried, M., and H. Broeshart. 1967. The soil plant system in relation to inorganic nutrition. Academic Press, N.Y.

----, and R. E. Shapiro. 1961. Soil-plant relationships in ion uptake. Annu. Rev. Plant Physiol. 12:91–112.

Greenland, D. J. 1979. The physics and chemistry of the soil root interface: Some comments. *In* J. L. Harley and R. S. Russell (ed.) The soil root interface. Academic Press, N.Y.

Hale, H. G., and L. D. Moore. 1979. Factors affecting root exudation: II 1970-1978. Adv. Agron. 31:93–124.

Hiltner, L. 1904. Arbeiten der Deutschen. Landwirtsch. Ges. 98:59–78.

Hodges, T. K. 1974. Ion absorption by plant roots. Adv. Agron. 25:163–207.

Huck, M. G. 1979. A photographic view of microscopic processes at the root-soil interface. *In* J. L. Harley and R. S. Russell (ed.) The soil root interface. Academic Press, N.Y.

----, B. Klepper, and H. M. Taylor. 1970. Diurnal variations in root diameter. Plant Physiol. 45:529–530.

Hughes, M. N. 1972. The inorganic chemistry of biological processes. John Wiley and Sons, London.

Jenny, H., and K. Grossenbacher. 1963. Root-soil boundary zones as seen in the electron microscope. Soil Sci. Soc. Am. Proc. 27:273–277.

Jungk, A., and S. A. Barber. 1974. Phosphate uptake rate of corn roots as related to the proportion of the roots exposed to phosphate. Agron. J. 66:554–557.

Kepert, D. G., A. D. Robson, and A. M. Posner. 1979. The effect of organic root products on the availability of phosphorus to plants. *In* J. L. Harley and R. S. Russell (ed.) The soil root interface. Academic Press, N.Y.

Koshland, D. E. 1970. Chapter 7. p. 341–396. *In* P. D. Boyer (ed.) The enzymes. 3rd ed. Vol. 1. Academic Press, N.Y.

Lauchli, A. 1976a. Apoplastic transport in tissues. *In* U. Luttge and M. G. Pitman (ed.) Encycl. plant physiol.: Transport in plants. 2B. Springer-Verlag, Berlin.

----. 1976b. Symplastic transport and ion release in the xylem. *In* I. F. Wardlaw and J. B. Passioura (ed.) Transport and transfer processes in plants. Academic Press, N.Y.

Leppard, G. G. 1974. Rhizoplane fibrils in wheat: demonstration and derivation. Science (Wash.) 185:1066–1067.

Levitski, A. 1978. Quantitative aspects of allosteric mechanisms. Springer-Verlag, N.Y.

Longeragan, J. F. 1979. The interface in relation to root function and growth. *In* J. L. Harley and R. S. Russell (ed.) The soil root interface. Academic Press, N.Y.

Lucas, W. L. 1979. Alkaline band formation in characorallina. Plant Physiol. 63:248–254.

Luttge, U., and M. G. Pitman. 1976. Transport in plants II. Part A Cells; Encycl. Plant Physiol. No. 5. Vol. 2. Springer-Verlag, Berlin.

Mengel, K., and E. A. Kirkby. 1978. Principles of plant nutrition. International Potash Institute, Bern.

Miller, M. H. 1974. Effects of nitrogen on phosphorus absorption by plants. p. 643–668. *In* E. W. Carson (ed.) The plant root and its environment. Univ. Press of Virginia.

Mokady, R. S., and P. F. Low. 1968. Simultaneous transport of water and salt through clays: I. Transport mechanisms. Soil Sci. 105(2):112–131.

Nagarajah, S., A. M. Posner, and J. P. Quirk. 1970. Competitive adsorption of phosphate with polygalacturonate and organic anions on kaolinite and oxide surfaces. Nature 228: 83–85.

Nissen, P. 1971. Multiphasic ion uptake in roots. p. 539–553. *In* W. P. Anderson (ed.) Ion transport in plants. Academic Press, N.Y.

Nye, P. H., and P. B. Tinker. 1977. Solute movement in the soil-root system. Univ. of California Press, Berkeley.

Oades, J. M. 1978. Mucilages at the root surface. J. Soil Sci. 29:1–16.

Olsen, S. R., and W. D. Kemper. 1967. Movement of nutrients to plant roots. Adv. Agron. 20:91–151.

Owusu-Bennoah, E., and A. Wild. 1979. Autoradiography of the depletion zone of phosphate around onion roots in the presence of vesicular-arbuscular mycorrhiza. New Phytol. 82:133–140.

Papahadjopoulus, D. 1971. Na^+-K^+ discrimination by pure phospholipid membranes. Biochem. Biophys. Acta. 241:254–259.

Pierce, W. S., and N. Higinbotham. 1970. Compartments and fluxes of K^+, Na^+ and cl^- in avena celoeoptile cells. Plant Physiol. 46:666–673.

Pitman, M. G. 1972. Uptake and transport of ions in barley seedlings. III. Correlation of potassium transport to the shoot with plant growth. Aust. J. Biol. Sci. 25:243–257.

----. 1977. Ion transport into the xylem. Annu. Rev. Plant Physiol. 23:71–88.

Rovira, A. P., R. C. Foster, and J. K. Martin. 1979. Origin, nature and nomenclature of the organic materials in the rhizosphere. *In* J. L. Haley and R. S. Russell (ed.) The soil root interface. Academic Press, N.Y.

Russell, R. S. 1977. Plant root systems. McGraw-Hill Book Co., London.

----, and D. Clarkson. 1976. Ion transport in root systems. *In* N. Sunderland (ed.) Perspectives in experimental biology. Vol. 2. Botany. Pergamon Press, Inc., Oxford.

Scott, F. M. 1978. Growth and structure of roots. p. 39–67. *In* Y. R. Dommergues and S. V. Krupa (ed.) Interactions between nonpathogenic soil microorganisms and plants. Elsevier Scientific Publishing Co., Amsterdam.

Skinner, M. F., and G. D. Bowen. 1974. The uptake and translocation of phosphate by mycelial strands of pine mycorrhizas. Soil Biol. Biochem. 6:53–56.

Slayman, C. L. 1970. Movements of ions and electrogenesis in microorganisms. Am. Zool. 10:377–392.

Smith, F. A. 1970. The mechanisms of chloride transport in Characean cells. New Phytol. 69:903–917.

Teorell, T. 1949. Membrane electrophoresis in relation to bioelectric polarization effects. Arch. Sci. Physiol. 3:205–219.

Tinker, P. B. 1980. Role of rhizosphere microorganisms in phosphorus uptake by plants. p. 617–654. *In* F. E. Khasawneh, E. C. Sample, and E. J. Kamprath (ed.) The role of phosphorus in agriculture. Am. Soc. of Agron., Crop Sci. Soc. of Am., and Soil Sci. Soc. of Am., Madison, Wis.

Turcheneck, L. W., and L. M. Oades. 1978. Organo-clay particles in soils. *In* W. W. Emerson and A. R. Dexter (ed.) Modification of soil structure. John Wiley and Sons, N.Y.

Ussing, H. H. 1949. The distinction by means of tracers between active transport and diffusion. Acta Physiol. Scand. 19:43–56.

Viets, F. G. 1944. Calcium and other polyvalent cations as acceleration of ion accumulation by excised barley roots. Plant Physiol. 19:466–480.

Chapter 13

Sorption and Movement of Pesticides and Other Toxic Organic Substances in Soils[1]

P. S. C. RAO AND R. E. JESSUP[2]

The use of various agrochemicals, such as fertilizers as supplemental crop nutrients and various pesticides (organic and inorganic) for managing crop pests, has contributed to continued high crop yields in the USA over the past two decades. Of about a million metric tons (1×10^6 t) of pesticides produced during 1979 in the USA, about a third (0.3×10^6 t) was exported, about 0.5×10^6 t were used in food and fibre production, while the remainder was consumed for industrial and urban uses. Therefore, the use of pesticides constitutes an important aspect of modern agriculture.

The efficacy of pesticides and the potential for environmental pollution from pesticides and their residues is controlled by various processes and factors. For soil-applied pesticides, sufficient quantities must be present in the solution phase or vapor phase within the crop root zone in order to effect pest control. The amount present in soil solution is de-

---

[1] Contribution from the Soil Science Dep., Univ. of Florida. Approved for Publication as Florida Agricultural Exp. Stn. Journal Series No. 4307. Partial financial support for this work was provided by USEPA Grant No. R-805794 from the R. S. Kerr Environ. Res. Lab., USEPA.

[2] Associate professor and scientific programmer, respectively, Soil Sci. Dep., Univ. of Florida, Gainesville, FL 32611.

Copyright © 1983 ASA, SSSA, 677 South Segoe Road, Madison, WI 53711. *Chemical Mobility and Reactivity in Soil Systems.*

termined, in part, by pesticide partitioning between solid phase and solution phases; such a partitioning will be referred to here as sorption. The leaching of pesticide out of the root zone is influenced by the rates and amounts of water moving beyond this zone as well as by sorption. Finally, various photochemical, microbial, and chemical processes will lead to pesticide dissipation in soils. Therefore, pesticide fate and transport in agroecosystems are directly influenced by sorption, transformations, and transport phenomena in soils. An understanding of these processes is essential not only to design better management practices but also for quantifying their environmental fate and transport required in assessment of environmental exposure and hazard.

Several books and monographs dealing with various aspects of environmental dynamics of pesticides in aquatic and terrestrial ecosystems have been published. Also, several excellent reviews covering sorption and transport of pesticides in soils and sediments are also available; these are cited in later sections. Therefore, emphasis is placed on an examination of most recent developments in quantitative descriptions of pesticide sorption and transport in soils. Because the same processes influence the behavior of other toxic organic substances (TOS) in soils, the discussions here are applicable to TOS in general. The major problems remaining in predicting pesticide and TOS sorption and transport under controlled laboratory conditions will be pointed out. Finally, the criteria for further work in measuring and modelling pesticide and TOS at a field scale will be discussed.

I. ADSORPTION-DESORPTION IN SOILS

It has long been recognized that sorption of pesticides on soils not only influences their efficacy as agrochemicals but also determines pesticide fate and transport in the environment. Thus, a considerable amount of data has been collected over the past two decades to characterize pesticide sorption on soils and sediments. These data have been summarized and critically evaluated in several review articles (Bailey and White, 1970; Helling et al., 1971; Hamaker and Thompson, 1972; Weber, 1972; Green, 1974; Weed and Weber, 1974; Farmer, 1975; Rao and Davidson, 1980; Lyman, 1982; Calvet, 1980; Kenega and Goring, 1980; McCall et al., 1981; Rao et al., 1982c).

With the increasing concern for the environmental fate of nonpesticidal, but related, TOS as well as with U.S. Environmental Protection Agency's publication of the list of 129 priority pollutants, data are being collected on sorption of several TOS on soils and sediments. Reinbold et al. (1979) reviewed the published data on sorption of energy-related organic pollutants, a variety of polycyclic aromatic compounds. Hassett et al. (1980) presented extensive experimental data on the sorption of several energy-related organic pollutants on a number of sediments.

Because of the availability of a large number of excellent review articles, no attempt is made here to review or summarize sorption data or to discuss sorption mechanisms. Rather, principal findings and the state-of-the-art in providing quantitative relationships useful in predictive models will be summarized. Particular attention will be given to equilibrium and kinetic sorption of pesticides and TOS.

A. Equilibrium Sorption

Among the large number of models proposed for describing equilibrium sorption (c.f., van Genuchten and Cleary, 1978; Travis and Etnier, 1981) the linear and Freundlich isotherm models have been the most commonly used for pesticides and TOS sorption on soils and sediments. These equations can be stated as follows:

$$S = KC \quad [1]$$

and

$$S = K_f C^N; \quad N < 1, \quad [2]$$

where S and C are, respectively, adsorbed-phase and solution-phase concentrations at equilibrium, while the sorption coefficients K, K_f, and N are empirical constants specific to a sorbent-sorbate combination. Note that for N = 1, Eq. [2] reduces to Eq. [1]. The values of K, K_f, and N are usually obtained by curve-fitting Eqs. [1] or [2] to measure data. Green et al. (1980) reviewed the advantages and limitations of various experimental methods available for measuring pesticide sorption. Koshinen (1980)[3] discussed the influence of various experimental variables on measured sorption coefficients, while Green and Yamane (1970), and Dao et al. (1982) discussed the propagation of errors and precision of measurements in experimentally determining the sorption coefficients.

The value of the partition coefficient (K or K_f) is a measure of the extent of pesticide sorption on soils. The value of K (or K_f) for a given pesticide or TOS may vary by one order of magnitude or more among several soils and sediments. Multiple regression analyses of K (or K_f) with several soil physical-chemical properties suggest that soil organic carbon content (% OC) may be the single best predictor of pesticide sorption coefficients for nonionic and polar pesticides. For ionic or ionizable compounds, the effect of pH and pK_a on speciation between ionic and molecular forms as well as surface charge characteristics of the soil, must be considered. Most researchers reported that for a given pesticide, the sorption coefficient normalized with respect to soil organic carbon content was essentially independent of soil type. This normalized sorption coefficient, designated as K_{oc}, is defined as follows:

$$K_{oc} = \frac{K}{\% \, OC} \times 100 \text{ for linear isotherms} \quad [3]$$

and

$$K_{oc} = \frac{K_f}{\% \, OC} \times 100 \text{ for nonlinear isotherms.} \quad [4]$$

The values of K_{oc} for a broad range of pesticides and other TOS may be found in reviews by Hamaker and Thompson (1972), Rao and Davidson (1980), Kenega and Goring (1981), and Karickhoff (1981). Recent work suggests that a K_{oc} value for a given pesticide is not only independent of soil type but also of particle-size fractions within or among different soils and sediments (Karickhoff et al., 1979; Rao et al., 1982c). All of these findings suggest that organic carbon is the principal sorbent of pesticides and TOS in soils and sediments. In no way does this imply, however, that inorganic soil constituents do not act as sorbents. Sorption on clay minerals is important for cationic pesticides (e.g., herbicides paraquat and diquat). However, since the soil organic matter and clay minerals exist in soils largely as clay-metal-organic complexes (Stevenson, 1976), it may be difficult to separate the independent contributions of clay minerals and organic matter to pesticide and TOS sorption. It is best to recognize that the K_{oc} concept provides a practical and useful simplification for sorption of nonionic pesticides and TOS on a broad range of soils. However, the use of K_{oc} to estimate pesticide sorption coefficients K or K_f in soils with either very low or very high organic carbon contents may be prone to considerable errors (c.f., Hamaker and Thompson, 1972). For predicting the sorption of organic cations, Brown (1983) proposed an empirical exchange equation based on the two-component competitive Langmuir relationship. In this derivation, the concentrations of all the competing cations are lumped together to give the concentration of a single equivalent cation, while the organic cation is treated as the other competing ion. Brown (1983) presented calibration and verification of such a model with experimental data for sorption of methylacridinium ion on a series of soils and sediments.

Although the K_{oc} value for a given pesticide is fairly constant (within a factor of two) among different soils or sediments, the K_{oc} values for different pesticides may vary over several orders of magnitude. Helling and Dragun (1981) note that molecular properties of pesticides and TOS affecting K_{oc} values include: (i) structure and conformation, (ii) acidic or basic dissociation constants, (iii) aqueous solubility, (iv) charge status, (v) polarity and polarizability, and (vi) molecular size. They further state that K_{oc} value reflects the relative affinity of soil organic carbon for the pesticide and the pesticide-water interactions. It is generally true that for nonionic pesticides, aqueous solubility, and K_{oc} are inversely correlated. The reader is referred to a recent paper by Karickhoff (1981) for further discussions of the relationship between molecular properties and K_{oc} values.

B. Non-Equilibrium Sorption

A large data base exists for characterizing pesticide and TOS partitioning at equilibrium between the soil and soil solution. By comparison, very few researchers have investigated the rate at which sorption equilibrium is attained. Under conditions of steady or transient water flow (in laboratory soil columns or field soil profiles), the contact time available may be insufficient to achieve sorption equilibrium between solid- and solution-phases. This fact has prompted several, if not many, researchers to measure the kinetics of pesticide sorption in soils. On the basis of published data available from "batch slurry" experiments, Rao and Davidson (1980) concluded that about 60 to 80% of the sorption reaction may be complete in less than a minute (i.e., essentially instantaneous). While the remainder is completed in one to a few hours. In spite of such apparent rapid sorption in short-term batch slurry experiments, for some pesticide-soil combinations sorption may continue at a very low rate for days or weeks (Hamaker and Thompson, 1972). Similar results for sorption of certain TOS on sediments were reported by Karickhoff (1979).

Although most batch experiments indicate essentially instantaneous or very rapid sorption of pesticides, flow experiments with packed soil columns suggest that this assumption may not be valid. This aspect and various mathematical models devised to explain nonequilibrium sorption during flow are further discussed in Section II of this chapter.

II. MATHEMATICAL MODELS FOR DESCRIBING SORPTION AND TRANSPORT IN SOILS

Combining Darcy's law and the continuity equation, the partial differential equation for water flow in soils can be derived. The equation for one-dimensional (vertical) transient water-flow with simultaneous soil-water extraction by plant roots is:

$$\frac{\partial \theta}{\partial t} = \frac{\partial}{\partial x}\left[D(\theta)\frac{\partial \theta}{\partial x}\right] - \frac{\partial}{\partial x}K(\theta) - U(x,t) \qquad [5]$$

where, θ is volumetric soil-water content, t is time, x is soil depth, $D(\theta)$ is soil-water diffusivity, $K(\theta)$ is soil hydraulic conductivity, and $U(x,t)$ is a sink term to account for plant uptake of water.

The equation for describing convective-diffusive-dispersive transport of pesticides during transient water flow is:

$$\frac{\partial}{\partial t}(\theta C + \varrho S) = \frac{\partial}{\partial x}\left[D_h \theta \frac{\partial C}{\partial x}\right] - \frac{\partial}{\partial x}(v\theta C) - \sum_{i=1}^{n} Q_i \qquad [6]$$

where ϱ is soil bulk density, D_h is hydrodynamic dispersion coefficient, v is average pore-water velocity, Q_i are various sink terms to account for pesticide losses (degradation and plant uptake), and other terms are as defined in earlier sections. The reader is referred to the papers by Bolt (1979), Biggar and Nielsen (1980), and Davidson et al. (1983) for a detailed discussion of the physical significance of D_h, and to the papers by Goring et al. (1975) and Kaufman (1983) for a treatment of pesticide degradation in soils. As the principal objective of this paper is to review sorption and transport of pesticides, Q_i terms are ignored here.

For transient water flow conditions occurring during infiltration and redistribution of water during and following a rainfall/irrigation event, Eqs. [5] and [6] must be solved simultaneously. For steady water flow conditions (i.e., $\partial\theta/\partial t = 0$). Eq. [6] simplifies (with $Q_i = 0$) to:

$$\frac{\partial C}{\partial t} + \frac{\varrho}{\theta}\frac{\partial S}{\partial t} = D_h \frac{\partial^2 C}{\partial x^2} - v\frac{\partial C}{\partial x} \qquad [7]$$

Functional relationships between the solution-phase (C) and the adsorbed-phase (S) concentrations must be specified in order to solve Eq. [7] subject to given initial and boundary conditions. When the sorption reactions are instantaneous, equilibrium exists between the solution-phase and adsorbed-phase concentrations. The equilibrium relationship between C and S is then specified by the adsorption isotherm (see Section I). For cases where sorption is not instantaneous, various rate laws have been proposed depending upon how the processes responsible for sorption non-equilibrium were conceptualized. These models, the conceptual basis for their formulation, and their verification with experimental data are examined in the following sections. Since attention is focused on the models for sorption, the mathematical treatment is easier for the steady water flow equation as in Eq. [7] rather than the transient water flow model as in Eq. [6]. However, conclusions regarding the sorption processes are generally applicable to both transient and steady water flow conditions.

A. Equilibrium Models

As indicated in Section I, linear and Freundlich isotherms have been the most common models used for describing pesticide and TOS sorption on soils. For a linear isotherm, given $S = KC$, Eq. [7] can be restated as:

$$R_T \frac{\partial C}{\partial t} = D_h \frac{\partial^2 C}{\partial x^2} - v \frac{\partial C}{\partial x} \qquad [8]$$

where, $R_T = \left[1 + \frac{\varrho K}{\theta}\right].$ [9]

R_T in Eq. [9] is referred to as the retardation factor (Hashimoto et al., 1964; Davidson and Chang, 1972). By defining $D_h^* = (D_h/R_T)$ and $v^* = (v/R_T)$, Eq. [8] becomes,

$$\frac{\partial C}{\partial t} = D_h^* \frac{\partial^2 C}{\partial x^2} - v^* \frac{\partial C}{\partial x}. \qquad [10]$$

For a nonadsorbed solute $K = 0$, hence $R_T = 1$. Thus, the average porewater velocity of a nonadsorbed solute is equal to v. For adsorbed solutes, such as pesticides, $K > 0$, $R_T > 1$, and $v^* < v$. From Eq. [10], note that with increasing adsorption, i.e., increasing K and R_T, the adsorbed solute pore-water velocity (v^*) decreases, hence the use of the expression retardation factor. For a nonlinear isotherm, $S = K_f C^N$, the retardation function, $R_T(C)$, is given by:

$$R_T(C) = \left[1 + \frac{\varrho N K_f C^{(N-1)}}{\theta}\right]. \qquad [11]$$

Note from Eqs. [9] and [11] that for a linear isotherm R_T is a constant and for a nonlinear isotherm $R_T(C)$ is concentration-dependent. Because for most pesticides $N \leq 1$ (c.f., Hamaker and Thompson, 1972; Rao and Davidson, 1980), $R_T(C)$ in Eq. [11] decreases with increasing concentration and increasing nonlinearity, (i.e., larger C and smaller N values). Over the range of solution concentrations associated with agricultural applications, however, linear sorption isotherms may be adequate (Rao and Davidson, 1980). Hence, for some applications the retardation factor (R_T) may be considered to be concentration-independent. However, for applications dealing with pesticide waste disposal on land, the assumption of linear sorption isotherms and constant R_T may lead to serious underestimation of pesticide leaching in soils (Davidson et al., 1978; Rao and Davidson, 1979; Davidson et al., 1980b, 1980c).

In the aforementioned models and others to be discussed in Section II of this paper, the adsorption-desorption isotherms are assumed to be reversible and single-valued (i.e., singular). A number of authors have presented experimental data demonstrating that adsorption-desorption isotherms may be non-singular; hence, different isotherm parameters need to be used in Eqs. [9] and [11], depending upon whether adsorption or desorption was the predominant process. A sound and well-accepted physical-chemical basis has yet to be proposed and the reasons for the experimentally observed sorption isotherm nonsingularity continue to be debated (Mukhtar, 1976[3]; Rao and Davidson, 1980; Koskinen, 1980[4]).
Empirical relationships, developed from measured data for specific soil-pesticide combinations, have been incorporated in the transport models to take into account nonsingular sorption isotherms (Hornsby and Davidson, 1973; van Genuchten et al., 1974, 1977). Influence of nonsingular isotherms on pesticide behavior may be of much less significance compared to the effects of sorption nonequilibrium during flow (van Genuchten et al., 1977). Therefore, a much greater emphasis must be placed on understanding and modelling nonequilibrium conditions for pesticide sorption during flow in soils. This is the subject of the next section of this chapter.

B. Non-Equilibrium Models

Various mathematical models have been proposed for describing nonequilibrium adsorption-desorption of pesticides during flow (Boast, 1973; van Genuchten and Cleary, 1978; Travis and Etnier, 1981). These models can be grouped into two classes: (i) chemical-process models, and (ii) physical-process models. In the first group of models, the sorption nonequilibrium is attributed to the time-dependence of sorption reactions

---

[3] Mukhtar, M. 1977. Desorption of adsorbed ametryne and diuron from soils and soil components in relation to rates, mechanisms, and energy of adsorption reactions. Ph.D. Diss. Univ. of Hawaii. Available from: Univ. Microfilms, Ann Arbor, Mich. (Diss. Abstr. 41:20-B).

[4] Koshkinen, W. C. 1980. Evaluation of the batch equilibrium method for characterization of adsorption-desorption of 2,4,5-T in soils. Ph.D. Diss. Univ. Microfilms, Ann Arbor, Mich. (Diss. Abstr. 41:20-B). 113 p.

at the soil-solution interfaces. In the second group of models, the sorption reactions at the soil-solution interfaces are assumed to be instantaneous; however, the rate of pesticide sorption is controlled by the rate at which pesticide is transported to (and form) the soil surfaces. For example, access to the soil surfaces may be controlled by pesticide diffusion through stagnant water films or by intra-particle diffusion. Selected examples of both physical- and chemical-nonequilibrium models are examined here.

The failure of simple one-site linear or nonlinear kinetic models (Hornsby and Davidson, 1973; Davidson and McDougal, 1973) to describe pesticide transport led Selim et al. (1976) and Cameron and Klute (1977) to propose a two-site sorption model. In their model, the pesticide sorption was considered to occur on two types of sites; sorption on type-1 sites was instantaneous, while sorption on type-2 sites was described as a first-order reversible process. For this model, the ($\partial S/\partial t$) term in Eq. [7] can be stated as:

$$\frac{\partial S}{\partial t} = FK \frac{\partial C}{\partial t} + k\{(1-F)KC - S_2\} \qquad [12]$$

where F is the fraction of total adsorption sites assigned to be type-1, K is the sorption partition coefficient, k is the first-order rate coefficient, and S_2 is the amount sorbed on type-2 sites.

In order to examine the inter-relationships between model parameters (i.e., system variables) and the mathematical similarities with other nonequilibrium models, it is convenient to express the transport equations in dimensionless form. Following van Genuchten (1981), we define:

$$T = (vt/L); Z = (x/L); P = \left(\frac{vL}{D_h}\right) \qquad [13]$$

$$R_T = \left[1 + \frac{\varrho K}{\theta}\right]; R_1 = \left[1 + \frac{\varrho FK}{\theta}\right] \qquad [14]$$

$$\omega = (kL/v) \qquad [15]$$

$$c_1 = \left[\frac{C - C_i}{C_o - C_i}\right]; c_2 = \frac{S_2 - [(1-F)KC_i]}{[(1-F)K(C_o - C_i)]} \qquad [16]$$

where T is dimensionless time or pore volumes, L is column length, P is Peclet number, c_1 and c_2 are dimensionless concentrations, C_i and C_o are the initial and applied solution concentration, respectively, and others are as defined earlier. Using these dimensionless variables, the two-site model can be now stated as:

$$\frac{\partial c_1}{\partial T} + (R_T - R_1)\frac{\partial c_2}{\partial T} = \frac{1}{P}\frac{\partial^2 c_1}{\partial Z^2} - \frac{\partial c_1}{\partial Z} \qquad [17]$$

$$\frac{\partial c_2}{\partial T} = \omega(c_1 - c_2). \qquad [18]$$

Van Genuchten (1981) summarized analytical solutions to the above equations for a variety of boundary conditions.

It is evident from Eqs. [17] and [18] that the two-site model can be expressed in terms of four dimensionless parameters, where P represents the physical process of hydrodynamic dispersion, and the parameters R_T and R_1 designate partitioning between sorption on total and type-1 sites, while the parameter ω is an index for the nonequilibrium sorption on type-2 sites. Note from Eq. [15] that ω, the index for the degree of nonequilibrium, is determined not only by the sorption rate coefficient (k) but also by the mean column residence time ($t_r = L/v$). Therefore, for a given k and L value, increasing pore-water velocity leads to a smaller ω, which in turn implies an increased nonequilibrium. As k value decreases (slower reaction rate), the mean column residence time (t_r) must be increased either by longer columns or by smaller porewater velocities to achieve sorption equilibrium.

Based on the theoretical analysis of van Genuchten and Cleary (1979) and Rao et al. (1983) it may be stated that for $\omega \geq 5$ near-equilibrium conditions exist for pesticide sorption. Hence, an equilibrium model, such as Eq. [8] or Eq. [11], may be used to describe pesticide transport in soils. Based on the discussions and data cited in Section I, and experimental conditions (i.e., L and v) employed by most authors, the near-equilibrium condition that $\omega \geq 5$ would appear to be valid. Several authors (Kay and Elrick, 1967; Elrick et al., 1966; Green et al., 1968; Hornsby and Davidson, 1973; Davidson and McDougal, 1973; van Genuchten et al., 1974; Rao et al., 1979; De Camargo et al., 1979; Hoffman and Rolston, 1980) reported that equilibrium conditions did not exist even at low porewater velocities. These observations apparently contradict the batch measurements of kinetic rate coefficients (Rao et al., 1979). Model parameters also needed to be adjusted for different experiments in the same soil column. Experimental results such as these prompted the development of alternate conceptualizations of the sorption phenomenon in soils during water flow.

In the models based on physical-nonequilibrium, the soil water is partitioned into mobile and stagnant regions. Convective-diffusive-dispersive solute transport is limited to the mobile regions only, while the stagnant regions serve as diffusive sinks/sources. Although pesticide sorption is instantaneous, the actual rate of sorption is controlled by the rate of solute transfer through the stagnant regions. Van Genuchten et al. (1976, 1977) presented a model based on the above concept. Their model can be stated as follows:

$$[\theta_M + f\varrho K]\frac{\partial C_M}{\partial t} + [\theta_A + (1-f)\varrho K]\frac{\partial C_A}{\partial t}$$
$$= \theta_M D_M \frac{\partial^2 C_M}{\partial x^2} - v_M \theta_M \frac{\partial C_M}{\partial x} \quad [19]$$

and

$$[\theta_A + (1-f)\varrho K]\frac{\partial C_A}{\partial t} = \alpha(C_M - C_A), \quad [20]$$

where the subscripts M and A designate parameters for the mobile and stagnant regions, respectively, f is the fraction of sorption sites residing in the mobile region, and α is a transfer coefficient for solute exchange between mobile and stagnant regions.

As before, Eq. [19] and [20] are expressed in dimensionless form using the following variables:

$$T = (v_M t\phi/L); Z = (x/L), \qquad [21]$$

$$\theta_S = (\theta_M + \theta_A); \phi = (\theta_M/\theta_S), \qquad [22]$$

$$R_1 = \left(\phi = \frac{f\varrho K}{\theta_S}\right); R_T = \left(1 + \frac{\varrho K}{\theta_S}\right), \qquad [23]$$

$$P = (v_M L/D_M), \qquad [24]$$

$$\beta = \left(\frac{\alpha L}{v_M \theta_M}\right), \qquad [25]$$

and

$$c_1 = \left(\frac{C_M - C_i}{C_o - C_i}\right); c_2 = \left(\frac{C_A - C_i}{C_o - C_i}\right). \qquad [26]$$

Using Eqs. [21] to [26], the solute transport equations based on physical nonequilibrium can be restated in a nondimensional form as follows:

$$R_1 \frac{\partial c_1}{\partial T} + (R_T - R_1) \frac{\partial c_2}{\partial T} = \frac{1}{P} \frac{\partial^2 c_1}{\partial Z^2} - \frac{\partial c_1}{\partial Z} \qquad [27]$$

and

$$(R_T - R_1) \frac{\partial c_2}{\partial T} = \beta(c_1 - c_2). \qquad [28]$$

In Eq. [27] and [28], hydrodynamic dispersion in the mobile regions is specified by P, the total retardation factor and that for the mobile regions are given by R_T and R_1, respectively, while ϕ is the fraction of porewater within the mobile regions and the diffusive mass transfer between mobile and stagnant regions is represented by β. Thus, β is an index of physical nonequilibrium for mass transfer and hence sorption. It may be noted from Eq. [25] that β is dependent upon α, the actual mass transfer coefficient, and upon the mean column residence time, $t_r = (L/v_M)$. Comparing Eqs. [28] and [18], we note that $\beta = [\omega/(R_T - R_1)]$. Based on the analysis of DeSmedt (1979)[5] and Rao and Jessup (1982a), $\omega > 5$ represents nearequilibrium conditions for diffusive mass transfer and Eqs. [27] and [28] can be reduced to Eq. [8].

A comparison of Eqs. [17] and [18] with Eqs. [27] and [28] reveals that these two models—one based on chemical nonequilibrium and the

---

[5] DeSmedt, F. 1979. Theoretical and experimental study of solute movement through porous media with mobile and immobile water. Doctoral Diss. Vrije Univ., Brussels, Belgium. June 1979.

other on physical nonequilibrium—are mathematically identical and that the solution to these models would be the same for given initial and boundary conditions. It is important to note that two models based on completely different conceptualizations of the system processes are in fact represented by the same mathematical solution. Van Genuchten (1981) developed a single computer program to estimate the model parameters from measured effluent breakthrough curves for both models. In a recent experimental evaluation, Nkedi-Kizza et al. (1983) confirmed that parameters of these two models are indeed related. Therefore, if all the model parameters are not obtained or estimated independent of the experimental data being simulated, the choice between chemical-process or physical-process models is impossible. Unfortunately, this fact has often been overlooked in the earlier literature.

In attempting to verify the physical-process nonequilibrium model, Eqs. [19] and [20], or equivalently Eqs. [27] and [28], α value was found not to be a constant but to vary with v_M and other experimental variables (van Genuchten et al., 1977; Gaudet et al., 1977). Rao et al. (1980b) presented theoretical analysis and experimental data to show that the α-dependence on system variables was indeed expected. They concluded that because in Eq. [20] the specific geometry of the stagnant regions was not explicitly taken into account, α must be varied. For the case when the stagnant regions can be represented by spherical geometries (as in aggregated media), Rao et al. (1980a) derived an analytic expression for relating α to physical constants of the system. In a later paper, Rao et al. (1980b) used this analytic expression to estimate independently the α value to describe successfully measured nonadsorbed solute effluent breakthrough curves from aggregated porous media over a wide range of pore-water velocities.

The work of Rao et al. (1980a, 1980b) was extended and generalized by Rao et al. (1983). Their physical conceptualization of the system is essentially similar to that of van Genuchten and Wierenga (1976, 1977), except that solue transfer between mobile and stagnant regions was described using Fick's second law for diffusion and the intra-aggregate pore-regions were considered to be stagnant. While Eq. [27] was used to describe solute transport in the mobile regions, Eq. [28] was replaced by the following expression:

$$3 (R_T - R_M) \frac{\partial c_\mu}{\partial T} = \gamma \left[\frac{1}{\eta^2} \left(\frac{\partial}{\partial \eta} \eta^2 \frac{\partial c_\mu}{\partial \eta} \right) \right], \qquad [29]$$

where, $\eta = (r/a)$ is the dimensionless radial distance within the stagnant regions (aggregates), a is aggregate radius, and c_μ is radially-varying concentration inside the aggregates. Also, c_2 in Eqs. [27] and [28] is defined as the volume-averaged relative concentration within the stagnant regions, i.e., for spherical aggregates,

$$c_2 (Z, T) = 3 \int_0^1 C_\mu (\eta, T; Z) \eta^2 \, d\eta \qquad [30]$$

Finally, the physical nonequilibrium index, γ, is defined as:

$$\gamma = \left[\frac{3D_\mu L (1-\phi)}{a^2 v_M \phi} \right], \qquad [31]$$

where, D_μ is effective molecular diffusion coefficient within the aggregates, and other parameters are as defined earlier.

Based upon their theoretical analysis and comparisons with experimental data, Rao et al. (1983) concluded that for cases when $\gamma > 1$, near-equilibrium conditions exist for diffusive solute transfer between the mobile and stagnant regions. Since sorption at the soil-solution interfaces was considered to be instantaneous, $\gamma > 1$ indicates that the more complex and explicit models (Eqs. [27], [28], and [29]) can be reduced to the simpler equilibrium models. That this can indeed be done was demonstrated by Rao et al. (1983) for adsorbed solutes and earlier by Passioura and Rose (1971) for nonadsorbed solutes using experimental data.

Most laboratory column experiments are conducted with sieved soils (a < 0.1 mm). Assuming that a = 0.1 cm, D_μ = 0.01 cm^2 h^{-1}, ϕ = 0.5, and L = 30 cm, and using these values in Eq. [31] physical nonequilibrium conditions are expected only when v_M > 90 cm h^{-1}. That is, at pore-water velocities less than 90 cm h^{-1}, the equilibrium model is expected to describe pesticide breakthrough curves from soil columns. The results of Hornsby and Davidson (1973), Davidson and McDougal (1973), van Genuchten et al. (1974), Rao et al. (1979), and Davidson et al. (1980a) all demonstrate that while the extent of sorption nonequilibrium decreased with decreasing pore-water velocity, the equilibrium models did not adequately predict the measured data except at low flow rates. Hamaker (1975) presented a comparison of laboratory and field measurements of pesticide sorption and leaching in soils and evaluated these data in terms of "real-world" field situations.

The above findings suggest that the rate-limiting step for pesticide sorption during flow may not be simple intra-particle diffusion or kinetics for sorption at the soil surfaces as has often been suggested. Sorption reactions at soil surfaces are apparently essentially instantaneous, while molecular diffusion within soil aggregates (< 0.1 cm radius) is also quite rapid. The extent of sorption nonequilibrium observed in soil column studies suggests that pesticide diffusion through certain regions of high diffusion resistance (i.e., small γ) may control the rates of pesticide sorption. Because soil organic matter is the principal sorbent of most pesticides, diffusion within the matrix of the soil organic matter can be suggested as a possible rate-limiting factor. However, additional basic research needs to be carried out to investigate further this possibility. Our own laboratory work in this area suggests that "batch slurry" methods are insensitive and may be unsuitable for investigating sorption kinetics. Sophisticated flow-equilibration and chromatographic techniques (e.g., ultra-filtration cells, high-pressure liquid chromatography, etc.) may help solve this puzzle.

III. SORPTION AND MOVEMENT UNDER FIELD CONDITIONS

Several models with varying degrees of complexity and conceptualization of the system processes can be developed. However, numerical solutions of such complex models require large amounts of computer time. Methods to measure independently the model input parameters are also inadequate. Use of such comprehensive simulation models at a field scale is confounded by two major problems. First, the soil physical, chemical, and biological characteristics vary spatially as well as temporally even within a single field. For example, soil-hydraulic conductivity, solute dispersion coefficient, average pore-water velocity, and similar flow intensity parameters are lognormally distributed (Nielsen et al., 1973; Biggar and Nielsen, 1976; Nielsen et al., 1983). Therefore, estimates of these model parameters are prone to considerable errors unless a large number of samples are taken (cf. Warrick et al., 1977). Most simulation models consider the model parameters to be deterministic and do not accommodate their stochastic nature. Second, the field-measured values of model output parameters (e.g., pesticide concentration distribution in the soil profile) also vary considerably owing to soil heterogeneity. Usage of simulation models at a field scale, requires the acceptance of the limitations imposed by uncertainties in the measured data used for model verification as well as the uncertainty in model input parameters and the associated confidence limits that should be placed on the model output.

Under field conditions, pesticide behavior is determined by a multitude of dynamic processes which occur simultaneously. A comprehensive field-scale simulation model should couple the interrelationships between these processes. However, time-varying boundary conditions (e.g., rainfall, pesticide applications) cannot be precisely specified under field conditions. The very complexity of the field problem and the soil heterogeneity suggest that much less accuracy is to be expected in simulations of field experiments compared to laboratory experiments. Experiences with development and testing simulation models for describing N dynamics in the crop root zone (Rao et al., 1976, 1981) suggest that simple models may be able to provide sufficiently accurate predictions of the fate of pesticides for field-scale applications.

A simple model for pesticide dynamics in soils should, at the minimum, include the following processes: (i) water and solute transport, (ii) sorption, and (iii) degradation. Fairly simple approaches, based on the piston displacement concepts, could be successfully used to describe water and pesticide transport (e.g., Rao et al., 1976, 1981). Approximate analytical solutions are also available to compute pesticide concentration distribution profiles during water infiltration and redistribution (DeSmedt and Wierenga, 1978). Techniques for approximating transient soil-water and solute transport by a steady-state flow model were discussed by Wierenga (1977). The simplest sorption model is the equilibrium linear isotherm given by Eq. [1]. Flow velocities encountered under field conditions are usually much smaller than those used in laboratory soil column experiments. Thus, with increased contact times between pesticide molecules and the soil surfaces (sorption sites), the assumption of equilibrium conditions may be more acceptable. However, for strongly aggregated or structured soils, physical nonequilibrium caused by pesticide diffusion into (and from) aggregate and peds may need to be explicitly accounted for. Addiscott (1977) presented a simple model for describing leaching in structured soils. Degradation of pesticides can be described fairly well by simple first-order kinetics, where the rate coefficient combines the rates of different processes responsible for pesticide dissipation. A constant value for this global rate coefficient for degradation appears to be satisfactory (Rao and Davidson, 1980; Rao et al., 1982c). Degradation rate coefficient values can be made a function of soil temperature and soil-water potential following the approach proposed by Walker (1976a, 1976b). An important process, not included in the above discussion is the plant uptake of pesticides by weed and crop species.

From the large number of field studies conducted on pesticide persistence, only a limited amount of data could be used for model verification. Therefore, it is difficult to assess whether or not simplified models are acceptable. The movement, retention, and degradation of propyzamide herbicide during a 100-day period in field plots was described by Leistra et al. (1974) using a simple flow model with linear equilibrium adsorption (Eq. [8]) and first-order degradation rate.

Leistra and associates (Leistra, 1975; Leistra and Dekkers, 1976; Leistra et al., 1976; Smelt et al., 1977; Leistra, 1978, 1979; Leistra et al., 1980; Bromilow and Leistra, 1980; Leistra and Smelt, 1981) have extended this work with other soils and pesticides. Similar experiments should be carried out to provide additional data for testing simple process-oriented models. The reader is referred to Green (1983) for further discussions on field-techniques for measuring parameters needed in simple predictive models.

The very accuracy with which laboratory measurements can be made and the mathematical ability to devise complex simulation models seem to be diverting our attention from the development and testing of simple models for field-scale application. Such models are not only easy to design, but necessary input data can be provided from existing data bases. Rao et al. (1982a) proposed various criteria for selection and use of simulation models. They suggested that the intended use of the model determines the conceptual completeness and complexity (or level of detail) of a simulation model. Recognizing the importance of spatial heterogeneity of soil properties, Rao et al. (1982a) also pointed out that any simulation model should not be expected to predict the system behavior, such as responses to imposed perturbations, any better than our ability to measure it. Thus, the judicious use of Okham's razor in the development of models for field-scale applications seems warranted. The importance of continued basic laboratory research on processes determining pesticide dynamics in soil water systems, however, cannot be overemphasized.

ACKNOWLEDGEMENTS

The authors are indebted to their colleagues at the Univ. of Florida, especially Drs. J. M. Davidson, P. Nkedi-Kizza, and A. G. Hornsby, for many helpful and thought-provoking discussions and for a critical review of this manuscript. Section III of this chapter is substantially based on Rao and Jessup's (1982) paper in Ecological Modeling.

LITERATURE CITED

Addiscott, T. M. 1977. A simple computer model for leaching in structured soils. J. Soil Sci. 28:554-563.

Bailey, G. W., and J. L. White. 1970. Factors influencing adsorption, desorption, and movement of pesticides in soil. Residue Rev. 32:29-92.

Biggar, J. W., and D. R. Nielsen. 1976. Spatial variability of the leaching characteristics of field soils. Water Resour. Res. 12:78-84.

----, and ----. 1980. Mechanisms of chemical movement in soils. p. 213-217. In A. Banin and U. Kafkafi (ed.) Agrochemicals in soils. Pergamon Press, N.Y.

Bolt, G. H. 1979. Movement of solutes in soils: Principles of adsorption/exchange chromatography. p. 285-348. In G. H. Bolt (ed.) Soil chemistry: Part B. Physico-chemical models. Elsevier Scientific Publishing Co., N.Y.

Boast, C. W. 1973. Modelling the movement of chemicals in soils by water. Soil Sci. 115:224-230.

Bromilow, R. H., and M. Leistra. 1980. Measured and simulated behavior of aldicarb and its oxidation products in fallow soils. Pestic. Sci. 11:389-395.

Brown, D. S. 1983. A model for predicting sorption of organic cations in soils and sediments. J. Environ. Qual. (in Review).

Calvet, R. 1980. Adsorption-desorption phenomenon. p. 1-30. In R. J. Hance (ed.) Interactions between herbicides and the soil. Academic Press, N.Y.

Cameron, D. R., and A. Klute. 1977. Convective-dispersive solute transport with a combined equilibrium and kinetic adsorption model. Water Resour. Res. 13:183-188.

Dao, T. H., D. B. Marx, T. L. Lavy, and J. Dragun. 1982. Effect and statistical evaluation of soil sterilization on aniline and diuron adsorption isotherms. Soil Sci. Soc. Am. J. 46:963-969.

Davidson, J. M., and R. K. Chang. 1972. Transport of picloram in relation to soil physical conditions and pore-water velocity. Soil Sci. Soc. Am. Proc. 36:357-261.

----, and J. R. McDougal. 1973. Experimental and predicted movement of three herbicides in water-saturated soil. J. Environ. Qual. 2:428-433.

----, L. T. Ou, and P. S. C. Rao. 1978. Adsorption, movement, and biological degradation of high concentrations of selected pesticides in soils. p. 233-244. In Proc. of 4th Annual Res. Symp. "Land Disposal of Hazardous Wastes", EPA-600/9-78-016. Cincinnati, Ohio.

----, P. S. C. Rao, R. E. Green, and H. M. Selim. 1980a. Evaluation of conceptual models for solute behavior in soil-water systems. p. 241-251. In A. Banin, and U. Kafkafi (ed.) Agrochemicals in soils. Pergamon Press, N.Y.

----, ----, and P. Nkedi-Kizza. 1983. Physical processes influencing water and solute transport in soils. p. 35-47. In D. W. Nelson, K. K. Tanji, and D. E. Elrick (ed.) Chemical mobility and reactivity in soil systems". Am. Soc. of Agron. and Soil Sci. Soc. Spec. Pub. No. 11, Am. Soc. of Agron., Madison, Wis.

----, ----, and L. T. Ou. 1980b. Movement and biological degradation of large concentrations of selected pesticides in soils. p. 93-107. In Proc. of 6th Annual Res. Symp. "Disposal of Hazardous Waste", EPA-600/9-80-010. Cincinnati, Ohio.

----, ----, ----, W. B. Wheeler, and D. F. Rothwell. 1980c. Adsorption, movement, and biological degradation of large concentrations of selected pesticides in soils. EPA-600/2-80-124, 111 p. Cincinnati, Ohio.

De Camargo, O. A., J. W. Biggar, and D. R. Nielsen. 1979. Transport of inorganic phosphorus in an alfisol. Soil Sci. Soc. Am. J. 43:884–890.

DeSmedt, F., and P. J. Wierenga. 1978. Approximate analytical solution to solute flow during infiltration and redistribution. Soil Sci. Soc. Am. J. 42:407–412.

Elrick, D. E., K. T. Erh, and H. K. Krupp. 1966. Application of miscible displacement techniques to soils. Water Resour. Res. 2:717–727.

Farmer, W. J. 1975. A literature survey of benchmark pesticides. George Washington Univ. Medical Center, Washington, D.C. 222 p.

Gaudet, J. P., H. Jegat, G. Vachaud, and P. J. Wierenga. 1977. Solute transfer, with exchange between mobile and stagnant water, through unsaturated sand. Soil Sci. Soc. Am. J. 41:665–671.

Goring, C. A. I., D. A. Laskowski, J. W. Hamaker, and R. W. Meikle. 1975. Principles of pesticide degradation in soil. p. 135–173. In R. Haque and V. H. Freed (ed.) Environmental dynamics of pesticides. Plenum Press, N.Y.

Green, R. E. 1974. Pesticide-clay-water interactions. p. 3–37. In W. D. Guenzi (ed.) Pesticides in soil and water. Soil Sci. Soc. Am., Madison, Wis.

----. 1983. Forecasting pesticide mobility in soils: Dispersion and adsorption consideration. In Proc. of Symp. USA-USSR Symp. "Prediction of Pesticide Behavior in the Environment", Yerevan, USSR (In press).

----, J. M. Davidson, and J. W. Biggar. 1980. An assessment of methods for determining adsorption-desorption of organic chemicals. p. 73–82. In A. Banin and U. Kafkafi (ed.) Agrochemicals in soils. Pergamon Press, N.Y.

----, and V. K. Yamane. 1970. Precision in pesticide adsorption measurements. Soil Sci. Soc. Am. Proc. 34:353–354.

----, ----, and S. R. Obien. 1968. Transport of atrazine in a latosolic soil in relation to adsorption, degradation, and soil water variables. Trans. 9th Int. Soil Sci. Congr. 1:195–204.

Hamaker, J. W. 1975. The interpretation of soil leaching experiments. p. 115–133. In R. Haque and V. H. Freed (ed.) Environmental dynamics of pesticides. Plenum Press, N.Y.

----, and J. M. Thompson. 1972. Adsorption. Vol. 1, p. 49–143. In C. A. I. Goring and J. W. Hamaker (ed.) Organic chemicals in the environment. Marcel Dekker Inc., N.Y.

Hashimoto, I., K. B. Deshpande, and H. C. Thomas. 1964. Peclet numbers and retardation factors for ion-exchange columns. Ind. Eng. Chem. Fund. 3:213–218.

Hassett, J. J., J. C. Means, W. L. Banwart, and S. G. Wood. 1980. Sorption properties of sediments and energy-related pollutants. EPA-600/3-80-041. 133 p. Athens, Ga.

Helling, C. S., and J. Dragun. 1981. Soil leaching tests for toxic organic chemicals. p. 43–88. In Protocols for environmental fate and movement of toxicants. Proc. of a Symp., AOAC 94th Annual Meeting, 21–22 Oct. 1980, Washington, D.C.

----, P. C. Kearny, and M. Alexander. 1971. Behavior of pesticides in soils. Adv. Agron. 23:147–240.

Hoffman, D. L., and D. E. Rolston. 1980. Transport of organic phosphate in soil as affected by soil type. Soil Sci. Soc. Am. J. 44:46–52.

Hornsby, A. G., and J. M. Davidson. 1973. Solution and adsorbed fluometuron concentration distribution in water-saturated soil: Experimental and predicted evaluation. Soil Sci. Soc. Am. Proc. 37:823–828.

Karickhoff, S. W. 1979. Sorption kinetics of hydrophobic pollutants in natural sediments. In Processes involving contaminants and sediments. Proc. of ACS National Symp., Honolulu, HI, April 1979.

----. 1981. Semi-empirical estimation of sorption of hydrophobic pollutants on natural sediments and soils. Chemosphere 10:833–846.

----, D. S. Brown, and T. Scott. 1979. Sorption of hydrophobic pollutants on natural sediments. Water Res. 13:241-248.

Kay, B. D., and D. E. Elrick. 1967. Adsorption and movement of lindane in soils. Soil Sci. 104:314-322.

Kenega, E. E., and C. A. I. Goring. 1980. Relationship between water solubility, soil sorption, octanol-water partitioning, and concentration of chemicals in biota. p. 78-115. *In* J. G. A. Eaton, P. R. Parrish, and A. C. Hendricks (ed.) Aquatic toxicology. Am. Soc. Testing and Materials, Spec. Tech. Pub. No. 707.

Leistra, M. 1975. Computed leaching of pesticides from soil as influenced by high rainfall, plant growth, and time of application. Agric. Environ. 2:137-146.

----. 1978. Computed redistribution of pesticides in the root zone of an arable crop. Plant Soil 49:569-580.

----. 1979. Computing the movement of ethoprophos in soil after application in spring. Soil Sci. 128:303-311.

----, ----, and J. J. T. I. Boesten. 1980. Measured and simulated behavior of oxamyl in fallow soils. Pestic. Sci. 11:379-388.

----, and W. A. Dekkers. 1976. Computed leaching of pesticides from soil under field conditions. Water Air Soil Pollut. 5:419-500.

----, and J. H. Smelt. 1981. Movement and conversion of ethoprophos in soil in winter: 2. Computer simulation. Soil Sci. 131:296-302.

----, ----, and Th. M. Lexmond. 1976. Conversion and leaching of aldicarb in soil columns. Pestic. Sci. 7:471-482.

----, ----, J. G. Verlaat, and R. Zandvoost. 1974. Measured and computed concentration patterns of propyzamide in field soils. Weed Res. 14:87-95.

Lyman, W. J. 1982. Adsorption coefficients for soils and sediments. p. 1-33. *In* Research and development of methods for estimating physicochemical properties of organic compounds of environmental significance. Final Report of Phase II, Contract No. DAMO-17-18-C-8073, U.S. Army Medical Res. and Dev. Command, Ft. Detrick, Md. (In press).

McCall, P. J., D. A. Laskowski, R. L. Swann, and H. L. Dishburger. 1981. Measurement of sorption coefficients of organic chemicals and their use in environmental fate analysis. p. 89-109. *In* Test protocols for environmental fate and movement of toxicants. Proc. of a Symp., AOAC 94th Annual Meeting, Oct. 21-22, 1980, Washington, D.C.

Nielsen, D. R., J. W. Biggar, and K. T. Erh. 1973. Spatial variability of field-measured soil-water properties. Hilgardia 42:215-259.

----, P. J. Wierenga, and J. W. Biggar. 1983. Spatial soil variation and mass transfers from agricultural soils. p. 65-78. *In* D. W. Nelson, K. K. Tanji, and D. E. Elrick (ed.) Chemical mobility and reactivity in soil systems. Am. Soc. of Agron. and Soil Sci. Soc. of Am. Spec. Pub. No. 11, Am. Soc. of Agron., Madison, Wis.

Nkedi-Kizza, P., J. W. Biggar, H. M. Selim, M. Th. van Genuchten, P. J. Wierenga, J. M. Davidson, and D. R. Nielsen. 1983. On the equivalence of two conceptual models for describing ion exchange during transport through an aggregated Oxisol. Water Resour. Res. (submitted).

Passioura, J. B., and D. A. Rose. 1971. Hydrodynamic dispersion in aggregated media: II. Effects of velocity and aggregate size. Soil Sci. 111:345-351.

Rao, P. S. C., and J. M. Davidson. 1979. Adsorption and movement of selected pesticides at high concentrations in soils. Water Res. 13:375-380.

----, and ----. 1980. Estimation of pesticide retention and transformation parameters required in nonpoint source pollution models. p. 23-67. *In* M. R. Overcash, and J. M. Davidson (ed.) Environmental impact of nonpoint source pollution. Ann Arbor Science Publishing Inc., Ann Arbor, Mich.

----, ----, and L. C. Hammond. 1976. Estimation of non-reactive and reactive solute front locations in soils. p. 235–241. *In* Proc. Hazardous Wastes Research Symp., EPA-600/9-76-015, Tucson, Ariz.

----, ----, and R. E. Jessup. 1981. Simulation of nitrogen behavior in the root zone of cropped land areas receiving organic wastes. p. 81–95. *In* M. J. Frissel and J. A. Van Veen (ed.) Simulation of nitrogen behavior in soil-plant systems. Pudoc, Wageningen, The Netherlands.

----, and R. E. Jessup. 1982. Development and verification of simulation models for describing pesticide dynamics in soils. Ecol. Modeling 16:67–75.

----, ----, J. M. Davidson, and H. M. Selim. 1979. Evaluation of conceptual models for describing nonequilibrium adsorption-desorption of pesticides during steady flow in soils. Soil Sci. Soc. Am. J. 43:22–28.

----, R. E. Jessup, and A. G. Hornsby. 1982. Simulation of nitrogen in agro-ecosystems: Criteris for model selection and use. Plant Soil 67:35–43. Proc. Int. Workshop "Nitrogen Cycling in Ecosystems of Latin American and the Caribbean," Cali, Colombia, March 16–21, 1981.

----, ----, D. E. Rolston, J. M. Davidson, D. P. Kilcrease. 1980a. Experimental and mathematical description of nonadsorbed solute transfer by diffusion in spherical aggregates. Soil Sci. Soc. Am. J. 44:684–688.

----, ----, and M. Th. van Genuchten. 1983. Non-adsorbed and adsorbed solute transport in water-saturated aggregated porous media: Theory, and model parameter analysis. Water Resour. Res. (In Review).

----, P. Nkedi-Kizza, and J. M. Davidson. 1982c. Retention and transformations of pesticides in relation to nonpoint source pollution from croplands. *In* G. W. Bailey, and F. Schaller (ed.) Agricultural management and water quality. Iowa State Univ. Press (In press).

----, D. E. Rolston, R. E. Jessup, and J. M. Davidson. 1980b. Solute transport in aggregated porous media: Theoretical and experimental evaluation. Soil Sci. Soc. Am. J. 44:1139–1146.

Reinbold, K. A., J. J. Hassett, J. C. Means, and W. L. Banwart. 1979. Adsorption of energy-related pollutants: A literature review. EPA-600/3-79-086, 170 p. Athens, Ga.

Selim, H. M., J. M. Davidson, and R. S. Mansell. 1976. Evaluation of a two-site adsorption-desorption model for describing solute transport in soils. p. 444–448. *In* Proc. Summer Computer Simulation Conf., Washington, D.C., July 12–14, 1976. Sponsored by AICHE, ISA, SHARE, SCS, AMS, AIAA, IEEE, AGU, BMES, and IAMCS.

Smelt, J. H., M. Leistra, and S. Voerman. 1977. Movement and rate of decomposition of ethoprophos in soil columns under field conditions. Pestic. Sci. 8:147–151.

Stevenson, F. J. 1976. Bound and conjugated pesticide residues. p. 180–207. *In* D. D. Kaufman, G. G. Still, and G. D. Paulson (ed.) Am. Chem. Soc. Symp. Series No. 29, Am. Chem. Soc., Washington, D.C.

Travis, C. C., and E. C. Etnier. 1981. A survey of sorption relationships for reactive solutes in soil. J. Environ. Qual. 10:8–17.

van Genuchten, M. Th. 1981. Non-equilibrium transport parameters from miscible displacemnet experiments. USDA Res. Rep. No. 119. 88 p.

----, and R. W. Cleary. 1982. Movement of solutes in soil: Computer simulated and laboratory results. p. 349–386. *In* G. H. Bolt (ed.) Soil chemistry; B. Physico-chemical models. (2nd rev. ed.). Elsevier Scientific Publishing Co., N.Y.

----, J. M. Davidson, and P. J. Wierenga. 1974. An evaluation of kinetic and equilibrium equations for the prediction of pesticide movement in porous media. Soil Sci. Soc. Am. Proc. 38:29–35.

----, and P. J. Wierenga. 1976. Mass transfer studies in sorbing porous media: I. Analytical solutions. Soil Sci. Soc. Am. J. 40:473–480.

----, ----, and G. A. O'Connor. 1977. Mass transfer studies in sorbing porous media: III. Experimental evaluations with 2,4,5-T. Soil Sci. Soc. Am. J. 41:278–285.

Walker, A. 1976a. Simulation of herbicide persistance in soil: I. Simazine and prometryne. Pestic. Sci. 7:41–49.

———. 1976b. Simulation of herbicide persistance in soil: II. Simazine and linuron in long-term experiments. Pestic. Sci. 7:50–58.

Warrick, A. W., G. J. Mullen, and D. R. Nielsen. 1977. Predictions of soil-water flux based upon field-measured soil-water properties. Soil Sci. Am. Proc. 41:14–19.

Weber, J. B. 1972. Interaction of organic pesticides with particulate matter in aquatic and soil systems. p. 55–120. *In* R. F. Gould (ed.) Fate of organic pesticides in the aquatic environment. Adv. Chem. Series No. 111, Am. Chem. Soc.,

Weed, S. B., and J. B. Weber. 1974. Pesticide-organic matter interactions. p. 39–66. *In* W. D. Guenzi (ed.) Pesticides in soil and water. Soil Sci. Soc. Am., Madison, Wis.

Wierenga, P. J. 1977. Solute distribution profiles computed with steady-state and transient water movement models. Soil Sci. Soc. Am. J. 41:1050–1054.

Wood, A. L., and J. M. Davidson. 1975. Fluometuron and water content distributions during infiltration: Measured and calculated. Soil Sci. Soc. Am. Proc. 39:820–826.

Chapter 14

Mobility of Radionuclides in Soil

G. W. Gee, Dhanpat Rai, and R. J. Serne[1]

Recent public interest in nuclear power production and waste disposal has led to increased awareness and concern about the biological hazards of radionuclide cycling in soil. Various radionuclides found in soils are shown in Table 1 and can be seen to originate from numerous sources including: radioisotopes formed during earth genesis, cosmic irradiation, fallout from atmospheric testing, uranium mill tailings, phosphate mill wastes, nuclear and coal fired power plants, defense activities, medical, industrial and research uses (Haury and Schikarski, 1977; Alexander and Bloemke, 1978; Beck et al., 1980; Roessler et al., 1980; U.S. Dep. of Energy, 1980).

Table 1. Sources of environmentally important radionuclides present in soils.

Source	Radionuclide
Naturally occurring	^{40}K, ^{222}Rn, ^{226}Ra, ^{230,232}Th, ^{235,238}U
Cosmic irradiation	^{3}H, ^{7}Be, ^{14}C, ^{22}Na
Fallout-weapons test	^{3}H, ^{90}Sr, ^{137}Cs, ^{239,240}Pu
Mining wastes—Uranium, phosphate, coal	^{222}Rn, ^{226}Ra, ^{230,232}Th, ^{235,238}U
Industrial wastes— Nuclear power plants, defense, and miscellaneous sources, including medical and research wastes	^{59,63}Ni, ^{60}Co, ^{90}Sr, ^{93,99}Zr, ^{99}Tc, ^{107}Pd, ^{129}I, ^{137}Cs, ^{144}Ce, ^{151}Sm, ^{152,154}Eu, ^{237}Np, ^{239,240,242}Pu, ^{241,243}Am

[1] Staff scientists, Geosciences Research and Engineering Dep., Pacific Northwest Lab., Richland, WA 99352.

Copyright © 1983 ASA, SSSA, 677 South Segoe Road, Madison, WI 53711. *Chemical Mobility and Reactivity in Soil Systems.*

The radionuclides that are present in soils as a result of natural distribution, cosmic irradiation and fallout generally do not present significant health hazards because of their low concentration resulting from widespread (global) distribution. Uranium mill tailings and nuclear fission wastes are potentially hazardous, but are present in concentrated forms in relatively few areas. Major commitments to nuclear energy and nuclear defense depend upon our ability to safely dispose of these wastes. Prediction of the potential mobility of actinides (e.g., U, Np, Pu, Am) and other important decay and fission products (e.g., Ra, Rn, Tc, Cs, Sr) requires knowledge of water flow and transport processes that control both leach rates and chemical interaction of leachates with the geologic media.

The mathematical framework for transport models needed to describe radionuclide mobility in soils is the same as that used for other soil trace constituents (Selim et al., 1977; Couchat et al., 1980). However, chemical transport models which adequately handle either trace soil constituents or radionuclide transport phenomena are often limited in scope and seldom include solute speciation, precipitation-dissolution reactions, and other factors influencing the adsorption-desorption processes of radionuclides.

Understanding how the chemical, physical, and biological characteristics of the soil interact with radionuclides provides insight into safe disposal methods for radiochemicals. The soil will likely be the final repository for most low-level radioactive wastes generated by the nuclear fuel cycle or from fallout. High level wastes are to be placed in deep-geological repositories isolated from the biosphere. However, complete isolation can not be proven for the time periods required (10^4 to 10^6 years), hence potential movement of these materials through soils and porous rock materials to the biosphere must be considered in safety assessments.

The mobility of the radionuclides in the soil system will depend on the physical and chemical properties of the soil, the chemistry of the soil solution and the characteristics of the radionuclide and the chemical nature of the waste itself (Fig. 1). Most radionuclides move in the liquid phase hence the prediction of radionuclide movement also relies on an understanding of the hydrology of the flow regime and adequate measurement of the hydrologic properties of the flow system (Yeh and Tamura, 1982; Nelson et al., 1980, 1983). However, ignoring chemical interactions in the flow system can lead to overly conservative estimates of transport and in some cases may result in unnecessary, detailed hydrologic modeling efforts to compensate for neglecting chemical interaction. Optimally, the combination of detailed geochemistry and hydrologic characterization would provide the most useful data base upon which to evaluate radionuclide mobility, hence proper technical assessment of radioactive waste repositories.

The objectives of this review are to summarize the factors that affect radionuclide mobility in soil, including adsorption-desorption, precipitation-dissolution reactions for liquid phase radionuclides as well as the transport of gaseous radionuclides in soils. Areas where more information is needed are also discussed. The mobility of selected radionuclides including isotopes of Sr, Cs, Co, Tc, I, Se, Pu, Am, Np, and Rn are discussed in additional detail. Two brief examples of waste management systems to control radionuclide mobility are presented.

RADIONUCLIDE ADSORPTION-DESORPTION ON SOILS

Before reviewing radionuclide adsorption-desorption, we wish to clarify terminology. As used in this paper, adsorption refers to surface processes wherein radionuclides in solution become incorporated onto soil component surfaces. The term includes ideal ion exchange with constant charge substrates such as three-layer (2:1) phyllosilicates as well as less well defined and variable surface reactions with two layer (1:1) phyllosilicates, amorphous soil coatings and organic matter, but excludes precipitation-crystallization of and structural substitutions in minerals or amorphous solids. Ideal cation-exchange reactions in soils are reviewed in Sposito (1981) and Thomas (1977) and the more general adsorption process is discussed by Mehlich (1981) and numerous authors in a recent monograph (see Anderson and Rubin, 1981).

There are two approaches currently used to study radionuclide adsorption, empirical studies that rely on distribution coefficient (Kd) measurements and mechanistic studies that strive to identify, differenti-

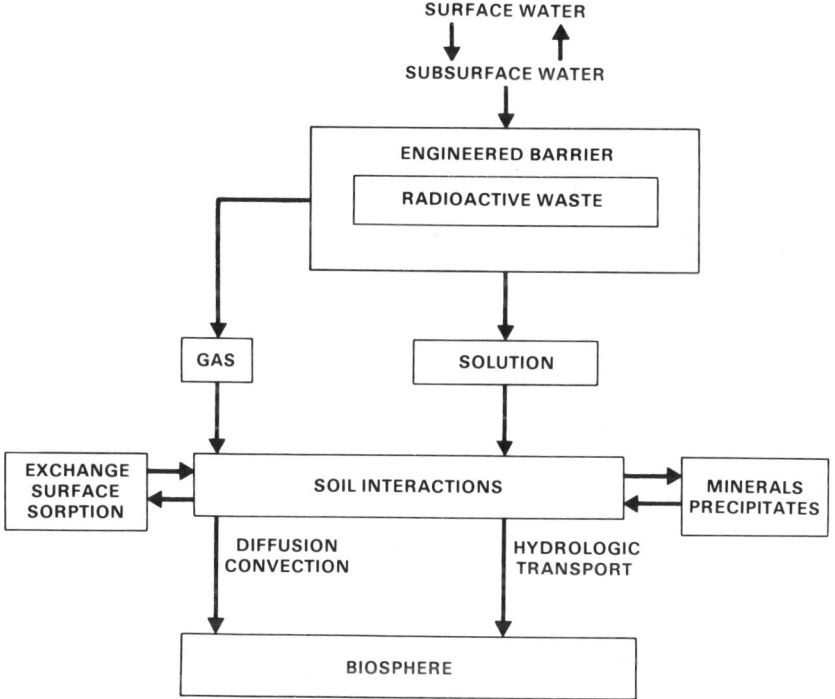

Fig. 1. Schematic of liquid and gaseous transport of radionuclide waste through soils and geologic media to the biosphere.

ate and quantify the physicochemical processes that control the observed adsorption.

THE Kd APPROACH

This approach allows the rapid determination of the relative affinity of nuclides for solid substrates without concern for the cause or controlling processes involved. The empirical distribution coefficient, often called "Kd," "D," or "Rd" is the ratio of the radioactivity or mass of the radionuclide adsorbed on the solid to the radioactivity or mass of the nuclide remaining in solution, i.e.,

$$Kd = \frac{\text{amount of radionuclide sorbed on soil}}{\text{amount of radionuclide remaining in solution}}, \text{ml/g} \qquad [1]$$

and is readily measured by laboratory experimentation. The Kd is often converted into a quantitative estimate of nuclide retardation relative to ground water, using the equation

$$R_f = 1 + (\text{bulk density} \times Kd/\text{porosity}) \qquad [2]$$

where R_f is the retardation factor, i.e., the velocity of the water divided by the velocity of the radionuclide (e.g., Inoue and Kaufmann, 1963; Freeze and Cherry, 1979). In its simplest form, use of this equation requires the assumption that Kd remains constant over time and space. On the other hand, through the use of computerized mass transport codes and discretized soil cells and time steps, variations in Kd, bulk density, porosity, and groundwater velocity over space and time can be accommodated. Almost every existing mass transport computer code used for safety assessment and environmental fate analyses relies on the distribution coefficient concept. Although Kd has no ultimate mechanistic rationale, it does represent a very convenient and useful statistic to describe the partitioning of radionuclide between the adsorbed and aqueous phases.

Until recently, almost all of the work reported in the literature on radionuclide retardation has been empirical in nature with measurement of distribution coefficients as the primary objective. The Kd has been used by some scientists much more selectively to refer to (i) the special case of ideal binary ion exchange where the element of interest is present in such trace concentrations that it does not appreciably affect concentration of the macro-constituent adsorbed on the solid or the activity coefficient ratio of the mass action exchange (Meyer, 1979; Rouston and Serne, 1972) or (ii) a special case of a Freundlich adsorption isotherm where adsorption is linear (Travis, 1978). Linear adsorption is implied by the use of the term, Kd, and typically occurs only at trace adsorbate concentrations in solution and trace loading onto the solid adsorbent surfaces. In both instances, the adsorption refers to equilibrium conditions. Kd values from literature sources (for example, Onishi et al., 1981) have been used rather indiscriminately to assess radionuclide mobility. Sometimes, and perhaps too often, the Kd values used in assessment of waste repositories are not appropriate for the expected repository conditions. A rather large effort has been initiated by the Department of Energy (Hostetler et al., 1980) to document and provide easy retrieval of information on adsorption-desorption data, including details on specific test conditions (soil, rock types, solution chemistry, contact times, experimental methods, etc.). This information will be useful in elucidating adsorption mechanisms and provide safety assessment modelers a more definitive and comprehensive data set from which they can determine if a selected Kd is appropriate for a given modeled repository scenario.

Recent studies of radionuclide adsorption have explored the relationship between the observed distribution coefficient and changes in soil and groundwater type. Variables such as soil mineralogy, particle size, cation-exchange capacity, organic matter content, groundwater composition, pH, and Eh have been studied. Contact time is also a variable which is commonly varied to collect kinetic information. A few studies have explored the effect of temperature, radionuclide concentration, and species distribution. Reversibility studies, measurement of both adsorption and subsequent desorption, are also becoming more common. Recently, comparisons among numerous laboratory methodologies have been attempted (Serne and Relyea, 1981). Field studies using isotopic tracers (Pickens et al., 1981) have been used successfully in determining reactive solute mobility and in measuring distribution coefficients, but this has been done under rather ideal conditions.

IDEAL EXCHANGE AND ITS LIMITATIONS

Although an exhaustive review of available data has not been performed, the general conclusions offered herein are from published reviews and annotated bibliographies (e.g., Serne and Relyea, 1981; Ames and Rai, 1978; Onishi et al., 1981). Ideal ion-exchange theory which adequately describes the adsorption of most macro cations, such as Ca^{2+}, Mg^{2+}, K^+ and Na^+, in soils is of limited utility for predicting the adsorption of most radionuclides. Aside from Cs, Sr, Ba and Ra, few radionuclides exhibit ideal cation-exchange behavior in soil adsorption studies. The above mentioned ideal radionuclides in general show adsorption that decreases in the presence of increased concentrations of competing cations, correlates with cation-exchange capacity and soil surface area.

For example, Sr and Cs have relatively simple aqueous chemistry and for many soils and clays their retention has been successfully described by mass action ion exchange constructs and/or by empirical functions of the solution concentration of competing cations. Routson and Serne (1972) and Hawkins and Short (1965) discuss the latter approach while Wahlberg and Fishman (1962) and Wahlberg et al. (1965) discuss the mass action ion exchange approach for Cs and Sr adsorption on clays.

Strontium adsorption studies by Baker and Beetem (1961), Baetsle et al. (1964), Cohen and Gailledreau (1961), and Wahlberg and Dewar (1965) and Cs adsorption studies by Ames and Hajek (1966) and Routson (1973) substantiate the predictive capabilities of the mass action exchange and empirical competing cation approaches. Conversely, Sr adsorption on soils with high organic matter, high carbonate contents or high hydrous oxide contents can show more complex adsorption behavior, possibly explained by adsorption on pH sensitive organic and hydrous oxide sites or precipitation (solid solution reactions) with calcite (see Tamura, 1964; Jenne and Wahlberg, 1968; Francis, 1978; Spaulding, 1980 for details).

Mica-like minerals such as illite tend to fix Cs (Sawhney, 1964; Tamura and Jacobs, 1960). When expanded layer silicates are collapsed to a 10 Å mica spacing by K saturation and heating, increased Cs selectivity results (Tamura and Jacobs, 1961; Tamura, 1964; Coleman et al., 1963). Some researchers have considered the exchange of trace Cs on micaceous minerals to be irreversible (Klechkovski and Gulyakin, 1958; Spitsyn et al., 1963), but the reverse rate may only be much slower than the adsorption rate (Routson, 1973). Cesium like many exchangeable radionuclide shows a strong dependence of adsorption on the soluble competing cations. Figure 2 illustrates the solution concentration dependence of the Kd for Cs on montmorillonite clay.

Aside from special cases such as those just mentioned, these ideal radionuclides follow classical exchange relationships for adsorption and desorption in the presence of competing cations.

For many radionuclides, the adsorption affinity as measured by a distribution coefficient is quite sensitive to the concentration of the radionuclide used in the experiment thus defying one of the basic assumptions

EFFECT OF SALT CONCENTRATION AND pH ON SORPTION OF CESIUM

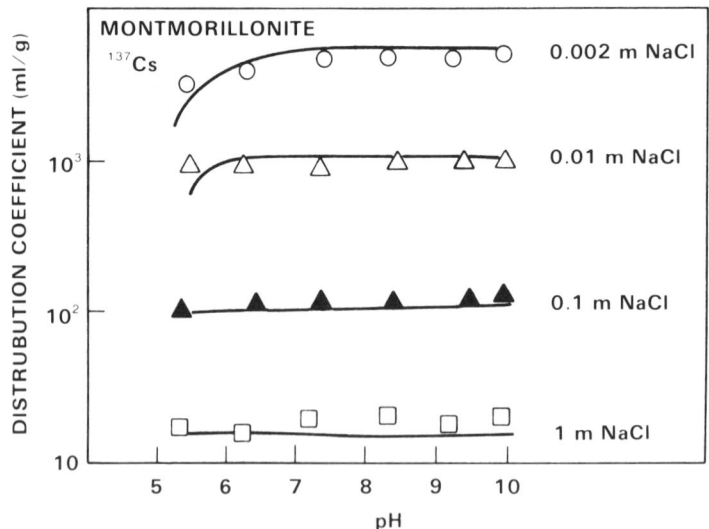

Fig. 2. Kd vs. pH for Belle Fourche montmorillonite clay. Symbols are experimental data and solid lines are predicted values using MINEQL a geochemical equilibrium model (after Relyea and Silva, 1981).

for using the Kd approach. For experiments performed at trace radionuclide concentrations below solubility constraints, Kd often increases as the radionuclide concentration decreases resulting in nonlinear isotherms. In addition, the steady-state distribution ratio measured from desorption experiments often are up to an order of magnitude larger than the distribution ratios measured from adsorption experiments which again violates a commonly used assumption for the Kd approach, reversibility. In general, surface adsorption reactions reach near-steady-state conditions within a few days. For most radionuclides, adsorption is sensitive to groundwater pH, soil hydrous oxide, and organic content. For multivalent radionuclides, the oxidation-reduction potential is often a key variable.

GENERAL ADSORPTION CONSIDERATIONS FOR TRANSITION METALS AND LANTHANIDES

Of the first and second transition series metals, released in reactor effluents and defense wastes, only ^{60}Co, ^{59}Ni, ^{63}Ni, ^{65}Zn, ^{93}Zr, ^{99}Tc, ^{106}Ru, ^{107}Pd, and ^{110m}Ag have long enough half-lives and decay modes and energies to be of environmental interest. Lanthanide series metals have radioactive isotopes for every element but only ^{144}Ce, ^{147}Pm, ^{151}Sm, ^{152}Eu, and ^{154}Eu have long enough half-lives and decay modes and energies to be of environmental interest. These elements, except Tc and perhaps Ru, consistently have very large distribution coefficients for typical near-neutral pH groundwaters. The ionic strength and competing cation composition cause only slight reductions in adsorption even at concentrations high enough to be considered as brines or bitterns (see Rhodes, 1957; Benson, 1960). Desorption experiments show that a significant proportion of these radionuclides do not exchange reversibly (Serne and Relyea, 1981). The type of soil mineralogy does not appear to significantly influence the adsorption affinity of these metals, although there appears to be a slightly enhanced adsorption by smaller particle-sized samples and high organic content soils which probably reflects increased active surface area and organic chelation, respectively. Adsorption of the transition elements is most sensitive to solution pH and may reflect chemisorption or specific adsorption of metal cations or hydrolized species onto soil oxide surfaces and condensed organic acids (Schwertmann and Taylor, 1977; Kinniburgh et al., 1976, 1977; James and Healy, 1972; Tewari et al., 1972). In some of the experiments it is possible that precipitation of insoluble metal hydroxides occurred and that the observed distribution coefficients may include a precipitation contribution.

The adsorption of technetium (Tc) is highly dependent on the oxidation-reduction potential as the predominate solution species and solubility of Tc are very sensitive to redox potentials. Under oxidizing conditions typical of near-surface and unsaturated soil systems, Tc exists predominately as the pertechnetate anion, TcO_4^-. This anion is quite soluble and mobile in most soil environments. In highly organic soils, significant adsorption and/or redox mediated precipitation has been observed (McFadden, 1980). Landa et al. (1977) also observed that Tc became bound in the solid phase in soil solutions with high microbiologic activity. Bondietti and Francis (1979) reported that TcO_4^- adsorbs on rocks, with large reducing potential (structural Fe^{2+}) such as basalt and some granites and shales. Under reducing conditions, thermodynamic predictions place solution Tc predominately as cationic Tc^{4+}, Tc^{2+} or hydrolysis complexes which would be at low concentrations because of the low solubility of technetium oxides (Bondietti and Francis, 1979).

COMPLEXATION WITH ORGANIC LIGANDS

Complexation of the transition metal ions with soluble organic ligands can decrease Kd and hence increase mobility. As an example, Means et al. (1978) observed a dramatic increase in the mobility of ^{60}Co from shallow land burial waste sites at Oak Ridge, Tenn. which they attributed to the presence of EDTA and perhaps fulvic and humic acids. EDTA, however, does not remain stable in soils indefinitely. Swanson (1982) and Jones et al. (1983) report time dependent decreasing mobility of ^{60}Co in the presence of EDTA in soils from Savannah River, SC and Hanford, Wash. which they attributed to the breakdown of the EDTA-cobalt complex with time and the high adsorption affinity of the free cobalt.

OTHER ENVIRONMENTALLY IMPORTANT RADIONUCLIDES

Iodine—Iodine-129 is released as I_2 gas from reactor operations and exists in waste streams and soil solutions predominately as the anion iodide, I^-, and under highly oxidized conditions iodate, IO_3^-. Where soils exhibit anion exchange at low pH (tropical soils, for example), adsorption of I^- or IO_3^- is expected. However, the anion exchange capacity of most temperate soils is small over the pH range 5 to 9, hence, the adsorption of iodide and iodate should be minimal. Several investigations with soils (Wildung et al., 1975; Gee and Campbell, 1980) and basalt rocks (Salter et al., 1981) in fact, show iodide to be a very mobile species under oxidizing conditions. The knowledge that I^- is strongly adsorbed by a few zeolites and forms highly insoluble compounds (such as AgI) can be used to immobilize iodide in nuclear waste packages (Bird and Lopata, 1980; Strickert et al., 1980). However, more common rocks, minerals, and soils offer little retardation capability for I^-.

Selenium—Selenium adsorption on soils has been reported by Singh et al., 1981; John et al., 1976, and Rajan and Watkinson, 1976. In relative terms Se in its two most common forms, selenite (SeO_3^{2-}) and selenate (SeO_4^{2-}), is adsorbed moderately by soils. (John et al., 1976 reports an average of 41% Se adsorption by 66 different soils). More selenate (SeO_4^{2-}) is usually adsorbed than selenite (SeO_3^{2-}) for a given soil and solution. High organic carbon and hydrous iron oxide containing soils immobilize Se best. Saline and alkaline (high pH) soils adsorb the least Se. The decrease in Se adsorption with the increase in pH is a typical anionic adsorption behavior resulting from the decrease in anionic exchange capacity with the increase in pH.

Tin—Very little data are available for the mobility of trace amounts of Sn in soil environments but from thermodynamic data we hypothesize that the low solubility of SnO_2 would limit Sn mobility in solution (Pourbaix, 1966).

ACTINIDES AND THEIR DAUGHTER PRODUCTS

Finally, the actinide series, Th through Cm represent potential environmental hazards. The important isotopes that exist in nature from Th and U ores or that are man-generated include ^{210}Pb, ^{222}Rn, ^{226}Ra, $^{229,230,232}Th$, ^{231}Pa, $^{233,234,235,236,238}U$, ^{237}Np, $^{238,239,240,241,242}Pu$, $^{241,243}Am$, and $^{242,244,245}Cm$. Of these, the largest hazards to man exist for ^{210}Pb, ^{226}Ra, ^{237}Np, $^{239,240,242}Pu$, $^{233,234,236,238}U$, and $^{241,243}Am$ (Serne and Relyea, 1981).

Transuranics—A significant data base exists in the referenced reviews on the expected environmental fate of the transuranics. In general, most transuranics exhibit moderate to high adsorption affinity for soils. Also, especially under reducing conditions, the oxides and hydroxides of these elements are relatively insoluble in neutral to alkaline pH waters. The solubility aspects of radionuclide mobility, particularly for the transuranics, are discussed in the next section.

Uranium and Daughters—There are numerous radionuclides produced by U, Th, and transuranic decay. Several of these exhibit long half-lives and exist in significant activities that cause potential environmental concern, such as ^{210}Pb, ^{226}Ra, and ^{230}Th daughter products of natural U. In most common soil environments, lead is adsorbed quite readily by hydrous oxides of Al, Fe, and probably Mn (e.g., Kinniburgh et al., 1976). Soil solution concentrations of lead may also be constrained by low solubility carbonate and phosphate minerals (Lindsay, 1979). Thorium hydroxides are also very insoluble (Baes and Mesmer, 1976). In recent studies with U mill tailings, Gee et al. (1980), Landa (1980), Sherwood and Serne (1982), and Serne et al. (1983) have demonstrated that ^{230}Th and ^{210}Pb are immobilized in neutral or slightly alkaline sediments but ^{238}U and ^{226}Ra can be complexed with solution carbonates and partially mobilized in neutral and alkaline groundwaters.

RECENT TRENDS IN ADSORPTION STUDIES

Recent radionuclide adsorption studies are tending towards mechanism identification, (e.g., Rai et al., 1980a), and development of adsorption mathematical models along the concepts originally described by James and Healy (1972). The models treat metal adsorption as the formation of a complex with the solid surface and rely on thermodynamic frameworks. Recent examples of this approach for the radionuclides, Cs and U, can be found in Silva et al. (1980, 1981) and Relyea and Silva (1981) (Fig. 2). These more mechanistic and mathematical modeling approaches will significantly increase our understanding of and predictive capability of trace constituent adsorption and clarify some of the uncertainties emanating from the empirical studies that are dependent upon specific soil/groundwater/radionuclide properties.

PRECIPITATION-DISSOLUTION REACTIONS IN SOIL SYSTEMS

In general, radionuclides exist in soils in trace quantities ranging from parts per million to less than parts per billion and thus can be treated as trace elements. Direct identification of trace element compounds in soils is difficult because these compounds, if present, comprise far less than 1% of the total mass and hence most research tools such as X-ray diffraction of bulk samples cannot be used for identification. Experimental evidence as to the exact nature of the solid phases of most trace elements present in soils is not available. Experiments which identify the nature of the solid phases, and quantify individual as well as complex solution species are urgently needed for predicting geochemical behavior of many radionuclides and trace elements. In the absence of direct methods, two indirect methods can be used for identifying the solubility controlling phases: (i) use of thermodynamic data as a basis for extrapolation and (ii) comparison of ion activity products with solubility products.

Thermodynamic data of different solids and solution species of an element can be used to predict the most stable phase and the total solution concentration of an element along with the nature of the solution species. Although thermodynamic data can be used to determine the direction of reactions and the ultimate equilibria, it does not provide information on rates of reactions. Therefore, in the absence of actual knowledge of solid phases in the soil and of kinetics of transformations of solids, thermodynamic data on the solid phases that have fast precipitation kinetics (such as amorphous oxides and hydroxides and salts) and those predicted to be the most stable can be used to set a range for the maximum solution concentration of an element in different environments. This assumes, of course, that accurate thermodynamic data on all possible solid phases, including solid solutions, and solution species are available. However, for most trace metals and radionuclides there are gaps in the data base and the accuracy of some of the available data is questionable. Recently, some excellent reviews of available information on solid phase-solution equilibria of major and some trace elments important in soil systems have been published (Lindsay, 1979, 1981; Mattigod et al., 1981). In addition, thermodynamic data for actinides and other radioisotopes have been summarized (Rai and Serne, 1977; Rai and Serne, 1978; Cleveland, 1979; Lemire and Tremaine, 1980; Allard et al., 1980; Burney and Harbour, 1974; Schulz, 1976; Baes and Mesmer, 1976; Langmuir, 1978; Fuger and Oetting, 1976). These literature data show that the reported solubility product values for many of the given solids differ by several orders of magnitude and hence accurate solubility data are not often available. In addition, the estimated errors even in the hydrolysis constants, especially of M(IV) oxidation states, are ±4 orders of magnitude (Lemire and Tremaine, 1980). The data on M(III) and M(IV) carbonate complexes are either not available or not reliable. Therefore additional experimental data are needed on different compounds that may be present in the soils and the solubilities of these compounds.

The second method which is often used by soil chemists to infer the presence of controlling solids, compares the ion activity products in solutions contacting soils with the solubility products of different solid phases (Lindsay, 1979; Jenne et al., 1980; Santillan-Medrano and Jurinak, 1975). In this method solid phases whose solubility products are similar to the observed ion activity products are assumed to be present in soils and hence to be the solids that control the trace metal concentrations in these solutions. Factors that limit the ion activity product approach include (i) unavailability of solubility products of some potentially important solid phases, (ii) inaccuracies in the existing thermodynamic data, and (iii) difficulties in measuring the nature and thermodynamic activity of solution species. However, from a practical standpoint this approach provides at least experimental information on solubilities and solution species in equilibrium with a solid, even though the nature of the solid phase is not known. For example, the ion product (solubility) data have been successful in predicting the concentration of certain elements in solutions, as evidenced by results for Si (Elgawhary and Lindsay, 1972), Fe (Norvell and Lindsay, 1982), and Zn and Cu (Norvell and Lindsay, 1969, 1972) in equilibrium with solid phases within soils. Several other studies also suggest that the solution concentrations of trace elements (Pb, Cd, Zn, Cu) may be solubility limited (Jenne et al., 1980; Santillan-Medrano and Jurinak, 1975; Norvell and Lindsay, 1969, 1972).

ACTINIDES

Table 1 lists the sources of actinides, such as Pu, Am, Np, and Cm and indicates that these radionuclides will be introduced into the soil or other geologic media primarily through contact with high level wastes or from radioactive fall-out. Recent studies have evaluated the solubility of actinides in different solids, such as Pu/Am contaminated soils and Pu and Np-doped borosilicate glasses (Rai and Strickert, 1980; Rai and Swanson, 1981; Rai and Ryan, 1982; Strickert and Rai, 1982; Rai et al., 1980a, 1980b, 1980c, 1981, 1982, 1983). Laboratory results have been combined with thermodynamic data to determine the nature of actinide compounds and their solubilities. Several studies on Pu, Np, and Am are briefly summarized below.

Plutonium—Predictions from thermodynamic data (Rai and Serne, 1977) indicate that crystalline PuO_2 is more stable than the hydrous Pu oxides; in addition, for oxidizing or reducing conditions, PuO_2 is predicted to be the most stable Pu compound among the simple oxides, hydroxides, carbonates, and phosphates for which information is available. Despite such a prediction, uncertainties in predicted Pu concentrations of up to five orders of magnitude resulted from variations in reported thermodynamic values for different Pu species (Rai and Serne, 1978).

The steady state solubility of different Pu solids in air-equilibrated solutions (Rai et al., 1980b, 1980c; Rai and Swanson, 1981) does indeed show that PuO_2 is more stable than Pu(IV) hydrous oxide (Fig. 3). Information on kinetics of dissolution and transformation of Pu(IV) solids is also needed to determine Pu concentrations to be expected in soils. Rai et al. (1980b) and Rai and Swanson (1981) have shown that Pu(IV) hydrous oxide has very fast precipitation kinetics (several minutes) and that the dissolution kinetics of Pu(IV) oxide, hydrous oxide, and polymer are also relatively fast (several days). Thermodynamic considerations dictate that compounds with high free energy would eventually convert to compounds with low free energy [e.g., $PuO_2(c)$]. However, low free energy compounds of radioactive materials can convert to higher free energy compounds due to radiolytic effects. The aging process, in which dehydration and crystallization are opposed by radiolytic effects, will convert these solids to a steady state material having properties between those of crystalline oxide and amorphic hydrous oxide (Rai and Ryan, 1982). In the case of ^{239}Pu, Rai and Ryan (1982) have shown that Pu(IV) hydrous oxide decreased in solubility over a 3.5 year period. The resultant solubility was comparable to that of $^{239}PuO_2(lc)$, a form of PuO_2 but with a less well-defined X-ray diffraction pattern. The results discussed above do indicate that $PuO_2(lc)$ is the most stable among simple ^{239}Pu compounds and if present in soils would control Pu concentrations.

Rai et al. (1980b) studied the solubility of ^{239}Pu-contaminated soils and compared these solubilities with the experimental solubility of PuO_2 and Pu(IV) hydrous oxide (see Fig. 3). The observed Pu solubilities of the soil were similar to the $PuO_2(lc)$ solubility, thus consistent with choice of $PuO_2(lc)$ as the solubility controlling phase. The presence of PuO_2 in these soils was also confirmed by x-ray diffraction of the Pu particles isolated from these sediments.

Rai and Strickert (1980) studied the solubility of Pu-doped glass, a possible waste form for high level nuclear waste disposal. Their results showed that Pu concentrations in solution contacting these glasses were similar to the concentrations maintained by $PuO_2(lc)$. Both experiments show that maximum Pu concentrations in soils and waste glass leachates may be estimated from the solubility of $PuO_2(lc)$.

Americium—Rai et al. (1981) studied the solubility of Am-contaminated soil samples to develop a predictive capability for Am concentrations in solutions and to determine Am solid phases that may form in soils. The Am concentrations were found to be controlled by a solid phase and fit the following reaction (Eq. [3])

$$Am(soil) + H^+ \leftrightarrows Am^+ (aq.\ complex) \qquad [3]$$

with a log equilibrium constant of -4.12. Their initial evaluations of Am^+ (aq. complex) based upon thermodynamic data reported in the literature suggested the likelihood of this species being $Am(OH)_2^+$. However, recent experimental data on Am hydrolysis species (Rai et al., 1983) showed that $Am(OH)_2^+$ could not have been the Am species. The Am solid present in these contaminated soils has not yet been identified, but it is not $Am(OH)_3$. As the above reaction indicates, the steady-state Am concentrations in solutions contacting the Am(soil) solid were found to decrease about 10-fold with one unit increase in pH. Further, Rai et al. (1981) reviewed adsorption experiments with a large number of rocks, soils, and minerals in which initial Am solution concentrations exceeded the solubility-limited concentration (Eq. [3]) and found that the final Am solution concentrations were similar to Am solutions contacting Am(soil). Despite the variation in rock or mineral types used in the experiments, the final observed Am concentration followed Eq. [3] suggesting a solubility control. The dissolution of Am(soil) and Am(OH)$_3$ (Rai et al., 1981; Rai et al., 1983) and AmO$_2$(c) is very rapid and steady state Am solution concentrations are observed in a few days.

Neptunium—No Np contaminated soils or sediments were available to study Np solid phases which may potentially form in soils, nor was there any experimental information on the solubilities of different Np solids. Thermodynamic data (largely estimated) for Np solution species and solid phases are available (Allard et al., 1980; Burney and Harbour, 1974). Predictions (Rai and Serne, 1978) based on the available thermodynamic data show that NpO$_2$(c) is the most stable compound among the Np oxides and hydroxides, at all pH and Eh values encountered in groundwaters (Baas Becking et al., 1960).

The solubility of Np in Np-doped borosilicate glass (a potential waste form) has been studied to predict the concentrations of Np leached from

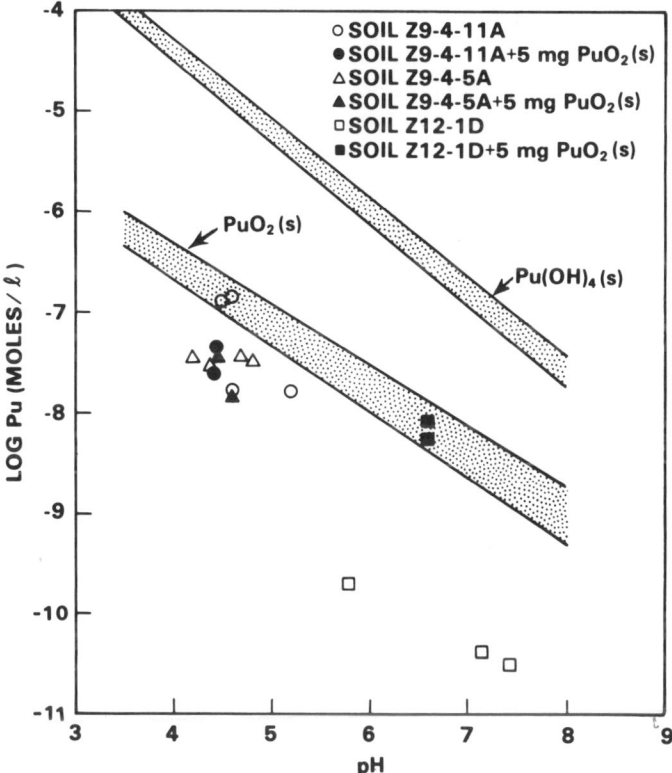

Fig. 3. Solubility-controlling plutonium solids in contaminated soils (after Rai et al., 1980b).

nuclear waste repositories in geologic environments and to determine the nature of possible Np solid phases that may control Np concentration (Rai et al., 1982). The concentrations of Np in solutions contacting the crushed doped glass were found to be controlled by a Np solid phase. Analogously to Pu, Np-doped glass was similar in solubility to $NpO_2(lc)$ (Strickert and Rai, 1982). Thus, the maximum concentration of the Np leached from this waste form can be predicted from the solubility of $NpO_2(lc)$. This conclusion is based upon similar Np concentrations in solutions contacting Np-doped glass, Np-doped glass plus $NpO_2(c)$, and $NpO_2(c)$ alone, under controlled redox potentials (pH + pe = 11.8) and a range of pH values.

In conclusion, the Pu, Am, and Np studies reported above were done under oxidizing conditions (in equilibrium with quinhydrone (pe + pH $\cong$ 11.82) and/or air) where Pu(V), Np(V), and Am(III) were the primary oxidation states in solution. The concentrations of these transuranic-elements in equilibrium with various solid phases under oxidizing conditions can reasonably be predicted from the studies reported above. Under reducing conditions, Pu and Np would be present in lower oxidation states [Pu as Pu(III) and/or Pu(IV) and Np primarily as Np(IV)], whereas the oxidation state of Am would not change. Because the nature of Pu and Np species change in reducing conditions, information on the solubilities of actinide compounds under these conditions is needed. However, using the experimental and available thermodynamic data, Wood and Rai (1982) predicted the solubilities under reducing conditions and compared them with Maximum Permissible Concentrations (MPC). Their analysis showed that for both oxidizing and reducing conditions, the solubility limited concentrations are very low and are either near or below the MPC. If subsequent studies verify the data, particularly for expected deep-geologic repository reducing conditions, it will further emphasize the importance, and in certain cases dominance, of solution chemistry in determining the suitability of soils and other geologic media for the containment of radioactive wastes.

GASEOUS RADIONUCLIDE TRANSPORT IN SOILS

Diffusion mechanisms are considered to be primarily responsible for the transport of gaseous radionuclides through soils. Gas transport by diffusion through soils has been the subject of comprehensive reviews (Currie, 1970; Tanner, 1980; Kimball, 1983) and diffusion coefficients have been determined for a number of soil gases including radon, ^{222}Rn. Strong dependence of gas diffusion on air-filled porosity has been observed for soil gases and empirical relationships have been useful for such gases as O_2 and CO_2 to relate the diffusion coefficient in soils to that observed in free air (Buckingham, 1904; Currie, 1960, 1961). For most soil gases the diffusion coefficient in water is as small as 1/10 000 of that in air at the same temperature and pressure, hence increasing soil water contents can cause significantly reduced gas flow rates. Efforts during the past 20 years have focused on analyzing the effects of moisture and pore size distributions on gas diffusion coefficients. Gas solubility in water and gas diffusion through small pores (Knudsen diffusion) also influence the diffusion of gas in soils, but these factors have not been studied in detail. It is clear, however, that estimates of radionuclide gas transport will require measurements of gas diffusion coefficients which in turn may be empirically related to soil moisture content and pore size distributions.

Radon-222, ^{3}H, and ^{14}C can be transported in the gas phase in soils. Since there is little evidence that tritium or carbon-14 behave differently than their nonradioactive counterparts (water vapor and CO_2), the major area of research in recent years has been on the transport of radon in soils.

RADON DIFFUSION THROUGH EARTHEN COVERS

Radon-222 is the major radiological hazard associated with uranium mill wastes. Radon movement through U tailings and into soil covers has been the subject of some recent intensive studies (Cohen, 1979; Rogers et al., 1980; Mayer et al., 1981; Zettwoog et al., 1982; Nielson et al., 1982). Radon gas is inert and hence is not expected to be chemically adsorbed on soils but can be solubilized to an appreciable degree by the soil solution. Recently, Nielson and Rogers (1982) have computed radon diffusion coefficients of soil cover materials using an analysis based on pore size distribution. Their analysis also included considerations of radon solubility in water and Knudsen diffusion which accounts for diffusion in very fine pores (<1 μm). Good agreement was found between calculated and experimental radon diffusivity values in soils with water contents in the 0 to 0.5 saturation range. At higher water contents (lower air-filled porosities) measured values tended to be lower than calculated. Silker and Rogers (1981) and Nielson et al. (1982) have measured radon diffusion coefficients that vary by more than an order of magnitude for similar material at similar air-filled porosities but packed to different moisture contents and densities. These data suggest that factors other than air-filled porosity, perhaps pore geometry, and pore blockage can influence the diffusion of radon through soils.

In the field, several nondiffusive mechanisms operate to enhance radon transport. Soil gases, including radon, have been observed to respond to wind turbulence near the soil surface (Kraner et al., 1964; Kimball, 1973) and to barometric pressure fluctuations (Clements and Wilkening, 1974). The results of these studies suggest that flux changes on the order of 20 to 60% can be obtained in deep, dry soils for barometric pressure changes of 1 to 2% per day and that wind turbulence can cause increased gas exchange ranging from 25 to 200%. Little effect of turbulence is observed, however, when the surface soil is fine-textured or wet (Kimball, 1983).

Observation of increased radon concentration in soil gas prior to major earthquakes (King, 1978) suggests that radon gas also may be mobilized by ultrasonic vibrations or by increased emanation in underlying rock fractures due to earthquake activity. However, it is not clear how the vibrations directly affect soil gas concentrations. Water content and barometric pressure changes in most cases should mask out perturbations created by ultrasonic vibrations, making it difficult to sort out effects. Wilkening (1980) concedes that enhanced gas transport is possible under certain circumstances but argues that radon gas transported in soils is largely controlled by diffusion.

Radon flux from a covered U tailings pile can vary by as much as 200% due to seasonal variations in precipitation and evaporation. These climatic variations influence the radon by directly affecting the water content in the soil profile (Mayer et al., 1981). However with adequate soil moisture and sufficient cover thickness radon exhalation rates can be controlled to low levels approaching natural background.

Control of ^{222}Rn by earthen covers takes advantage of the short half-life (3.8 d) of radon. Wet or thick earthen covers can create a sufficiently tortuous or long diffusion path that the radon can decay to 1/100 or less of its initial value, hence drastically reducing the surface radon flux. Radon will be produced for thousands of years due to the long-lived parent of radon, ^{226}Ra ($t_{1/2}$ = 1620 years), so radon control by an earthen cover ultimately will depend upon the long-term moisture content of the cover, as affected by climate and vegetation, and the ability of the cover to withstand erosional forces of wind and water, which may remove the cover over a long period of time.

WASTE MANAGEMENT EXAMPLES

Transport of radionuclides in soils is a major concern in waste management design. Numerous engineered barriers have been proposed to confine the waste and to retard or minimize migration rates from burial sites. Both physical and chemical barriers have been proposed. Two brief examples of waste management systems will be discussed.

ARID SITE DISPOSAL

In arid region disposal sites, engineering modifications of the soil physical system have been proposed. The purpose is to alter the hydrologic properties of the system so as to minimize water contact with the waste forms. Winograd (1974, 1981) has proposed hydraulic isolation by using coarse gravel around solid waste packages to keep the radioactive waste from direct contact with liquid water (Fig. 4). If unsaturated conditions persist, water would contact the waste material primarily through vapor transport. Water flow rates in soils at arid sites have been calculated to be less than 1 mm year^{-1} (Winograd, 1981). Transport of radionuclides would be correspondingly slow. Any chemical transport by liquid would be at almost immeasurable rates.

WET SITE DISPOSAL

In the eastern USA and in Canada, radioactive waste sites are located in regions where the water table is near or in the burial zone. Cherry et al. (1978) proposed that radioactive wastes in Canada be buried in dense, unfractured clay deposits below the water table (Fig. 5). The advantage of this type of burial scheme is the slow rate at which the radioactive waste would diffuse through the clay and into the groundwater. Assuming a

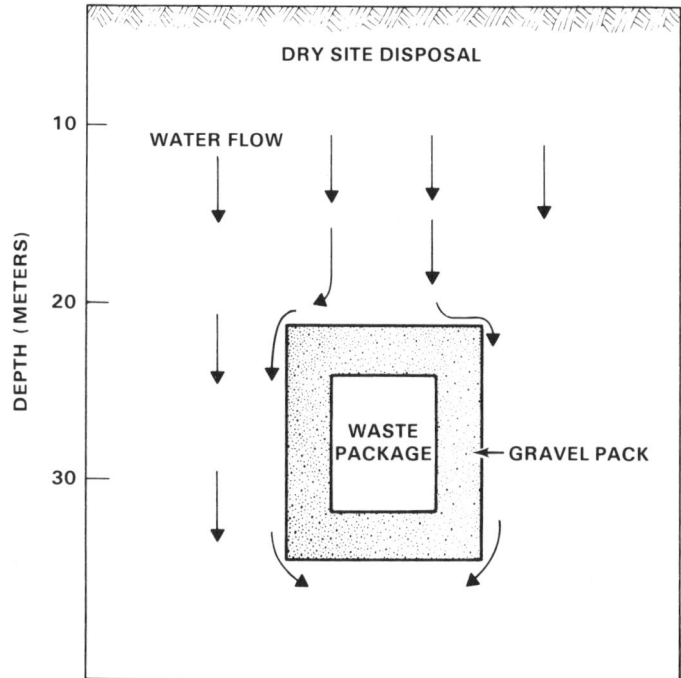

Fig. 4. Dry site waste burial scheme (after Winograd, 1974).

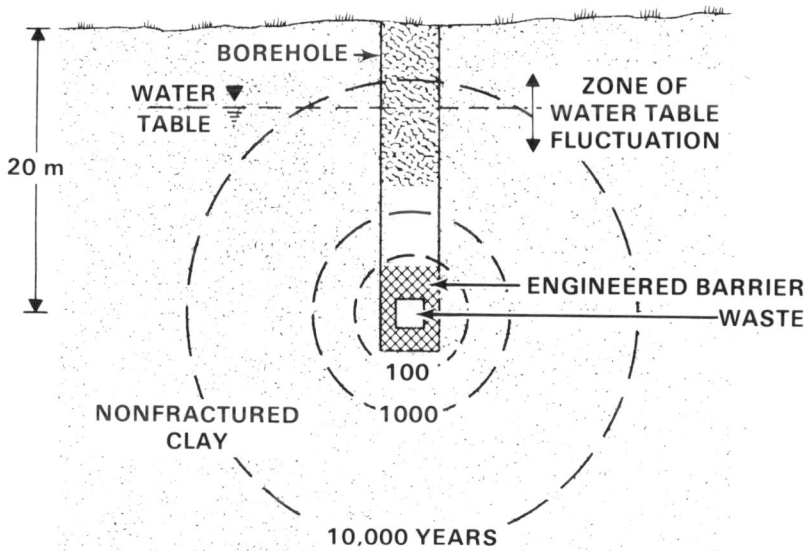

Fig. 5. Wet site waste burial scheme (after Cherry et al., 1978). Dashed circles with times indicate radionuclide diffusion rate from waste form into saturated clay materials.

typical diffusion coefficient of 10^{-6} cm^2 s^{-1} and no adsorption, calculations show that the relatively short-lived radionuclides such as ^{60}Co, ^{3}H, ^{90}Sr, and ^{137}Cs would not migrate more than a few meters before their activity is lost to decay. On the otherhand, ^{226}Ra would have some potential for diffusion to the biosphere if adsorption does not occur to an appreciable extent. However because of the combined effects of radioactive decay and the slow rate of diffusion, the total amount of ^{226}Ra migration would be very small compared to the initial ^{226}Ra concentration. If adsorption of ^{226}Ra occurs, time for increased radioactive decay could reduce the activity to a very low level before the diffusion front reaches the biosphere. Carbon-14 would migrate but will likely be retarded significantly by soil reactions with carbon dioxide.

Selection of specific sites where tight unfractured clay occur can be made by using water analysis for ^{2}H, ^{3}H, ^{14}C and ^{18}O. Water that originated during the most recent glacial episode (10 000 to 100 000 years ago) should contain no detectable, or low concentrations of these radionuclides. Measurements of extremely low concentrations in the interstitial water of clays would suggest that little water movement is occurring in these sediments. The combination of tracer measurements, with hydrologic site evaluation and transport modeling, provides the data base necessary to properly assess the suitability of such sites for storage of radioactive contaminants.

SUMMARY

This paper reviews existing literature and current research on the mobility of radionuclides in soils. Water chemistry and soil interactions play major roles in determining radionuclide mobility. The adsorption of radionuclides on soil materials can be described by empirical distribution coefficients but these describe adsorption affinity for only the specific soil/groundwater/nuclide concentration used in the experiment. Indiscriminate use of available distribution coefficients for adsorption estimates for soil/groundwater/nuclide conditions not studied has led to erroneous predictions and cast serious doubts on the validity of distribution coefficient approach. Although valid and practical for the exact conditions studied, the strict specificity of empirical distribution coefficients presents serious limitations in their use for generalized radionuclide mobility predictions. Therefore, current thrusts to understand radionuclide adsorption include mechanistic approaches such as mass action, site-binding, and triple layer models. Hopefully these approaches will enhance our predictive capabilities in determining radionuclide adsorption and clarify uncertainties originating from the misuse of empirical distribution coefficient studies.

The low solubility of many radionuclides particularly those in the transition metals and lanthanide and actinide series requires consideration of the chemical reactions that control solid phase existence or formation in the waste form or in the soil in the presence of groundwater. Often, the solid phase, which controls the radionuclide solubility is not known and the kinetics of dehydration and crystalization processes are not well documented for many of the radionuclides. In spite of these uncertainties, examples are given where measurements have identified likely solids (e.g., PuO_2 and NpO_2) that control the solubility of environmentally hazardous radionuclides such as Pu and Np in waste glass and contaminated soils. Further, although not identified, it appears that some solid phase may be controlling the fate of Am in soils.

Solubility constraints based on literature data and on experiments described herein have been used successfully to make preliminary estimates of the maximum concentrations of many important radionuclides in soil environments. To improve these preliminary calculations and provide more defensible information, additional solubility measurements and general improvement in the measured thermodynamic constants are needed for many radionuclides including Am, Pu, and Np.

Gaseous transport of radionuclides through soils is influenced by a number of factors including soil compaction, soil moisture content, barometric pressure, and air turbulence. In general, however, gas transport of radionuclides, particularly radon, can be considered a diffusional process, with rates controlled by gas concentration gradients and moisture content and density dependent diffusion coefficients.

Radioactive waste sites (repositories) are designed to isolate the waste from the biosphere. Conceptual designs for both wet and dry (climate) sites appear feasible, but control of transport, particularly at wet sites, will be strongly influenced by chemical controls that govern radionuclide solubility and the adsorption of radionuclides on soil.

ACKNOWLEDGMENTS

This work was funded by the U.S. Dep. of Energy through the Office of Nuclear Waste Isolation and the Uranium Mill Tailings Remedial Action Program and by the U.S. Nuclear Regulatory Agency through the Office of Nuclear Regulatory Research. These programs have supported Pacific Northwest Laboratory in extensive nuclear waste management research activities.

LITERATURE CITED

Alexander, C. W., and J. O. Bloemke. 1978. Origin and characteristics of low-level non-transuranic waste from the nuclear fuel cycle. p. 55–78. In M. W. Carter, A. A. Moghissi, B. Kahn, (ed.) Management of low level radioactive waste. Vol. 1. Pergamon Press, Inc., N.Y.

Allard, B. H. Kipatsi, and J. O. Liljenzin. 1980. Expected species of uranium, neptunium, and plutonium in neutral aqueous solutions. J. Inorg. Nucl. Chem. 42:1015–1027.

Ames, L. L., Jr., and B. F. Hajek. 1966. Statistical analysis of cesium and strontium sorption on soils. BNWL-CC-539. Pacific Northwest Lab., Richland, Wash.

————, and Dhanpat Rai. 1978. Radionuclide Interactions with Soil and Rock Media, Vol. 1: Processes influencing radionuclide mobility and retention, element chemistry and geochemistry, conclusions and evaluation. Vol. 2. Annotated bibliography. EPA 52016-78-007a and EPA 520/6-78-0076 available from NTIS as PB-292 460.

Anderson, M. A., and A. J. Rubin (ed.) 1981. Adsorption of inorganics at solid-liquid interfaces. Science Publishers, Inc., Ann Arbor, Mich.

Baas Becking, L. G. M., I. R. Kaplan, and D. Moore. 1960. Limits of the natural environment in terms of pH and oxidation reduction potentials. J. Geol. 68:243–284.

Baes, C. F., Jr., and R. F. Mesmer. 1976. The hydrolysis of cations. John Wiley and Sons, N.Y.

Baetsle, L. H., P. Dejonghe, W. Maes, E. S. Simpson, J. Souffriau, and P. Staner. 1964. Underground radionuclide movement. EURAEC-703.

Baker, J. H., W. A. Beetem, and J. S. Wahlbert. 1964. Adsorption equilibria between earth materials and radionuclides, Cape Thompson, Alaska. TID-20638. U.S. Geologic Survey, Washington, D.C.

Beck, H. L., C. V. Cogolak, K. M. Miller, and W. M. Lowder. 1980. Perturbations on the natural radiation environment due to utilization of coal as in energy source. p. 1521–1558. *In* Natural radiation environment III. CONF-780422. Available from NTIS, Springfield, VA.

Bensen, D. W. 1960. Review of soil chemistry research at Hanford, HW-67201.

Bird, G. W., and V. J. Lopata. 1980. Solution interaction of nuclear waste anions with selected geological materials. p. 419–426. *In* Scientific basis for waste management, Vol. 2. Plenum Press, N.Y.

Bondietti, E. A., and C. W. Francis. 1979. Geologic migration potentials of Tc-99 and Np-237. Science 203:1337–1340.

Buckingham, E. 1904. Contributions to our knowledge of the aeration of soils. U.S. Dep. of Agric. Bureau of Soils Bull. 25. 52 p.

Burney, G. A., and R. M. Harbour. 1974. Radiochemistry of neptunium. National Academy of Sciences—National Research Council Nucl. Sci. Ser., NAS-NS-3060.

Cherry, J. A., R. W. Gillham, and G. E. Grisak. 1978. A concept for long-term isolation of solid radioactive waste in fine-grained deposits. p. 1021–1040. *In* M. W. Carter, A. A. Moghissi, and b. Kahn (ed.) Management of low-level radioactive waste. Pergamon Press, N.Y.

Clements, W. E., and M. H. Wilkening. 1974. Atmospheric effects on ^{222}Rn transport across the earth-air interface. J. Geophys. Res. 79:5025–5029.

Cleveland, J. M. 1979. Critical review of plutonium equilibria of environmental concern. *In* E. A. Jenne (ed.) Chemical modeling in aqueous systems. ACS Symposium Series 93, Am. Chem. Soc., Washington, DC.

Cohen, B. L. 1979. Methods for predicting the effectiveness of uranium mill tailings covers. Nucl. Instrum. and Methods 164:595–599.

Cohen, P., and C. Gailledreau. 1961. Preliminary investigations of the adsorption of radiostrontium in saclay soil. Three-Component System. TID-7621, p. 90–99.

Coleman, N. T., R. J. Lewis, and D. Craig. 1963. Sorption of cesium by soils and its displacement by salt solutions. p. 290–294. *In* Proc. Soil Sci. Soc. of Am.

Couchat, Ph., F. Brissaud, and J. P. Gayraud. 1980. A study of strontium-90 movement in a sandy soil. Soil Sci. Soc. Am. J. 44:7–13.

Currie, J. A. 1960. Gaseous diffusion in porous media. Part 2: Dry granular materials. Br. J. Appl. Phys. 11:318–324.

———. 1961. Gaseous diffusion in porous media. Part 3: Wet granular materials. Br. J. Appl. Phys. 12:275–281.

———. 1970. Movement of gases in soil respiration. p. 152–171. *In* Sorption and transport processes in soils. Monograph Society Chemical Ind. No. 37.

Elgawhary, S. M., and W. L. Lindsay. 1972. Solubility of silica in soils. Soil Sci. Soc. Am. Proc. 36:439–442.

Francis, C. W. 1978. Radiostrontium movement in soils and uptake in plants. TID-27564 Tech. Inf. Ctr. USDOE, 131 p. Oak Ridge National Laboratory, Oak Ridge, Tenn.

Freeze, R. A., and J. A. Cherry. 1979. Groundwater contamination. p. 384–462. *In* Groundwater. Prentice-Hall, Inc., N.J.

Fuger, J., and F. L. Oetting. 1976. The chemical thermodynamics of actinide elements and compounds, Part 2: The actinide aqueous ions. Int. Atomic Energy Agency, Vienna.

Gee, G. W., and A. Campbell. 1980. Monitoring and physical characterization of unsaturated zone transport-laboratory analysis. PNL-3304, Pacific Northwest Lab., Richland, Wash.

----, ----, D. R. Sherwood, R. G. Strickert, and S. J. Phillips. 1980. Interaction of uranium mill tailings leachate with soils and clay liners: Laboratory analysis/progress report. NUREG/CR-1494, Pacific Northwest Laboratory, Richland, Wash.

Haury, G., and W. Schikarski. 1977. Radioactive inputs into the environment, comparison of natural and man-made inventories. p. 165–188. In Werner Stumm (ed.) Global chemical cycles and their alterations by man. Dahlem Kon Ferenzen, Berlin.

Hawkins, D. B., and H. L. Short. 1965. Equations for the sorption of cesium and strontium on soil and clinoptilolite. IDO-12046. INEL, Idaho Falls, Id.

Hostetler, D. D., R. J. Serne, A. J. Baldwin, G. M. Petrie. 1980. Status report on SIRS. Sorption information retrieval system. PNL-3528. Pacific Northwest Laboratory, Richland, WA.

Inoue, Y., and W. J. Kaufmann. 1963. Prediction of movement of radionuclides in solution through porous media. Health Phys. 9:705–715.

James, R. O., and R. W. Healy. 1972. Adsorption of hydrolyzable metal ions at the oxide/water interface. I. Co(II) adsorption on SiO_2 and TiO_2 as model systems, II. Charge reversal of SiO_2 and TiO_2 colloids by adsorbed Co(II), La(III), and Th(IV) as model systems, III. A thermodynamic model of adsorption. J. Colloid Interface Sci. 40:42–81.

Jenne, E. A., J. W. Ball, J. M. Buchard, D. V. Vivit, and J. H. Barks. 1980. Geochemical modeling: Apparent solubility controls on Ba, Zn, Cd, Pb, and F in waters of the Missouri tri-state mining area. p. 353–361. In D. D. Hemphill (ed.) Trace substances in environmental health-XIV. Univ. of Missouri, Columbia, Mo.

----, and J. S. Wahlberg. 1968. Role of certain stream-sediment components in radioion sorption. U.S.G.S. Professional Paper 433-F.

John, M. K., W. M. H. Saunders, and J. W. Watkinson. 1976. Selenium adsorption by New Zealand soils: 1. Relative adsorption of selenite by representative soils and the relationship to soil properties. N.Z. J. Agric. Res. 19:143–151.

Jones, T. L., G. W. Gee, J. L. Swanson, and R. R. Kirkham. 1983. A laboratory and field evaluation of the mobility of cobalt-60/EDTA. In R. G. Post (ed.) Waste management 1983. Proc. of the Symposium on Waste Management at Tuscon Arizona February 28–March 3, 1983. Univ. Arizona Press, Tuscon.

Kimball, B. A. 1973. Water vapor movement through mulches under field conditions. Soil Sci. Soc. Am. Proc. 37:813–818.

----. 1983. Canopy gas exchange: Gas exchange with soil. p. 215–225. In H. M. Taylor, W. R. Jordan, and T. R. Sinclair (ed.) Limitations to efficient water use in crop production. Am. Soc. of Agron., Madison, Wis.

King, C. Y. 1978. Radon emanation on San Andreas fault. Nature 271(5645):516–519.

Kinniburgh, D. G., M. L. Jackson, and J. K. Syers. 1976. Adsorption of alkaline earth, transition, and heavy metal cations by hydrous oxide gels of iron and aluminum. Soil Sci. Soc. Am. J. 40:796–799.

----, K. Sridhar, and M. L. Jackson. 1977. Specific adsorption of zinc and cadmium by iron and aluminum hydrous oxides. p. 231–239. In Biological implications of metals in the environment. CONF-750929, available from NTIS.

Klechkovski, V. M., and I. V. Gulyakin. 1958. The behavior in soils and plants of traces of strontium, cesium, ruthenium, and zirconium. Soviet Soil Sci. 3:219–230.

Kraner, H. W., G. L. Schraeder, and R. D. Evans. 1964. Measurements of the effects of atmospheric variables on radon-222 flux and soil-gas concentrations. p. 191–215. In J. A. S. Adams and W. M. Lowder (ed.) The natural radiation environment. Univ. Chicago Press.

Landa, E. R. 1980. Isolation of uranium mill tailings and their component radionuclides from the biosphere—some earth science perspectives. U.S. Geol. Survey Circular 814, Arlington, VA.

———, L. J. Thowig, and R. G. Gast. 1977. Effect of selective dissolution, electrolytes, aeration, and sterilization on Tc-99 sorption by soils. J. Environ. Qual. 6:181–187.

Langmuir, D. 1978. Uranium solution-mineral equilibria at low temperatures with applications to sedimentary ore deposits. Geochim. Cosmochim. Acta 42:547–569.

Lemire, R. J., and P. R. Tremaine. 1980. Uranium and plutonium equilibria in aqueous solutions to 200°C. J. Chem. Eng. Data 25:361–370.

Lindsay, W. L. 1979. Chemical equilibria in soils. John Wiley and Sons, N.Y.

———. 1981. Solid phase-solution equilibria in soils. p. 183–202. In R. H. Dowdy, J. A. Ryan, V. V. Volk, and D. E. Baker (ed.) Chemistry in the soil environment. Am. Soc. of Agron. and Soil Sci. Soc. of Am. Spec. Pub. No. 40, Am. Soc. of Agron., Madison, Wis.

Mattigod, S. V., G. Sposito, and A. L. Page. 1981. Factors affecting the solubilities of trace metals in soils. p. 203–221. In R. H. Dowdy, J. A. Ryan, V. V. Volk, and D. E. Baker (ed.) Chemistry in the soil environment. Am. Soc. of Agron. and Soil Sci. Soc. of Am. Spec. Pub. No. 40, Am. Soc. of Agron., Madison, Wis.

Mayer, D. W., C. A. Oster, R. W. Nelson, and G. W. Gee. 1981. Radon diffusion through multilayer earthen covers: Models and simulations. PNL-3989, Pacific Northwest Lab., Richland, Wash.

McFadden, K. M. 1980. The chemistry of technetium in the environment. PNL-2579, Pacific Northwest Laboratory, Richland, WA.

Means, J. L., D. A. Crerar, and J. O. Duguid. 1978. Migration of radioactive wastes: Radionuclide mobilization by complexing agents. Science 200:1477–1481.

Mehlich, A. 1981. Charge properties in relation to sorption and desorption of selected cations and anions. p. 47–76. In R. H. Dowdy, J. A. Ryan, V. V. Volk, and D. E. Baker (ed.) Chemistry in the soil environment, Am. Soc. of Agron. and Soil Sci. Soc. of Am. Spec. Pub. No. 40, Am. Soc. of Agron., Madison, Wis.

Meyer, R. E. 1979. Systematic study of metal ion sorption on selected geologic media. In R. J. Serne (ed.) Task 4 second contractor information meeting, Vol. 1. PNL-SA-6352, Vol. 1, Pacific Northwest Lab., Richland, Wash.

Nelson, R. W., P. R. Meyer, P. L. Oberlander, S. C. Sneider, D. W. Mayer, and A. E. Reisenauer. 1983. Model evaluation of seepage from uranium tailings disposal above and below the water table. NUREG/CR 3078 (PNL-4461) Pacific Northwest Laboratory, Richland, Wash.

———, A. E. Reisenauer, and G. W. Gee. 1980. Model assessment of alternatives for reducing seepage of contaminants from buried uranium mill tailings at the Morton ranch site in central Wyoming. NUREG/CR1495 (PNL-3378) Pacific Northwest Lab., Richland, Wash.

Nielson, K. K., D. C. Rich, and V. C. Rogers. 1982. Comparison of radon diffusion coefficients measured by transient-diffusion and steady-state laboratory methods. NUREG/CR-2875 (PNL-4370). NRC/GPO-USNRC, Washington, D.C.

———, and V. C. Rogers. 1982. A mathematical model for radon diffusion in earthen materials. NUREG/CR-2765. NRC/GPO-USNRC, Washington, D.C.

Norvell, W. A., and W. L. Lindsay. 1969. Reactions of EDTA complexes of Fe, Zn, Mn, and Cu with soils. Soil Sci. Soc. Am. Proc. 33:86–91.

———, and ———. 1972. Reactions of DTPA chelates of iron, zinc, copper, and manganese with soils. Soil Sci. Soc. Am. Proc. 36:778–783.

———, and ———. 1982. Estimation of the concentration of Fe^{3+} and the $(Fe^{3+})(OH^{-3})$ ion product from equilibria of EDTA in soil. Soil Sci. Soc. Am. J. 44:46:710–715.

Onishi, Y., R. J. Serne, E. M. Arnold, C. E. Cowan, and F. L. Thompson. 1981. Critical review: Radionuclide transport, sediment transport, and water quality mathematical modeling, and radionuclide adsorption/desorption mechanisms. NUREG/CR-1322 (PNL-2901), Pacific Northwest Laboratory, Richland, Wash.

Pickens, J. F., R. E. Jackson, and K. J. Inch. 1981. Measurement of distribution/coefficients using a radial injection dual-tracer test. Water Resour. Res. 17:529–544.

Pourbaix, M. 1966. Atlas of electrochemical equilibria in aqueous solutions. Pergamon Press, Inc., London.

Rai, Dhanpat, and J. L. Ryan. 1983. Crystallinity and solubility of Pu(IV) oxide and hydrous oxide in aged aqueous suspensions. Radiochim. Acta 30:213–216.

----, and R. J. Serne. 1977. Plutonium activities in soil solutions and the stability and formation of selected plutonium minerals. J. Environ. Qual. 6:89–95.

----, and ----. 1978. Solid phases and solution species of different elements in geologic environments. PNL-SA-8448, Pacific Northwest Lab., Richland, Wash.

----, ----, and D. A. Moore. 1980a. Interactions of plutonyl(VI) with soil minerals. PNL-SA-8448, Pacific Northwest Laboratory, Richland, Wash.

----, ----, and ----. 1980b. Solubility of plutonium compounds and their behavior in soils. Soil Sci. Soc. Am. J. 44:490–495.

----, R. J. Serne, and J. L. Swanson. 1980c. Solution species of plutonium in the environment. J. Environ. Qual. 9:417–420.

----, and R. G. Strickert. 1980. Maximum concentration of actinides in geologic media. Trans. Am. Nucl. Soc. 35:185–186.

----, ----, and G. L. McVay. 1982. Neptunium concentrations in solutions contacting actinide-doped glass. Nucl. Tech. 58:69–76.

----, ----, D. A. Moore, and J. L. Ryan. 1983. Am(III) hydrolysis constants and solubility of Am(III) hydroxide. Radiochim. Acta (In press).

----, ----, ----, and R. J. Serne. 1981. Influence of an americium solid phase on americium concentrations in solutions. Geochim. Cosmochim. Acta. 45:2257–2265.

----, and J. L. Swanson. 1981. Properties of plutonium(IV) polymer of environmental importance. Nucl. Tech. 54:107–112.

Rajan, S. S., and J. W. Watkinson. 1976. Adsorption of selenite and phosphate on an allophane clay. Soil Sci. Soc. Am. J. 40:51–54.

Relyea, J. F., and R. J. Silva. 1981. Application of a site-binding, electrical, double-layer model to nuclear waste disposal. PNL-3898, Pacific Northwest Lab., Richland, Wash.

Rhodes, P. W. 1957. The effect of pH on the uptake of radioactive isotopes from solution by a soil. Soil Sci. Soc. Am. Proc. 21:389–392.

Roessler, C. E., R. Kantz, W. E. Bolch, Jr., and J. A. Wethington, Jr. 1980. The effects of mining and land reclamation on the radiological characteristics of the terrestrial environment of Florida's phosphate region. p. 1476–1493. In T. F. Gessel and W. M. Lowder (ed.) Natural radiation environment III. CONF-780422. Available from NTIS Springfield, Va.

Rogers, V. C., R. F. Overmyer, K. M. Putzig, C. M. Jensen, K. K. Nielson, and B. W. Sermon. 1980. Characterization of uranium tailings cover materials for radon flux reduction. NUREG/CR-1081.

Routson, R. C. 1973. A review of studies on soil-waste relationships on the Hanford reservation from 1944 to 1967. BNWL-1464, Battelle Northwest Lab., Richland, Wash.

----, and R. J. Serne. 1972. One dimensional model of the movement of trace radioactive solute through soil columns: The PERCOL Model. BNWL-1718, Battelle, Pacific Northwest Lab., Richland, Wash.

Salter, P. F., L. L. Ames, and J. E. McGarrah. 1981. The sorption behavior of selected radionuclides on Columbia River Basalts. RHO-BWI-LD-48, Rockwell Hanford Operations, Richland, Wash.

Santillan-Medrano, J., and J. J. Jurinak. 1975. The chemistry of lead and cadmium in soil: Solid phase formation. Soil Sci. Soc. Am. Proc. 39:851–855.

Sawhney, B. L. 1964. Sorption and fixation of microquantities of cesium by clay minerals: Effects of saturating cations. Soil Sci. Soc. of Am. Proc. 28:183–186.

Schulz, W. W. 1976. The chemistry of americium. TID-26971. Technical Information Center, Energy Research and Development Administration, Washington, D.C.

Schwertmann, U., and R. M. Taylor. 1977. Iron oxides. p. 145–180. In Minerals in soil environments. Soil Sci. Soc. of Am., Madison, Wis.

Selim, H. M., J. M. Davidson, and P. S. C. Rao. 1977. Transport of reactive solutes through multilayered soils. Soil Sci. Soc. Am. J. 41:3–10.

Serne, R. J., S. R. Peterson, and G. W. Gee. 1983. Laboratory measurements of contaminant attenuation of uranium mill tailings by sediments and clay liners. NUREG/CR-3124, Pacific Northwest Laboratory, Richland, Wash.

----, and J. F. Relyea. 1981. The status of radionuclide sorption-desorption studies performed by the WRIT prgoram. PNL-3997, Pacific Northwest Lab., Richland, Wash.

Sherwood, D. R., and R. J. Serne. 1982. Evaluation of selected neutralizing agents for the treatment of uranium tailings leachates. NUREG/CR-3030, Pacific Northwest Lab., Richland, Wash.

Silker, W. B., and V. C. Rogers. 1981. Factors influencing radon attenuation by tailings covers. p. 265–274. In 4th Symposium on Uranium Tailings Management. Ft. Collins, Colo., October 25–26.

Silva, R. J., L. V. Benson, A. W. Yee, and G. A. Parks. 1980. Theoretical and experimental evaluation of waste transport in selected rocks. In J. F. Relyea (ed.) Waste isolation safety assessment program: Task 4 third contactor information meeting, Vol. 1. PNL-SA-8571, Pacific Northwest Laboratory, Richland, Wash.

----, A.F. White, and A. W. Yee. 1981. Sorption modeling studies and measurements. p. 204–214. In J. F. Relyea (ed.) Waste/rock interactions technology program FY-80 information meeting. PNL-3887, Pacific Northwest Lab., Richland, Wash.

Singh, M., N. Singh, and P. S. Relan. 1981. Adsorption and desorption of selenite and selenate selenium on different soils. Soil Sci. 132:134–141.

Spitsyn, V., V. C. Balukova, and T. A. Ermanova. 1963. Studies on sorption and migration of radioactive elements in soil. p. 569–577. In Treatment and storage of high-level radioactive wastes. IAEA, Vienna (in Russian).

Spaulding, B. P. 1980. Adsorption of radiostrontium by soil treated with alkali metal hydroxides. Soil Sci. Soc. Am. J. 44:703–709.

Sposito, G. 1981. Cation exchange in soils: A historical and theoretical perspective. p. 13–30. In R. H. Dowdy, J. A. Ryan, V. V. Volk, and D. E. Baker (ed.) Chemistry in the soil environment. Am. Soc. of Agron. Spec. Pub. No. 40. Am. Soc. of Agron., Madison, Wis.

Strickert, R. G., A. M. Friedman, and S. Fried. 1980. The sorption of technetium and iodine radioisotopes by various minerals. Nuclear Tech. 49:253–266.

----, and Dhanpat Rai. 1982. Solubility-limited neptunium concentrations in redox-controlled suspensions of NpO_2. PNL-SA-10590, Pacific Northwest Lab., Richland, Wash.

Swanson, J. L. 1982. Effect of organic complexants on the mobility of nickel and cobalt in soils: Status report. PNL-4389, Pacific Northwest Laboratory, Richland, Wash.

Tamura, T. 1964. Reactions of cesium-137 and strontium-90 with soil minerals and sesquioxides. ORNL-P-438.

----, and D. G. Jacobs. 1960. Structural implications in cesium sorption. Health Phys. 5: 149–154.

----, and ----. 1961. Improving cesium selectivity of bentonites by heat treatment. Health Physics. 5:149–154.

Tanner, A. b. 1980. Radon migration in the ground: A supplementary review. In T. F. Gessel and W. M. Lowder (ed.) The natural radiation environment III. CONF-780422 (Vol. 1). Available from NTIS Springfield, Va.

Tewari, P. H., A. B. Campbell, and W. Lee. 1972. Adsorption of cobalt (+2) by oxides from aqueous solution. Can. J. Chem. 50:1642–1648.

Thomas, G. W. 1977. Historical development in soil chemistry: Ion exchange. Soil Sci. Soc. Am. J. 41:230–238.

Travis, C. C. 1978. Mathematical description of adsorption and transport of reactive solutes in soil: A review of selected literature. ORNL-5403, Oak Ridge National Laboratory, Oak Ridge, Tenn.

U.S. Dep. of Energy. 1980. National low-level waste management program. Managing low-level radioactive wastes—A proposed approach, LLWMP-1 published by EG&G Idaho, Inc., Idaho Falls, Id.

Wahlberg, J. S., and R. S. Dewar. 1965. Comparison of distribution coefficients for strontium exchange from solutions containing one and two competing cations. U.S.G.S. Bull. 1140-D.

----, and M. J. Fishman. 1962. Adsorption of cesium on clay minerals. U.S.G.S. Bulletin 1140-A, Arlington, Va.

Wildung, R. E., R. C. Routson, R. J. Serne, and T. R. Garland. 1975. Pertechnetate, iodide and methyl iodide retention by surface soils. BNWL-1950, pt 2, p. 37–40. Pacific Northwest Lab., Richland, Wash.

Wilkening, M. 1980. Radon transport processes below the waters surface. p. 90–104. *In* T. F. Gessell and W. M. Lowder (ed.) Natural radiation environment III. CONF 780422, NTIS, Springfield, Va.

Winograd, I. J. 1974. Radioactive waste storage in the arid zone. Trans. Am. Geophys. Union 55(10):884–894.

----. 1981. Radioactive waste disposal in thick unsaturated zones. Science 212(4502):1457–1464.

Wood, B. J., and Dhanpat Rai. 1982. Nuclear waste isolation: Actinide containment in geologic repositories, PNL-SA-9549, Pacific Northwest Lab., Richland, Wash.

Yeh, G. T., and T. Tamura. 1982. Geohydrochemical considerations in land disposal of low-level wastes. Nucl. Sci. Eng. 82:206–219.

Zettwoog, P., N. Fourcade, F. E. Campbell, H. Caplan, and J. Haile. 1982. The radon concentration profile and the flux from a pilot-scale layered tailings pile. Health Phys. 43:428–433.

Chapter 15

Movement of Heavy Metals in Soils[1]

R. H. DOWDY AND V. V. VOLK[2]

Societal concerns about the quality of our air and water have lead to an increased interest in land application of waste materials, particularly those materials that may improve crop production directly or indirectly by improving the soil as a rooting medium (Dowdy et al., 1976). However, land application of waste materials can also degrade the environment with the introduction of potentially harmful substances such as trace or heavy metals into the soil. Heavy metals have been applied to soil with pesticides, as plant nutrients, atmospheric fallout, and as a constituent of waste products. Although applications of waste products to soils have added more metals to selected soils, pesticides and plant nutrient supplements are applied to large areas. Heavy metals that have received the most attention with regard to accumulation in soils, uptake by plants, and contamination of groundwaters include Cd, Cr, Cu, Hg, Ni, Pb, and Zn.

Sewage sludge contains all the above heavy metals and has been applied to croplands; thus, in the past decade, research on movement of heavy metals has largely centered around sludge. The heavy metals associated with sludge may pose a threat to the environment as a result of land spreading this material. Within this context, the objective of this paper is to review appropriate, and present new, research data that will help define soil conditions where leaching of sludge-borne metals might occur.

---

[1] Contribution from Soil and Water Management Research Unit, USDA-ARS and the Minnesota and Oregon Agric. Exp. Stns., Paper No. 1824-0, Scientific Journal Series.

[2] Research soil scientist, USDA-ARS and professor, Univ. of Minnesota, St. Paul, MN 55108; and professor, Dep. of Soil Sci., Oregon State Univ., Corvallis, OR 97331.

Copyright © 1983 ASA, SSSA, 677 South Segoe Road, Madison, WI 53711. *Chemical Mobility and Reactivity in Soil Systems.*

In the soil, heavy metals may: (i) occur on ion exchange sites, (ii) be incorporated into or on the surface of crystalline or noncrystalline inorganic precipitates, (iii) be incorporated into organic compounds, or (iv) be in soil solution. Most researchers have recognized that heavy metals are sparingly soluble and occur predominantly in a sorbed state or as part of insoluble inorganic or organic compounds. Because of their low solubility, movement of heavy metals in soils has generally been considered to be minimal.

Movement from zone of application could occur by diffusion, either as a free ion or as a complex, by mass flow with the water front, or by movement of metal-laden particulates through open channels in the soil. Movement of heavy metals through channels may be extensive in unique situations of rodent tunneling or extensive soil cracking during wetting and drying cycles. Generally, diffusion of metals occurs over short distances and will play a major role in metal uptake by plants (Barber, 1974). However, mass flow is probably the principle means by which heavy metals move appreciable distances within soils and will be the main emphasis of this chapter.

Movement of heavy metals with water in soils requires that the metal be in the soluble phase or associated with mobile particulates. Chemical parameters of the soil such as pH, ion sorption sites, ionic strength, and ligands which may form sparingly soluble precipitates affect the concentration of metal in solution (Corey et al., 1981). The mobility of Zn in coarse-textured soils was related to soil pH, organic matter, phosphate, and clay mineral content (Barrows et al., 1960). The formation of the ionic $Zn(OH)_4^=$ solution complex reduced Zn sorption by an alkaline soil and thus increased its mobility (Jurinak and Thorne, 1955). With most heavy metals, precipitation occurs with a pH increase, thus reducing ionic concentration in soil solution (Lindsay, 1979).

Organic matter and hydrous oxides, as ionic sorption sites, strongly affect metal mobility in the soil profile. Heavy metals have vacant d-orbitals such that they are often chelated or complexed with soil organic materials. These chelated compounds may be more soluble than inorganic precipitates. Several authors (Holtzclaw et al., 1978; Hodgson et al., 1965; Sposito et al., 1976) have related the solubility of heavy metals to soluble complexing organic matter present in the soil. Fulvic acid has been considered a most important organic constituent in relation to interactions with clay minerals and metal ions (Sposito et al., 1976, 1982). The concerns with organic matter and metal concentration in soil solution become very important when one considers that heavy metals are applied to soils as a constituent of sewage sludge, although the metals may not be associated with the sludge organic materials (Schaumberg et al., 1980).

Cadmium and Zn were leached by $CaCl_2$ from a Baltimore sludge more than Cu or Pb (Lagerwerff et al., 1976). The Cl anion appears to have acted as a competing complexing ligand for Cd and Zn. Ion pair complexes of Cd and Zn with Cl would be expected to increase the total solubility of the two metals, making them more mobile in the soil profile. The Cu and Pb could have formed relatively strong complexes with the solid organic matter phase.

EXPERIMENTAL OBSERVATIONS

Column Studies

Numerous mechanisms are simultaneously involved in the transport of heavy metals applied to soils with sewage sludge. To better define the leachability of sludge-borne trace metals, several investigators have measured the movement of metals in soil columns under controlled laboratory conditions. In one study, samples of each genetic horizon (to a depth of 230 cm) of a deep, Fayette silt loam loess (fine-silty, mixed, mesic Typic Hapludalf) were obtained from a non-cropped, non-eroded, and forested area (Dowdy, unpublished). The very thin A_1 horizon and undecomposed organic matter were discarded. Air-dried soil was packed by horizon in 15-cm diam columns to a density similar to that in the field. The columns were constructed so that each horizon occupied one-third the depth of the horizon in the undisturbed soil profile.

Anaerobically digested sewage sludge containing 15, 7190, 1780, and 1540 mg/kg of Cd, Cr, Cu, and Zn, respectively, was mixed into the top (A_2) horizon sample at a rate of 100 g/kg soil (dry weight basis). Except the high level of Cr, the metal concentrations were similar to most sewage sludges (Sommers, 1977). The soils treated with sludge and comparable non-treated soils were leached with 75 cm of distilled water over a 38-day period with essentially equally timed 2.5 cm increments of water addition. After leaching, the columns were sectioned by horizon and sampled. The samples (4 g, dry weight basis) were equilibrated with 20 ml 4 N HNO_3 for 20 h, decanted, and the supernatant analyzed for Cd, Cr, Cu, Ni, Pb, and Zn by atomic absorption spectroscopy utilizing continuum hydrogen lamp background correction where appropriate (Dowdy et al., 1978).

Soil pH in the A_2 zone of incorporation was 6.6 initially and dropped to 6.3 as a result of sludge application (Table 1). The pH throughout the rest of the profile ranged from 4.8 to 5.6 prior to sludge application, except for a pH of 7.8 in the zone of carbonate accumulation (C_3) which occurred at the 179-cm depth.

Essentially all sludge-borne Cd, and most of the Cr, Cu, and Zn were extracted by 4 N HNO_3 (Table 1). These metals did not appear to move into any soil horizons below the zone of incorporation, with the possible exception of Cu movement into the B_1 and B_{21} soil layers. In other studies Cu was the first heavy metal to be detected in run-off waters from land that received surface applications of sewage sludge (Dowdy et al., 1980).

To simulate removal of the A_1 and A_2 horizons by erosion, the identical experiment was repeated with sludge incorporated into the B_1 horizon. Such a situation closely simulates the field situation under cultivation. Again, no downward movement of trace metals was observed (Table 2). In both experiments, no movement of Ni or Pb was observed using the 4 N HNO_3 extraction as the criteria for movement (data not shown). Emmerich et al. (1982) carried out column studies that used a

Table 1. The extractable metal content (4.0 N HNO₃) of a Fayette soil after sludge application to the A2 horizon and subsequent leaching.

Soil				Metal concentration						
Horizon		pH‡		Cd		Cr		Cu		Zn
Desig-nation	Depth†	Initial	Final	Treatment§						
				0	100	0	100	0	100	0	100
	cm						mg/kg			
A2	2–25	6.6	6.3	0.2	1.5	4.3	670	4.2	150	14	125
B1	25–43	4.8	6.5	0.1	0.1	4.9	5.0	4.1	4.8	13	12
B21	43–76	5.0	6.9	<0.1	0.1	6.1	6.0	5.2	6.4	15	15
B22	76–97	5.1	6.5	<0.1	<0.1	6.2	6.6	6.6	6.9	17	17
B23	97–127	5.1	6.4	<0.1	<0.1	6.2	6.5	6.7	6.9	17	18
B3	127–152	5.2	6.1	0.1	0.1	6.2	6.4	6.3	6.4	18	22
C1	152–163	5.6	6.0	0.1	0.2	6.6	6.5	6.6	6.7	19	19
C2	163–179	5.6	6.0	0.2	0.2	7.2	6.5	6.3	6.4	18	19
C3	179–230	7.8	7.8	0.1	0.1	5.2	5.2	5.2	5.1	15	15

† Depth as it exists in the field.
‡ For the 100 g/kg treatment.
§ Represents the application of 100 g sludge/kg soil which added 1.4, 720, 180, and 155 mg/kg of Cd, Cr, Cu, and Zn, respectively, to the A2 horizon.

Table 2. The extractable metal content (4.0 N HNO₃) of a Fayette soil after sludge application to the B1 horizon and subsequent leaching.

Soil				Metal concentration						
Horizon		pH‡		Cd		Cr		Cu		Zn
Desig-nation	Depth†	Initial	Final	Treatment§						
				0	100	0	100	0	100	0	100
	cm						mg/kg			
B1	25–43	4.8	6.3	0.2	1.4	5.1	780	4.8	175	13	135
B21	43–76	5.0	6.7	<0.1	<0.1	6.5	6.2	5.4	5.7	15	12
B22	76–97	5.1	6.2	<0.1	<0.1	6.4	6.6	7.1	7.3	17	16
B23	97–127	5.1	6.2	<0.1	<0.1	6.4	5.9	6.7	6.7	18	17
B3	127–152	5.2	5.5	<0.1	<0.1	6.4	6.3	6.2	6.2	18	17
C1	152–163	5.6	5.9	0.2	0.1	6.5	6.3	6.7	6.8	20	19
C2	163–179	5.6	5.7	0.2	0.2	6.7	6.3	6.9	6.8	20	19
C3	179–230	7.8	7.8	0.1	0.1	5.0	5.0	5.6	5.6	17	17

† Depth as it exists in the field.
‡ For the 100 g/kg treatment.
§ Represents the application of 100 g sludge/kg soil which added 1.4, 720, 180, and 155 mg/kg of Cd, Cr, Cu, and Zn, respectively, to the B1 horizon.

similar experimental approach, but leached the soil with 5 m of Colorado River water over a 25-month period, and observed no metal movement.

The movement of metals through intact 50 cm long cores of five Indiana soils ranging in textures from a sand to a silty clay loam was studied by Sommers et al. (1979). Surface soil pH ranged from 4.7 to 6.3.

Table 3. Distribution of 4 M HNO₃ extractable Cu, Zn, Cd, Pb, and Ni in soils treated with 22.4 t ha⁻¹ of sewage sludge (Sommers et al., 1979).

Soil	Depth	Cu	Zn	Cd	Pb	Ni
	cm			mg/kg†		
Plainfield	0 –7.5	55	314	11.7	13.0	9.1
(sand)	7.5–15	3	3	0	0.7	−0.7
Maumee	0 –7.5	41	286	9.3	11.8	9.6
(fine sandy loam)	7.5–15	0	−3	−0.1	0.8	0.4
Parr	0 –7.5	36	310	10.5	14.5	8.8
(silt loam)	7.5–15	1	1	0	0	0.9
Chalmers	0 –7.5	45	366	12.3	17.0	10.3
(silty clay loam)	7.5–15	0	0	0	0	0
Fincastle	0 –7.5	36	268	9.9	6.5	8.6
(silt loam)	7.5–15	0	−5	−0.1	−3.5	−0.8

† Values are the difference between treated and untreated columns.

Table 4. Distribution of 4 M HNO₃ extractable Cu, Zn, Cd, Pb, and Ni in Fincastle soil incubated with Kokomo sludge (Sommers et al., 1979).

Rate	Depth	Cu	Zn	Cd	Pb	Ni
t/ha	cm			mg/kg‡		
22.4(1)†	0 –7.5	36	268	9.9	6.5	8.6
	7.5–15	0	−5	−0.1	−3.5	−0.8
44.8(1)	0 –7.5	78	572	23.0	18.5	18.5
	7.5–15	6	21	1.3	−3.5	1.1
89.6(1)	0 –7.5	133	822	41.0	29.5	29.5
	7.5–15	34	243	11.7	7.0	8.8
89.6(2)	0 –7.5	178	1151	44.0	37.0	32.0
	7.5–15	6	36	1.4	0	1.5
89.6(3)	0 –7.5	175	1179	46.0	36.5	33.5
	7.5–15	6	33	1.3	1.0	0.4
89.6(4)	0 –7.5	152	916	43.5	34.5	32.5
	7.5–15	3	3	0.6	−3.0	1.0

† Number of sludge applications.
‡ Values are the difference between treated and untreated columns.

A digested sludge containing 13 540 mg Zn, 2170 mg Cu, 540 mg Pb, 740 mg Ni, and 480 mg Cd/kg was incorporated into the upper 7.5 cm soil layer at a rate equivalent to 22.4 t ha⁻¹. A 22.4 t ha⁻¹ rate is within the application range recommended for sewage sludge on agronomic crops such as corn (*Zea mays* L.). Columns were leached monthly with ~3.2 cm of distilled water over a 1-year period. Metal concentrations in leachate samples did not increase with sludge addition. After 1 year, the columns were sectioned by depth and soil samples extracted with 4 M HNO₃. No sludge-borne metals were detected in soils below the 15-cm depth. In fact, essentially all added metals were retained within the zone of incorporation, even on the Plainfield sand (mixed, mesic Typic Udipsamment) with a pH of 4.9 (Tables 3 and 4). The slightly increased metal concentrations in the 7.5 to 15-cm soil layer when sludge was added at the 44.8 and 89.6 t

Table 5. Extractable metals (4.0 N HNO₃) in Bold silt loam soil 4 years following sludge application and cropped continuously with barley.

Soil Depth cm	\multicolumn{2}{c}{pH}		\multicolumn{2}{c}{Cd}		\multicolumn{2}{c}{Cu}		\multicolumn{2}{c}{Ni}		\multicolumn{2}{c}{Zn}
Treatment†	0	100	0	100	0	100	0	100	0	100
	\multicolumn{10}{c}{mg/kg}
0–15	5.4	6.4	0.3	7.2	5.8	44.0	7.8	25.0	18	145
23–38	5.9	5.9	<0.1	0.2	4.5	4.3	7.7	8.3	14	15
46–61	5.8	5.8	<0.1	<0.1	4.2	4.5	7.8	7.7	18	16
69–84	5.8	5.8	<0.1	<0.1	6.3	6.1	9.3	8.7	18	15
100–115	5.8	5.7	<0.1	<0.1	6.2	5.8	10.3	9.6	15	13

† Represents the application of 100 t ha⁻¹ of sludge which added 20, 110, 65, and 470 kg/ha of Cd, Cu, Ni, and Zn, respectively, to the soil.

ha^{-1} rates, probably occurred because ". . . some mixing of sludge with the 7.5 to 15-cm increment (of soil) could have occurred during sludge application."

Giordano and Mortvedt (1976) recognized the importance of soil pH associated with N fertilization and investigated mobility of heavy metals as a function of N and sewage sludge application to an Ennis sandy loam (fine-loamy siliceous, thermic Fluventic Dystrochrepts) soil. In their column studies, tall fescue (*Festuca arundinacea* Schreb.) was grown and the soil was leached with the equivalent of 12 cm of deionized water for 4, 8, and 12 weeks after fescue was planted. Metal concentrations in the leachate from sludge-treated soils did not exceed those measured in leachate from soils which received no sewage sludge. The N fertilizer, as urea, did not affect downward movement of Cd, Cr, Ni, Pb, or Zn in soil. Metal uptake by tall fescue did increase, but the increased uptake appeared to result from increased growth as a response to added N. The soil columns were sampled at 1-cm increments and the samples extracted with 0.5 N HCl. Mobility of trace metals applied as inorganic salts was greater under excessive leaching than when the metals were applied with sewage sludge.

Field Studies

A comprehensive study to investigate the effect of land spreading anaerobically digested sewage sludge from Chicago, Ill., was organized by personnel from approximately 15 Agricultural Experiment Stations, federal agencies, and municipalities. The sludge was applied at rates of 20 t ha^{-1} year^{-1} for 4 years or one application of 100 t year^{-1} at each location. Barley was grown as a test crop and soil samples were collected periodically. Although the study is still underway, preliminary results from Minnesota and Oregon indicate minimal movement of heavy metals as a function of waste application. In Minnesota, soil samples collected from the profile of a deep loess (Bold silt loam), pH = 6.2, to a depth of 115 cm were extracted with 4 N HNO₃ for 20 h. The Cd, Cr, Cu, Ni, Pb,

Table 6. Cadmium, Cu, Ni, and Zn distribution in a Willamette sandy loam soil after treatment with $(NH_4)_2SO_4$ and sewage sludge.

Depth	$(NH_4)_2SO_4$ (kg/ha)		Sewage sludge (t/ha)		
	0	160	20	40	100
cm	mg/kg				
	Cd				
0–15	1.29	0.94	3.55	6.81	2.84
15–30	0.28	0.28	1.70	1.36	2.20
30–60	0.16	0.11	0.13	0.35	0.15
60–75	0.11	0.10	0.12	0.26	0.12
	Cu				
0–15	10.2	8.4	22.2	40.1	20.1
15–30	4.1	4.2	12.4	10.9	16.0
30–60	3.9	3.2	4.2	5.6	4.4
60–75	4.4	4.1	4.7	5.7	4.5
	Zn				
0–15	35.2	28.8	71.7	129.4	59.8
15–30	16.5	17.1	42.3	40.2	49.1
30–60	10.9	10.0	11.6	16.4	10.8
60–75	12.0	11.0	14.2	16.1	11.5
	Ni				
0–15	6.8	6.1	12.4	21.5	10.7
15–30	4.0	4.3	7.7	7.7	9.2
30–60	3.5	3.0	3.7	5.2	2.9
60–75	4.1	3.4	4.6	5.1	3.2

† Extracted with $1 N HNO_3$.

and Zn concentrations in soil collected from the 0- to 15-cm horizon increased with sludge application. Below a depth of 23 cm no difference between the trace metal content in the soil which received sludge and the control soil was observed (Table 5). In Oregon, no movement of Cd, Cu, Ni, or Zn was observed in a Willamette sandy loam (fine-silty, mixed, mesic Pachic Ultic Argixeroll) soil (pH 6.0) (Table 6). The increase in metal concentration in the 15 to 30 cm soil depth was attributed to sludge incorporation during tilling operations. During the study, the soil received ~50 cm/year of rainfall.

Numerous authors (Sidle and Kardos, 1977; Sidle et al., 1976, 1977; Parker et al., 1978) have observed retention of about 95% of Cd, Cu, Pb, and Zn in the 0 to 30-cm depth after application of sewage sludge to soil. Kelling et al. (1977) extracted soils which had been previously treated with sewage sludge and found somewhat increased levels of Cd, Cu, and Ni at the 15 to 30-cm depth, but felt the increase was related to tillage incorporation. The Zn concentration did increase at depths greater than 30 cm so some movement of Zn apparently occurred.

A field study in Minnesota was also conducted to assess the potential for sludge-borne metal movement down the soil profile. A total of 180 t ha^{-1} of a high Cd, waste-activated sludge was incorporated into the top 20

Table 7. Extractable metals (1.0 N HNO₃) in Waukegan silt loam soil following 3 years of sludge application and cropped continuously with corn.

Soil		Metal concentration							
	pH		Cd		Cu		Ni		Zn
					Treatment†				
Depth	0	180	0	180	0	180	0	180	0	180
cm					mg/kg					
10–20	6.6	6.5	<0.1	8.2	3.3	44.6	4.2	19.4	8.5	91.9
40–50	5.6	5.5	<0.1	<0.1	4.0	3.8	4.4	4.5	10.0	9.0
70–80	5.8	5.8	<0.1	<0.1	3.2	3.4	3.7	3.4	9.1	8.4

† Represents the application of 180 t ha⁻¹ of sludge over a 3-year period, which added 25, 125, 43, and 345 kg/ha of Cd, Cu, Ni, and Zn, respectively, to the soil.

cm of a Waukegan (fine-silty over sandy or sandy skeletal, mixed, mesic Typic Hapludoll) silt loam soil pH ≅ 6.4, over a 3-year period. After three corn crops, soil samples were collected at three depths: 10 to 20 cm, 40 to 50 cm, and 70 to 80 cm. No metals had reached the 40-cm depth, as measured by a 1.0 N HNO₃ extraction (Table 7). The 25 kg Cd/ha loading in this study exceeded the maximum cumulative Cd application allowed by the Environmental Protection Agency (1979) for this soil (CEC = 20 meq/100 g) cropped to food-chain crops.

In Illinois, Hinesly and Jones (1977) measured the Zn, Cu, and Cd in drainage water from a sludge-amended Blount silt loam soil (pH ≅ 6.8). Liquid sludge was surface applied throughout the growing season for 5 years and tilled into the soil prior to corn planting. Accumulated sludge additions reached 251 t ha⁻¹ and corresponded to metal loadings of 1257, 270, and 64 kg/ha of Zn, Cu, and Cd, respectively. During years 3 through 5, drainage water collected from tile lines placed at an 86.4-cm depth was analyzed for trace metals. Geometric mean concentrations of Zn in drainage waters increased from 18 to 35 μg/L over the 3-year period. Similarly, both Cd and Cu concentrations in drainage waters increased from 5 to 10 μg/L as a result of sludge additions. These increases, although significant, were very small and, indeed, data collected from an identical set of plots showed no significant increase in Cd or Zn concentrations of drainage waters as a result of sludge applications. Analyses of HCl-HF digests of subsoil samples (30 to 76 cm) collected after 5 years of sludge additions showed no change in subsoil levels of Cd or Zn attributable to sludge applications (Hinesly et al., 1979).

Metal movement in a sludge-amended Hublersburg clay loam soil (clayey, illitic, mesic, Typic Hapludult) with a surface pH of 5 and 10.6% organic matter was also studied by analyzing water samples collected from 15 and 120-cm depths (Sidle and Kardos, 1977). After applications of up to 27 t ha⁻¹ of sludges, soil samples were collected to a depth of 120 cm. The Cd, Cu, and Zn concentrations in the percolate water collected from both soil depths increased with sludge application, but <1% and ~3% of the applied Cu and Zn, respectively, were leached below the 120-cm depth. From 4 to 7% of the applied Cd was believed to be lost with the percolate water. It appeared that most of the applied Cu was

Table 8. Comparison of heavy metal content of groundwater collected under control and sludge-treated soil (Ellis et al., 1981).

Treatment	Metal					
	Cd	Cr	Cu	Ni	Pb	Zn
	mg/kg					
Control	2	18	27	16	21	68
	(2)‡	(3)	(8)	(1)	(4)	(13)
Sludge†	2	29	70	94	26	715
	(0.5)	(12)	(16)	(45)	(14)	(355)

† Sludge applied was 99, 81, and 140 t ha^{-1} in year I, II, and III, respectively.
‡ Standard deviation between replications of the same treatment.

fixed in the upper 15 cm of soil. The 0.1 N HCl extractable Cu increased at all depths of this acid soil, with the greatest increases occurring in the upper profile. Since the 0.1 N HCl extractable Cd and Zn did not increase with depth, the authors speculated that soluble Cd, and to a certain extent, Zn, may be moving quickly through the soil via a series of interconnected channels.

Field studies at Muskegon, Mich. (Ellis et al., 1981) were conducted under conditions where the potential for leaching sludge-borne metals was great: (i) very sandy soil (CEC < 5 meq/100 g), (ii) very high sludge loadings (320 t ha^{-1} over a 3-year period), and (iii) high water application rates, as sewage effluent. Total metal loadings were 12.6, 770, 930, 3490, and 370 kg ha^{-1} for Cd, Ni, Cu, Zn, and Pb, respectively. The concentrations of Cu, Ni, and Zn in groundwater samples, at the 2 to 3-m depth, taken in July following final sludge additions, increased with sludge applications (Table 8). The movement of these metals through the soil profile was confirmed by elevated levels of DTPA extractable Cu, Ni, and Zn in the 15 to 30-cm soil layer. No appreciable leaching of Cd, Cr, or Pb occurred, even under conditions most conducive to metal movement.

Under high rainfall and acid soil pH of a Cecil sandy clay loam (clayey, kaolinitic, thermic, Typic Hapludult), King and Morris (1972) did not observe movement of Zn into the 15 to 30-cm soil horizon. As much as 40 cm of liquid sewage sludge had been applied over a 2-year period. An increased Mn content in lower soil horizons was considered a result of the pH decrease from 5.2 to 4.2 associated with the sludge application. Under similar climatic conditions of 176 cm of rainfall, Boswell (1975) noted little movement of 1.0 N HCl extractable Cd, Cr, Cu, Pb, and Zn beyond the 30-cm depth of a Davidson clay loam soil (pH = 5.3) amended with 16.8 t ha^{-1} of sludge.

Williams et al. (1980) added sewage sludge from two sources over 3 years to Dublin loam (pH 5.2 to 5.6) in California. While the soil pH was similar to studies in the southeast, annual rainfall in California was considerably less. Sludge was added at rates up to 225 t ha^{-1} and mixed into the surface 20 cm of soil. The sludge application increased the 4 N HNO$_3$ extractable metal content of the surface soil. Even though the soil acidity increased with sludge application, the metal movement in the soil was limited to a depth of 30 cm over the 3-year study period. Zinc and Pb moved downward about 10 cm from the zone of incorporation while Cd, Cr, Cu, and Ni movement was limited to a depth of only 5 cm below the zone of incorporation.

Most heavy metal movement studies have been 1 to 5 years in length. In a longer study, Andersson and Nilsson (1972) reported that As, Cd, Co, Cr, Cu, Hg, Mn, Ni, Pb, Se, and Zn applied with sewage sludge (84 t ha^{-1}) over a 12-year period remained in the surface 20 cm of soil. Recovery exceeded 100% for all elements except Mn, where 98% of that added was recovered (Page, 1974).

Finally, the most probable situation where metal movement may occur would be on coarse-textured soils on which sludge disposal ponds have been located (Lund et al., 1976). The concentration of 4 N HNO$_3$ extractable Cd, Cr, Cu, Ni, and Zn was greater in soils sampled under the disposal ponds than at an untreated site adjacent to the treatment plants. The increase in metal concentration occurred to depths as great as 3 m under some ponds. Since the measured chemical oxygen demand related closely to changes in metal concentration, the authors suggest that the metals moved as soluble metal-organic complexes.

The practice of entrenching sludge can also be considered a worst-possible case situation with respect to metal movement. Sikora et al. (1980) studied the movement of Cd, Cu, and Zn directly below sludge entrenched in a loamy sand soil. The soil was sampled with time at 5, 20, 40, 60, and 80 cm below the trench. Even though the sludge contained 18 mg Cd and 627 mg Cu/kg, no detectable movement of these metals was observed as determined by extraction of soils with DTPA. After 1.9 years, DTPA extractable Zn concentrations had not increased above background levels. However, by the 2.7 years sampling date, the Zn concentration in soil 80 cm under the trench was significantly higher than background levels (3.6 vs. 0.5 mg/kg). Similar findings were reported by Ritter and Eastburn (1978) where as much as 180 t ha^{-1} of sewage sludge was applied on the surface of 183 cm long columns of the same soil type, and subsequently leached with 102 cm of water. Cadmium and Cu were not detected in the leachate. They concluded that "hazardous amounts of Cd, Cr, Cu, and Zn would not be leached to the groundwater under the experimental conditions of this study when the tested sludges were applied to Coastal Plain soils at recommended rates."

CONCLUSIONS

Although some studies have noted movement of trace metals to layers below the soil surface or depth of incorporation, it appears that Zn is the element which most consistently has potential to move in soils. The mechanisms by which Zn moves, whether as Zn^{2+}, an anionic form, or as a chelated compound has not yet been qualified.

Heavy metal movement will most likely occur with large applications to a sandy, acid, low organic matter soil which receives high rainfall or irrigation. Even under these conditions the extent of movement will be limited. Movement of heavy metals may occur through open soil channels or cracks where the soil has had no opportunity to attenuate them.

LITERATURE CITED

Andersson, A., and K. O. Nilsson. 1972. Enrichment of trace elements from sewage sludge fertilizers in soils and plants. Am. Biol. 1:176–179.

Barber, S. A. 1974. Influence of the plant root on ion movement in soil. p. 522–569. *In* E. W. Carson (ed.) The plant root and its environment. Univ. Press of Virginia, Charlottesville.

Barrows, H. L., M. S. Neff, and N. Gammon, Jr. 1960. Effect of soil type on mobility of zinc in the soil and on its availability from zinc sulfate, to tung. Soil Sci. Soc. Am. Proc. 24: 367–372.

Boswell, F. C. 1975. Municipal sewage sludge and selected element applications to soil: Effect on soil and fescue. J. Environ. Qual. 4:267–272.

Corey, R. B., R. Fujie, and L. L. Hendrickson. 1981. Bioavailability of heavy metals in soil-sludge systems. p. 449–465. *In* Proc. 4th Ann. Conf. Applied Res. & Practices on Municipal & Industrial Waste. 28–30 Sept. 1981. Univ. of Wisconsin-Extension, Madison.

Dowdy, R. H., C. E. Clapp, D. R. Duncomb, and W. E. Larson. 1980. Water quality of snowmelt runoff from sloping land receiving annual sludge applications. p. 11–15. *In* Proc. Natl. Conf. Municipal and Industrial Sludge Utilization and Disposal. Information Transfer, Inc., Silver Springs, Md.

----, R. E. Larson, and E. Epstein. 1976. Sewage sludge and effluent utilization in agriculture. p. 138–153. *In* Proc. Land Application of Waste Materials Conf., March 15–18, 1976, Des Moines, Iowa, Soil Conserv. Soc. Am., Ankeny, Iowa.

----, W. E. Larson, J. M. Titrud, and J. J. Latterell. 1978. Growth and metal uptake of snap beans growth on sewage sludge-amended soil: A four-year study. J. Environ. Qual. 7:252–257.

Ellis, B. G., A. E. Erickson, L. W. Jacobs, J. E. Hooks, and B. D. Knezek. 1981. Cropping systems for treatment and utilization of municipal wastewater and sludge. Project Report, EPA 600/2-81-065, Robert Kerr USEPA Lab., Ada, Ok. p. 38–65.

Emmerich, W. E., L. J. Lund, A. L. Page, and A. C. Chang. 1982. Movement of heavy metals in sewage sludge-treated soils. J. Environ. Qual. 11:174–178.

Environmental Protection Agency. 1979. Criteria for classification of solid waste disposal facilities and practices. Fed. Regist. 44:53438–53468.

Giordano, P. M., and J. J. Mortvedt. 1976. Nitrogen effects on mobility and plant uptake of heavy metals in sewage sludge applied to soil columns. J. Environ. Qual. 5:165–168.

Hinesly, T. D., and R. L. Jones. 1977. Heavy metal content in runoff and drainage waters from sludge-treated field lysimeter plots. p. 27–44. *In* Proc. Natl. Conf. on Disposal of Residues on Land. 13–15 Sept. 1976. St. Louis, Mo. USEPA and Inf. Transf., Inc., Rockville, Md.

----, E. L. Ziegler, and G. L. Barrett. 1979. Residual effects of irrigating corn with digested sewage sludge. J. Environ. Qual. 8:35–38.

Hodgson, J. F., H. R. Geering, and W. A. Norvell. 1965. Micronutrient cation complexes in soil solution: Partition between complexed and uncomplexed forms by solvent extraction. Soil Sci. Soc. Am. Proc. 29:665–669.

Holtzclaw, K. M., D. A. Keech, A. L. Page, G. Sposito, T. J. Ganje, and N. B. Ball. 1978. Trace metal distribution among the humic acid, the fulvic acid, and precipitable fractions extracted with sodium hydroxide from sewage sludge. J. Environ. Qual. 7:124–127.

Jurinak, J. J., and D. W. Thorne. 1955. Zinc solubility under alkaline conditions in a zinc bentonite system. Soil Sci. Soc. Am. Proc. 19:446–448.

Kelling, K. A., D. R. Keeney, L. M. Walsh, and J. A. Ryan. 1977. A field study of the agricultural use of sewage sludge: 3. Effects of uptake and extractability of sludge-born metals. J. Environ. Qual. 6:352–358.

King, L. D., and H. D. Morris. 1972. Land disposal of liquid sludge: II. The effect of soil pH, manganese, zinc, and growth and chemical composition of rye (*Secale cereale* L.). J. Environ. Qual. 1:425–429.

Lagerwerff, J. V., G. T. Biersdorf, and D. L. Brower. 1976. Retention of metals in sewage sludge. I. Constituent heavy metals. J. Environ. Qual. 5:19–25.

Lindsay, W. L. 1979. Chemical equilibria in soils. John Wiley and Sons, N.Y.

Lund, L. J., A. L. Page, and C. O. Nelson. 1976. Movement of heavy metals below sewage disposal ponds. J. Environ. Qual. 5:330–334.

Page, A. L. 1974. Fate and effects of trace elements in sewage sludge when applied to agricultural lands. A literature review. U.S. EPA Rep. No. EPA-670/2-74-005. Office of Res. and Dev.; USEPA, Cincinnati, Ohio 45268. 108 p.

Parker, G. R., W. W. McFee, and J. M. Kelly. 1978. Metal distribution in forested ecosystems in urban and rural northwestern Indiana. J. Environ. Qual. 7:337–342.

Ritter, W. F., and R. P. Eastburn. 1978. Leaching of heavy metals from sewage sludge through Coastal Plains soils. Commun. Soil Sci. Plant Anal. 9:785–798.

Schaumberg, G. D., C. S. LeVesque-Madore, G. Sposito, and L. J. Lund. 1980. Infrared spectroscopic study of the water-soluble fraction of sewage sludge-soil mixtures during incubation. J. Environ. Qual. 9:297–303.

Sidle, R. C., J. E. Hook, and L. T. Kardos. 1976. Heavy metals application and plant uptake in a land disposal system for waste water. J. Environ. Qual. 5:97–102.

----, ----, and ----. 1977. Accumulation of heavy metals in soils influenced by extended wastewater irrigation. J. Water Poll. Control Fed. 49:311–318.

----, and L. T. Kardos. 1977. Transport of heavy metals in a sludge-treated forested area. J. Environ. Qual. 6:431–437.

Sikora, L. J., N. H. Franks, C. M. Murray, and J. M. Walker. 1980. Trenching of digested sludge. J. Environ. Eng. Div. ASCE. 106:351–361.

Sommers, L. E. 1977. Chemical composition of sewage sludges and analyses of their potential use as fertilizers. J. Environ. Qual. 6:225–232.

----, D. W. Nelson, and D. J. Silviera. 1979. Transformations of carbon, nitrogen, and metals in soils treated with waste materials. J. Environ. Qual. 8:287–294.

Sposito, G. K., K. M. Holtzclaw, and J. Baham. 1976. Analytical properties of the soluble, metal-complexing fractions in sludge-soil mixtures. 2. Comparative structural chemistry of fulvic acid. Soil Sci. Soc. Am. J. 40:691–697.

----, ----, C. S. LeVesque, and C. T. Johnston. 1982. Trace metal chemistry in arid-zone field soils amended with sewage sludge: II. Comparative study of the fulvic acid fraction Soil Sci. Soc. Am. J. 46:265–270.

Williams, D. E., J. Vlamis, A. H. Pukite, and J. E. Corey. 1980. Trace element accumulation, movement, and distribution in the soil profile from massive applications of sewage sludge. Soil Sci. 129:119–132.

Chapter 16

Movement of Phosphorus and Nitrogen in Soil Following Application of Municipal Wastewater[1]

J. E. HOOK[2]

The practice of land treatment of municipal wastewater provides an important case for movement of chemicals in soil. Wastewater (WW) applications increase soil water content during the normally driest season of the year. In many instances WW plus rainfall additions exceed soil and plant evaporative losses, thus resulting in net recharge to subsurface aquifers or surface water bodies. Further, WW contains chemicals naturally present in the original municipal water supply, and those added during use and treatment of the water. The movement of these added, as well as, natural soil constituents has held the attention of many scientists during the past 20 years. The purpose of this paper is to briefly review and summarize many of the significant observations of this research as they relate to movement of P and N through soils.

PHOSPHORUS

Phosphorus typically constitutes less than 1% of total dissolved solids in municipal WW. It causes no direct human health problems. However,

---

[1] Contribution of the Dep. of Agron., Univ. of Georgia, Coastal Plain Stn., Tifton, GA 31793.
[2] Assistant professor, Dep. of Agron.

Copyright © 1983 ASA, SSSA, 677 South Segoe Road, Madison, WI 53711. *Chemical Mobility and Reactivity in Soil Systems.*

it was the P contamination in lakes and streams which provided the impetus for land application of WW.

Phosphorus was identified as a limiting nutrient in the eutrophication processes of some lakes. Links were established between P discharges from municipal WW treatment facilities and subsequent algal blooms in receiving waters. Application of the WW to land, it was assumed, would utilize the soils well-known P-fixation capacity to prevent WW P from entering surface water. In testing this hypothesis researchers have used several criteria to evaluate the soil's ability to keep P from leaving the site. These include comparison of P concentration in the solution phase in soils of treated and untreated sites and measurement of P retained in the soil. In addition, they have examined the phenomena which affect P movement and retention and have developed several models to predict short and long-range behavior of P movement.

Increases in P concentration in subsoil solution following land application of WW are indicative of penetration of applied P and/or mobilization of existing soil P. Unfertilized soils typically contain 0.001 to 0.10 mg P L^{-1} in the solution phase of the subsoil (Kardos et al., 1974; Burton and Hook, 1979; Cole and Schiess, 1978; Nutter et al., 1979). By contrast, secondary treated WW contains 4 to 16 mg P L^{-1}, principally as orthophosphate ions (Pound and Crites, 1973). As this WW moves through the soil, changes in P concentration along the flow path indicate the degree of renovation brought about by the whole soil-plant system.

In many cropland sites irrigated with 50 to 500 cm of municipal WW per year, observed changes in P concentration of subsoil solutions have been small. Kardos et al. (1974) found P concentration at 120 cm increased slightly during 7 years of WW addition to corn (*Zea mays* L.) and other crops grown on a clay loam Typic Hapludalf. Mean annual concentrations increased slightly from 0.02 to 0.07 mg P L^{-1}. This increase was nearly identical to that observed under a fertilized, but unirrigated, control area. Larson (1979) reported mean P concentrations which varied from 0.02 to 0.06 mg P L^{-1} in subsoil water of a silt loam Typic Hapludoll where corn and forage crops had been irrigated with WW for 5 years. Fertilized control plots irrigated with well water had similar P concentrations in the subsoil. Hook and Tesar (1978) reported concentrations of 0.02 to 0.11 mg P L^{-1} in subsoil solution of a silt loam Typic Hapludalf during the 3rd year of WW additions to several crops. Varying the rates of WW additions from 2.5 to 7.5 cm/week had no effect on subsoil P concentrations in the 3rd year. Vaisman et al. (1981) found P concentration remained below the detection limit (0.5 mg P L^{-1}) in drainage from a grass covered sand dune irrigated with WW for 2 years. Ketchum and Vaccaro (1977) likewise reported P concentrations < 0.5 mg P L^{-1} in subsoil below forages irrigated with WW for 2 years.

In forested sites irrigated with WW, and even in nonirrigated control sites, wide variations in P concentration of subsoil water were common. Sopper and Kerr (1979) reported P concentrations in soil water at 120 cm varied from 0.03 to 0.80 mg P L^{-1} during 14 years of WW applications of up to 7.5 cm/week. Mean annual P concentrations within unirrigated, unfertilized control sites on this clay loam Typic Hapludalf varied from

0.02 to 0.13 mg P L^{-1} over that period. There was no consistent trend indicating increases in P concentration in subsoil solution over the 14 years. Urie (1973) found weekly variation in P concentration of 0.02 to 0.30 mg P L^{-1} in shallow wells (1 to 3 m) in both WW treated and nonirrigated control sites. Those sites were within a pine plantation located on a loamy sand Typic Haplorthod. Nutter et al. (1979) found a 4-year mean of 0.26 to 0.49 mg P L^{-1} in various depths of the subsoil of a WW irrigated forest on a sandy loam Typic Hapludult. A nearby nonirrigated site had 0.22 to 0.27 mg P L^{-1} in the subsoil. Burton and Hook (1979) found total P concentration in subsoil solution from two WW irrigated sites was similar to P concentration in a nearby control site. In that forest, which had a mixture of well to poorly drained soils, there was a tendency for P concentrations to increase toward the end of the 2nd year.

In most of the previously mentioned studies, the authors concluded that subsoil P concentrations changed negligibly over the period of study. That was not always the case, however. Sopper and Kerr (1979) reported that year-round WW application to a sandy loam Ultic Hapludalf in a forested site resulted in significant and regular increases in soluble P. Mean annual P concentrations at the 120 cm depth increased from 0.04 mg P L^{-1} in 1963 to 0.39 mg P L^{-1} 7 years later. Hortenstine (1976) found greater than 1.2 mg P L^{-1} at 60 cm in a fine sand Arenic Haplaquod after only a few months of WW applications. This was a fourfold increase over pretreatment levels at this site. Iskandar and Syers (1980) observed 7.3 mg P L^{-1}, 10 times higher than at a nearby untreated site, in soil water collected at 80 cm. The site which had a loamy coarse sand Typic Xeropsamment had received WW applications for 4 years. Phosphorus concentration increases in these WW treatment sites were related to soil properties and to specific application rates.

Because land treatment systems are WW treatment systems, mass balances have been computed for many sites to provide degree of renovation information. Even in those sites where WW application resulted in P concentration increases in the subsoil solution, as compared with nontreated sites, the balances indicated that the net movement of P out of the subsoil was usually equivalent to less than 5% of the annual P application. Kardos and Hook (1976) calculated that leaching losses from several land treatment sites were less than 3% over 10 to 11 years of WW addition regardless of soil type or of vegetative cover. Cole and Schiess (1978) and Johnson et al. (1979) reported P retention in the upper 10 cm alone was greater than 99% in the first 2 years of WW treatment, even though the soil was a gravelly, loamy sand Typic Haplorthod. Karlen et al. (1976) found P in tile drain water under a calcareous till Udollic Ochraqualf contained the equivalent of less than 0.7% of P applied even at the highest simulated WW rate of 200 cm year^{-1}. Leland et al. (1978) calculated that more P, the equivalent of 15% of the applied P, moved past 150 cm during winter of the 2nd year than during previous application periods when losses were less than 2% of the applied P. Over the 2 years of treatment P equivalent to 5% of the applied P moved past the 150 depth of this old field site. On a nearby old field site irrigated only during the growing season, P losses were less than 2% (Hook and Burton, 1978). Where mass balances have been computed, they indicate a minimum movement of P past the reference depth.

Most of the conclusions presented above concerned changes in P concentrations in soil solutions or P balances computed from those concentrations. Most researchers used porous, ceramic cup soil water samplers to obtain those solutions. Hansen and Harris (1975) reported that concentrations of various chemicals extracted from the soil would vary because of the nonuniformity of ceramic cups from sampler to sampler, and would vary because of differences in the degree of evacuation used. Van der Ploeg and Beese (1977) using a mathematical study of flow around porous cup water extractors, found that the volume of soil represented by a single sample was related to degree of evacuation. They noted, further, that the presence of an evacuated sampler distorted the flow path of percolating soil water. Given the transient nature of soil water flux, soil water content, and concentration of the chemical of interest, P concentrations in extracted soil solutions must be interpreted carefully.

As the soil approaches saturation, flow occurs in pores of larger and larger diameters. Anderson and Bouma (1977) indicated that soil to solution contact time can be restricted near saturation because flow occurs in large pores. They used the term "hydrodynamic dispersion" to describe water movement through relatively large pores around slowly permeable peds with stagnant water. Without hydrodynamic dispersion, water applied to a soil would displace all of the water previously in equilibrium with the soil before the applied solution would move out. In a soil without hydrodynamic dispersion, soil water samplers could obtain representative sampling from solutions moving by them provided sampling intervals were short enough to detect changes in concentration. In a soil with hydrodynamic dispersion, samplers obtain a mix of moving and stagnant water.

The importance of hydrodynamic dispersion has been noted (Quisenberry and Phillips, 1976; Thomas and Phillips, 1979; Shuford et al., 1977). These authors describe flow in large pores or along ped faces and cracks which resulted in rapid movement of chemicals deep into the soil. Shaffer et al. (1979) compared NO_3 and Cd concentration in soil water collected by porous cup samplers with that collected by catching water dripping from the ceiling of an excavation in a funnel apparatus. Successive irrigations to nearly saturated soil apparently resulted in flow in large pores and cracks. When these irrigations contained Cd or NO_3, the water collected from the drips or channelized flow at 112 cm had increases in either Cd or NO_3 within minutes of the addition to the surface. Cadmium which should have been held tightly on exchange sites increased to 500 μg L^{-1} in the funnel samples during the day of the addition and the concentration remained well above preapplication levels for 10 d. Water collected in the ceramic cup samples, however, contained Cd at near preapplication levels (0.25 to 0.50 mg ml^{-1}) for 10 d. Cadmium concentration in ceramic cup samples never reached that observed in the funnel samples.

If estimates of leached P are in doubt because of water sampling techniques or because of uncertain water recharge volumes, then corroborative evidence should be obtained by directly measuring P retained in the soil. Unfortunately, most measurements of soil P were aimed at accessing the P available to plants (Hook and Burton, 1978; Kardos et al., 1974). This available P is only a portion of the total soil P.

A few analyses of total soil P have been reported. Latterell et al. (1982) compared pretreatment with posttreatment concentrations of total-P in soils of the Minnesota studies at the end of 5 years of WW applications. They found total P decreased in the upper 60 cm in fertilized plots irrigated with well water. However, total P increased somewhat in the upper 60 cm in both the low and high rate WW application plots. Assuming the concentration increases over the 5 years were significant and accurate, the high rate treatment increased total-P by 435 kg ha^{-1} in the upper 60 cm. With 831 kg P ha^{-1} added and only 22% of that removed in corn harvests (Dowdy et al., 1978) then a net 649 kg P ha^{-1} could have been retained. The difference, 213 kg P ha^{-1}, or 25% of total-P added presumably moved past 60 cm. That represents considerably higher leaching losses than could be calculated from mass balances using soil water samples taken at 60 cm. Sommers et al. (1979) in a follow up study at the Pennsylvania site found even smaller differences between WW irrigated sites and nearby nontreated sites. Although the reed canarygrass site, for example, received 2800 kg P ha^{-1} over the 1964 to 1975 period, they found only 465 kg ha^{-1} more P in the upper 15 cm of the WW treated than the nonfertilized control site. Below 15 cm there was actually less total P in the effluent treated than control site. Crop harvests removed 550 kg P ha^{-1} during that time (Kardos and Hook, 1976). This left more than 60% of the added P unaccounted for in soil retention and crop removal. Mass balances of this type are helpful but suffer because of inherent problems in sampling soils and in measuring total P. Nonetheless they provide another perspective to P movement from land treatment sites.

Much of the effort in describing P movement in WW systems has been in predicting when P concentration in the leachate would become unacceptable. Because there is no single standard for P concentration, researchers have used various levels in their modelling attempts. Harter and Foster (1976) simulated WW additions by adding pulses of P (50 μg 25 ml^{-1}) to 0.5 g soil samples until the soil retained no additional P. A Webster soil (fine-loamy, mixed, mesic Typic Haplaquoll) with low P sorption capacity retained no additional P after only four pulses. A Venango, a soil with high P sorption capacity continued to adsorb P for 16 pulses. Polynomial curves relating the P added to the P sorbed were used to predict ideal movement of P resulting from successive additions to the surface. Predicted penetration of the P leaching front after 200 kg P ha^{-1} had been added was 17 cm for the Webster but only 3 cm for the Venango.

Ellis (1973) determined adsorption maxima for several soils of the Muskegon land treatment site using traditional treatment of the Langmuir adsorption isotherm. Because the maximum quantity of P would not be reached except for very high levels of P in solution, the percentage of the maximum which would be saturated in eqilibrium 10 mg P L^{-1} was calculated for each horizon. The Rubicon sand A_1 held the least P, 4 kg P ha^{-1} cm^{-1}. The Warsaw clay loam B$_{21}$ horizon with its Fe accumulation held the most, 71 kg P ha^{-1} cm^{-1}.

Tofflemire et al. (1978) similarly used Langmuir adsorption isotherm to classify for each horizon of 35 soils of New York for their ability to adsorb P. However, they chose a 5 mg P L^{-1} equilibrium solution rather than a 10 mg P L^{-1} as used by Ellis (1973). They found for purposes of rating various soils ability to retain P, grouping by parent material or by B horizon taxonomy was most useful. They also noted B horizon sorption capacity was often over 2.5 times that of the C horizon. Sawhney and Hill (1975) compared B horizon material of six Connecticut soils for their P sorption capacity in equilibrium with 6.4 mg P L^{-1}—the concentration of P in domestic water entering on-site seepage fields. Sorption capacities varied from 9.0 to 29.0 mg P/100 g soil. Samples from existing seepage fields in operation for 12 years indicated that the soil only 15 to 30 cm from the trench still retained about 1/10 of the original sorption capacity.

They suggested regeneration of P sorption sites occurs. This later hypothesis is consistent with findings of Sommers et al. (1979) who reported on sorption capacities of soils at the Pennsylvania State University WW site. Phosphorus sorption capacity remaining on WW irrigated soils was lower than that capacity of nontreated control sites. However, even at the forest site with its sandy loam topsoil, the soil (0 to 15 cm) was only in equilibrium with a 0.58 mg P L^{-1} solution, considerably less concentrated than the WW which had been added for the previous 10 years.

Use of P sorption capacity to predict retention or movement of P requires several assumptions. Laboratory determinations of P isotherms involve addition of P solutions, shaking for complete mixing, and timing to allow the adsorption equilibrium to be reached. In the field complete mixing does not necessarily occur. As was pointed out above, channelization can occur so that only the surfaces of the channels or faces of soil peds may come into contact with the added WW solution. In this case the mass of P moved below a soil layer would be greater than that with the assumed complete mixing. For similar reasons rapid flow in larger pores and cracks may exceed time necessary for equilibrium to be obtained. Data of Sommers et al. (1979) indicate that even though the surface soil (0 to 15 cm) is not in equilibrium with P concentration of applied WW, substantial amounts of P moved past that layer and accumulated in 15 to 30 and 30 to 60 cm layers of the sandy loam soil. In a strongly structured Typic Hapludalf soil near saturation, flow rate in the large pores was calculated to be 3.96 cm min^{-1}, and these pores accounted for 72% of the total flow through the soil (Shaffer et al., 1979). In that soil and similar soils, the assumption that there is sufficient reaction time for equilibration may not be met.

Several investigators (Enfield et al., 1976; Munns and Fox, 1976) have noted a slow sorption reaction which continues after an apparent eqilibrium when solutions remain in contact with soils for a long enough time. Enfield et al. (1976) showed that this reaction continued for over 300 h for soils which were initially in "equilibrium" (at 1 h) with solutions of greater than 1 mg P L^{-1}. Mansell et al. (1977) described this slow reaction as immobilization or fixation occurring as the weaker bonds of the initially adsorbed P undergo transformation to stronger chemical bonds. While this immobilization would not be directly involved in retention of P from a moving solution of WW, the transformation to stronger bonding may free adsorption sites for further rapid retention. Additionally, the backward reaction mobilization is also slow. Thus, immobilized P would not be rapidly released to low P solutions, such as rain water moving through the soil.

The precipitation of P from soil solution has not been as widely reported as the adsorption reactions. Lance (1977) found WW P additions to calcareous sand columns resulted in an initial low P concentration in the leachate. This was followed by a period increasing P concentration until a levelling off occurred. This latter steady state concentration was a function of the infiltration rate and could be increased or decreased by increasing or decreasing the flow rate. He suggested that, during the early period which had minimal P leaching, adsorption was occurring. After the adsorption capacity was exceeded, a time dependent immobilization or precipitation reaction accounted for most of the P retention. Because of the rapid response to changes in infiltration rate, Lance (1977) suggested that precipitation was responsible for continued P removal. With an infiltration rate of 5 cm d^{-1}, P retention continued even after 60 000 kg P ha^{-1} had been applied. When the flow rate exceeded 15 cm of WW d^{-1}, there was insufficient time for nucleation of Ca-P compounds, and P retention in the column decreased. Robbins and Smith (1977) also found P retention related to flow rates. Again precipitation of calcium phosphates was suggested. Enfield and Bledsoe (1975) reported the apparent sorption of P in lab samples was 1.5 to 3.0 times greater after 4 months equilibration than after 5 days. Barrow (1980) showed that the adsorption reactions described by Enfield and Bledsoe (1975) could be divided into two types based on soil reaction. The calcareous soils had continued reduction of P in solution until a concentration of 0.2 mg P L^{-1} was reached, regardless of the amount of P initially in the solution. This concentration corresponded to the solubility of octacalcium phosphate. Nonalkaline soils had lower rates of continued sorption with time, as would be obtained by the slow immobilization reaction (Mansell et al., 1977). When precipitation can occur, the formation of precipitates provides much higher ultimate P retention capacities than would be predicted by short-term adsorption alone.

Movement of P in land treatment systems is not only a function of soil properties and loading rates but also of vegetative management. Plants provide a sink for P through uptake from soil solution. This creates a dynamic situation related to vegetative growth rates and residue mineralization rates. When no harvests occur, recycling of plant residues should ultimately release P to the soil, although not necessarily into the rapid channelized flow. Where harvests occur, crop removal can significantly reduce net mobilization of P. Karlen et al. (1976), Palazzo (1981), Bole and Bell (1978), Kardos et al. (1974), Overman and Nguy (1975), and Dowdy et al. (1978) have reported that the amount of P which can be recovered in various crops is 14 to 40 kg ha^{-1} in corn grain, 16–58 kg ha^{-1} in corn silage, 23 to 34 kg ha^{-1} in alfalfa (*Medicago sativa* L.) hay, and 33 to 56 kg ha^{-1} in cool-season grasses. Crop harvest assures that that portion of added P will not leach from the soil.

NITROGEN

While WW P provided the impetus to use soil-plant systems treatment of WW, N renovation has received considerable attention. The existence of a drinking water quality standard, 10 mg NO$_3$-N L^{-1} (Environmental Protection Agency, 1976) provided a focus for evaluating the effectiveness of land treatment systems with respect to N. Secondary treated municipal WW typically contains 15 to 35 mg total N L^{-1} (Pound and Crites, 1973). Most of this N is in organic and NH$_4^+$ forms. Because these forms can be converted to NO$_3^-$-N by mineralization and nitrification, municipal WW potentially contains excessive (> 10 mg N L^{-1}) nitrate concentrations.

The N cycle in soil plant systems is complex. However, its components are well-known even though not all phenomena associated with them are fully understood (Allison, 1966; Keeney, 1973). Nitrogen retention in soil can occur when organic-N compounds are adsorbed or immobilized, when NH$_4^+$-N is held on exchange sites or fixed within clay and/or organic compounds and when inorganic-N forms are assimilated by microorganisms. The N which is not retained in soils can appear in one of three sinks—leaching loss, gaseous loss, and plant uptake.

Retention of N compounds has been examined for several land treatment sites. Kardos et al. (1974) found no differences in total Kjeldahl N (TKN) among cropland sites receiving 0, 2.5, and 5.0 cm WW week^{-1} for 6 years. Because more than 8500 kg TKN ha^{-1} was contained in the upper 150 cm of soil, most of the 180 kg N ha^{-1} added annual would have to be retained before detectable differences in TKN would occur. Sommers et al. (1979) did report slightly higher levels of TKN in soil of an old field, a hardwood forest, and a grass field on a clay loam Typic Hapludalf after 11 years of WW additions. The differences between treated and control occurred only in the upper 15 cm and were 700, 450, and 390 kg N ha^{-1} for the old field, hardwood and grass area, respectively. Latterell et al. (1982) compared total-N in fertilized control plot with that in a WW treated soils. While they found no significant differences in the upper 30 cm, the reported values for total-N were approximately 1900 kg ha^{-1} greater in the WW treated soil than the fertilized soil. The difficulty in detecting significant increases of total-N of soils above already high background levels makes calculation of N retention in WW treated soils difficult.

Rather than examining the soils directly, most researchers have concentrated their efforts on accessing the importance of the N sinks. Field studies of N in leachate from land treatment systems are numerous. Hook and Kardos (1977) examined nitrate levels in soil water from various WW treatment sites which had different vegetation or soil. Where WW was irrigated year round to a forested site, NO$_3$-N concentration varied from 10 to 60 mg L^{-1} in soil water at 120 cm. At an old field site, NO$_3$-N concentration at 120 cm did not exceed 10 mg L^{-1} when weekly additions were 5.0 cm during the growing season. However, when the application was increased to 7.5 cm, NO_3-N concentration increased to 10 to 25 mg L^{-1}. In a hay field, harvested three times annually, NO_3-N remained below 5 mg L^{-1} at 120 cm even with year-round irrigations with WW. However, with only growing season applications a corn field was not as effective in preventing NO_3-N concentrations from increasing to > 10 mg N L^{-1} in the subsoil. Clapp et al. (1977) reported NO_3^- + NH_4^+-N at 125 cm varied with crop cover and time of the growing season. With growing season applications of WW to a corn crop, mineral-N concentration exceeded 20 mg N L^{-1} during June and July and then decreased to less than 10 mg N L^{-1} in August to October. In contrast, mineral N did not exceed 4 mg N L^{-1} throughout the season with forages. Hook and Tesar (1978) reported mineral-N concentration at 150 cm was releated growth stage for annual forages, cutting management for perennial forages and WW application rate for both perennials and annuals.

Iskandar (1978) observed seasonal changes in NO_3^--N in leachate from closed lysimeter plots which received primary and secondary WW year round. Highest NO_3^--N concentrations (10 to 40 mg N L^{-1}) occurred during early summer. He explained that those high levels apparently resulted from nitrification of NH_4^+-N which had accumulated during winter applications. Breuer et al. (1979) found mean annual NO_3-N concentration in soil with no vegetative cover was 4.5 and 10.6 mg L^{-1} during the 1st and 2nd year of WW application. Where grass cover existed NO_3^--N was 1.0 and 4.4 mg L^{-1} during those 2 years. Nutter et al. (1979) found leachate NO_3^--N from a forested site irrigated with WW varied with N application rate. With preaeration of the WW total-N was lowered to 13.8 mg L^{-1} before irrigation. Leachate from the site contained less than 10 mg NO_3^--N L^{-1}. Without preaeration WW contained 24.4 mg total-N L^{-1}. Irrigation with this WW doubled the N loading and leachate NO_3^--N increased to more than 15 mg L^{-1}. These and other field studies point out the responsiveness of the leaching sink to seasonal changes in weather as well as plant growth and microbial activity.

Nitrate is only weakly adsorbed by soils, and most NO_3^- compounds are highly soluble. Nitrate consequently moves with the water when biological reactions do not or cannot occur (Hill, 1972). Because biological reactions essentially cease near freezing temperatures, NO_3^- added to very cold soils will move through the soil and enter receiving waters altered only by dilution (Leland et al., 1978). Likewise, NO_3^- moving from biologically active soil zones will not be further retained or converted to other forms during flow into and through nonreducing aquifers. Dilution and mixing with low NO_3^- water will lower the NO_3^- concentration. However, this is an unreliable means of meeting the water quality standard (10 mg NO_3^--N L^{-1}) because flow paths in aquifers are difficult to predict (Walker et al., 1973b; Ellis and Childs, 1973; Parizek, 1973).

Most of the N in municipal WW is in the NH_4^+ form (Pound and Crites, 1973). In the presence of nitrifying organisms in aerobic soil environments, NH_4^+ is very unstable. Walker et al. (1973a) for example, found rapid conversion of NH_4^+ to NO_3^- in the vacinity of seepage fields for dispersion of septic tank effluent. Eighty percent of the N reaching the seepage field was in the form of NH_4^+. Clogging of the gravel-soil interface in the field maintained anaerobic conditions within the gravel envelope but allowed aerobic, unsaturated flow from that interface to the underlying water table. Within 2 cm of that interface NH_4^+ was sorbed by cation exchange sites in the soil and converted to NO_3^-. Conversion was complete in a matter of hours.

While nitrification is a very rapid reaction, the predominance of NH_4^+ rather than NO_3^- in WW is still advantageous. To the extent that NH_4^+ in infiltrating water can be exchanged on soil surfaces, NH_4^+ will be retained in the biologically active zone as the water percolates through the soil. The absence of more than 1 mg NH_4^+-N L^{-1} in subsoil water at almost all field study sites for land treatment indicates that exchange and/or conversion of NH_4 prevents NH_4^+-N movement on all but the sandiest soils. Left in the biologically active zone by sorption and exchange NH_4^+-N and NO_3^--N derived from it will be available for microbial assimilation and crop uptake until the next rainfall or WW event. Where winter applications of WW are made, the NH_4^+ form is even more important. Hook and Kardos (1977) found that when NH_4^+ was the dominant N form in the WW, N leaching losses during winter were less than where NO_3^- was the major form.

With the dominance of biological reactions involving N it is not surprising that there is a strong seasonal effect on N movement (Hook and Kardos, 1977; Iskandar, 1978; Larson, 1979; Burton and Hook, 1979; Hook and Tesar, 1978). During winter, NH_4^+ from applied WW can be held on exchange sites in soil. As the soil warms in spring, nitrification of stores NH_4^+ begins. If plants such as cool-season grasses are active, much of the nitrified N could be taken up. However, spring applications of WW, as well as spring rainfall, and mineralization of previous years' plant residues increase potential N leaching just at the time that nitrification of stores NH_4^+ occurs. The result may be a flush of high NO_3^- water in spring (Karlen et al., 1976; Hook and Tesar, 1978; Iskandar, 1978).

The role of plants in preventing NO_3^- movement is well documented. Nitrogen is taken up and immobilized in plant tissue. However, without harvest of the plant, much of the annual uptake would be released in subsequent years. This has been evident where WW was applied to mature forests (Sopper and Kerr, 1979; Burton and Hook; 1979; Nutter et al., 1979). Annual nutrient requirements for biomass production in a mature forest are in equilibrium with the existing nutrients. They are effectively recycled. Mature forests may be able to use additional N and water to promote a more rapid annual production. However, after a few years application, a new steady state exists where N release from residues is sufficient to supply the annual requirements of trees and plants. With this N sink satisfied, subsequent N additions which are not retained in soil are lost through leaching or denitrification.

As with P, harvest of the plant biomass can assure that the N taken up will not be able to leach out. Nitrogen removal from the land treatment site can be quite high: 74 to 170 kg ha^{-1} for corn grain, 100 to 250 kg ha^{-1} for corn silage, 114 to 284 kg ha^{-1} for alfalfa, 100 to 472 kg ha^{-1} for cool-season grasses (Palazzo and McKim, 1978; Ellis et al., 1980; Kardos et al., 1974; Clapp et al., 1977; Overman and Nguy, 1975; Larson, 1979).

Uptake of N is not uniform throughout the season. This results in NO_3^- leaching within the season. Hook and Tesar (1978) and Larson (1979) reported NO_3^--N greater than 10 mg L^{-1} in the subsoil during establishment of the corn. Corn was efficient in removing N only for the 6 to 8 week period of rapid vegetative growth. Increased NO_3^- leaching was also observed shortly after each alfalfa harvest (Hook and Tesar, 1978). By applying the WW only as N is needed by the crop, it should be possible to minimize NO_3^- losses.

Denitrification serves as an important sink for N in soils of land treatment sites. The potential for NO_3^- leaching is immediately removed. Denitrification appears to occur in the root zones of rapidly growing plants. It has been suggested that the interior of peds in moist soils may become anaerobic even a short distance from aerobic pores (Broadbent, 1973). This would allos sites for denitrification. Root exudates could provide a reduced carbon energy source to enhance denitrification. Nitrate in subsoil water remains below 10 mg N L^{-1}. Where annual growth of nonwoody vegetation is rapid, NO_3^- leaching does not reach the steady state with most of added N appearing in the leachate as has been observed in mature forests. This has been reported for old field vegetation (Hook and Kardos, 1978; Hook and Burton, 1979), for volunteer vegetation following clear cutting (Sopper and Kerr, 1979) and for mowed and unmowed grasses (Hook and Burton, 1979).

Denitrification to remove NO_3^- from the solution phase and thereby reduce NO_3^- leaching has been incorporated specifically into two land treatment systems—overland flow and high rate groundwater recharge. In overland flow, lateral, downslope movement of WW results in zones within the field with sufficient aeration for nitrification. Downslope positions, however, are sufficiently anaerobic for denitrification (Peters and Lee, 1978). Additionally, vegetative cover on the slope may aid the denitrification process by producing organic substrates for denitrifiers and by removing O_2 through root respiration.

With deep recharge systems alternate periods of drying and flooding brings about nitrification of NH_4^+ retained during flood periods and denitrification of NO_3^- formed during drying. Net effectiveness of the process in preventing leaching of either NH_4^+ or NO_3^- depends upon lengths of periods of flooding and drying and upon infiltration rates during flooding (Lance and Whisler, 1972; Lance et al., 1976).

Nitrate moves readily in soil, but it must be controlled on-site because of the drinking water standard. The soil itself has less effect on N retention than the vegetation. The irrigation and vegetation at land treatment sites must be properly managed together for effective control of NO_3^- movement from the site.

SUMMARY

Field studies of land treatment of WW found little evidence of P penetration to the subsoil. However, questions concerning ability of soil water sampler to access the net P flowing past the sample depth cast some doubt on these observations. Mass balances rely upon these observations of P in subsoil water for estimates of leaching quantities and suffer accordingly. Soil sampling provides one method to verify estimates of leaching losses; however, soil sampling techniques and high levels of total P in pretreated soils also create uncertainties. Both physical and chemical properties of the soil have been related to P retention capacity. Management of WW addition, both in instantaneous rates of application and annual application rates, was related to the movement of P. Harvest of plant residues enhanced P renovation at land treatment sites.

Nitrogen was found to leach readily as NO_3^--N when vegetative uptake was minimal. Temporary retention of NH_4^+ on soil exchange sites during cold season, aided year-round renovation but in some instances it led to a flush of high NO_3^- water in spring. Applications of WW to actively growing vegetation resulted in the lowest N leaching losses, particularly when that vegetation was removed by crop harvests.

LITERATURE CITED

Allison, F. E. 1966. The fate of nitrogen applied to soils: A review. Adv. Agron. 18:219–258.

Anderson, J. L., and J. Bouma. 1977. Water movement through pedal soils. Soil Sci. Soc. Am. J. 41:413–418.

Barrow, N. J. 1980. Differences among some North American soils in the rate of reaction with phosphate. J. Environ. Qual. 9:644–648.

Bole, J. B., and R. G. Bell. 1978. Land application of municipal sewage wastewater: Yield and chemical composition of forage crops. J. Environ. Qual. 7:222–226.

Breuer, D. W., D. W. Cole, and P. Schiess. 1979. Nitrogen transformation and leaching associated with wastewater irrigation in Douglas fir poplar, grass and unvegetated systems. p. 19–33. *In* W. E. Sopper and S. N. Kerr (ed.) Utilization of municipal wastewater and sludge on forest and disturbed land. The Pennsylvania State Univ. Press, University Park, Pa.

Broadbent, F. E. 1973. Factors affecting nitrification-denitrification in soils. p. 232–245. *In* W. E. Sopper and L. T. Kardos (ed.) Recycling treated municipal wastewater and sludge through forest and cropland. The Pennsylvania State Univ. Press, University Park, Pa.

Burton, T. M., and J. E. Hook. 1979. A mass balance study of application of municipal wastewater to forests in Michigan. J. Environ. Qual. 8:589–596.

Clapp, C. E., D. R. Linden, W. E. Larson, G. C. Marten, and J. R. Nylund. 1977. Nitrogen removal from municipal wastewater effluent by a crop irrigation system. p. 139–150. *In* R. C. Loehr (ed.) Land as a waste management alternative. Ann Arbor Science Publishing, Inc., Ann Arbor, Mich.

Cole, D. W., and P. Schiess. 1978. Renovation of wastewater and response of forest ecosystems: The pack forest study. Vol. 1. p. 323–331. *In* H. L. McKim (coor.) State of knowledge in land treatment of wastewater. U.S. Army Cold Regions Res. and Eng. Lab., Hanover, N.H.

Dowdy, R. H., G. C. Marten, C. E. Clapp, and W. E. Larson. 1978. Heavy metal content and mineral nutrition of corn and perennial grasses irrigated with municipal wastewater. Vol. II. p. 175–182. *In* H. L. McKim (coor.) State of knowledge in land treatment of wastewater. U.S. Army Cold Regions Res. and Eng. Lab., Hanover, N.H.

Ellis, B. G. 1973. The soil as a chemical filter in recycling. p. 46–70. *In* W. E. Sopper and L. T. Kardos (ed.) Recycling treated municipal wastewater and sludge through forest and cropland. The Pennsylvania State Univ. Press, University Park, Pa.

----, and K. E. Childs. 1973. Nutrient movement from septic tank and lawn fertilization. Tech. Bull. No. 73-5. Dep. of Nat. Resources. Lansing, Mich.

----, A. E. Erickson, L. W. Jacobs, J. E. Hook, and B. D. Knezek. 1980. Cropping systems for treatment and utilization of municipal wastewater and sludge. Completion Report of Projects G-005292 01 and R805270010. Robert S. Kerr Environ. Res. Lab., Office of Res. and Dev., U.S. Environ. Protection Agency, Ada, OK 74820. 187 p.

Enfield, L. G., and B. E. Bledsoe. 1975. Kinetic model for orthophosphate reactions in mineral soils. USEPA Report (EPA-660/2-75-002).

Enfield, C. G., C. C. Harlin, Jr., and B. E. Bledsoe. 1976. Comparison of five kinetic models for orthophosphate reactions in mineral soils. Soil Sci. Soc. Am. J. 40:243–249.

Environmental Protection Agency. 1976. National interim primary drinking water regulations. USEPA Report (EPA-570/9-76-003). U.S. Government Printing Office, Washington, D.C.

Hansen, E. A., and A. R. Harris. 1975. Validity of soil-water samples collected with porous ceramic cups. Soil Sci. Soc. Am. Proc. 39:528–536.

Harter, R. D., and B. B. Foster. 1976. Computer simulation of phosphorus movement through soils. Soil Sci. Soc. Am. J. 40:239–242.

Hill, D. E. 1972. Wastewater renovation in Connecticut soils. J. Environ. Qual. 1:163–167.

Hook, J. E., and T. M. Burton. 1978. Land application of municipal effluent on old fields and on grass lands. p. 25–65. In T. M. Burton (ed.) The Felton-Herron Creek, Mill Creek, Pilot Watershed Studies, Env. Prot. Tech. Series, EPA-905/9-78-002.

----, and ----. 1979. Nitrate leaching from sewage irrigated perennials as affected by cutting management. J. Environ. Qual. 8:496–502.

----, and L. T. Kardos. 1977. Nitrate relationships in the Pennsylvania state living filter system. p. 181–198. In R. C. Loehr (ed.) Land as a waste management alternative. Ann Arbor Science Publishing, Inc., Ann Arbor, Mich.

----, and ----. 1978. Nitrate leaching during long term spray irrigation for treatment of secondary sewage effluent at woodland sites. J. Environ. Qual. 7:30–34.

----, and M.B. Tesar. 1978. Land application to croplands. p. 66–85. In T. M. Burton. The Felton-Herron Creek, Mill Creek, Pilot watershed studies. Environ. Protection Tech. Series EPA-905/9-78-002.

Hortenstine, C. C. 1976. Chemical changes in the soil solution from a spodosol irrigated with secondary treated sewage effluent. J. Environ. Qual. 5:335–338.

Iskandar, I. K. 1978. The effect of wastewater reuse in cold regions on land treatment systems. J. Environ. Qual. 7:361–368.

----, and J. K. Syers. 1980. Effectiveness of land application for phosphorus removal from municipal wastewater at Manteca, Calif. J. Environ. Qual. 9:616–621.

Johnson, D. W., D. W. Breuer, and D. W. Cole. 1979. The influence of anion mobility on ionic retention in wastewater-irrigated soils. J. Environ. Qual. 8:246–250.

Kardos, L. T., and J. E. Hook. 1976. Phosphorus balance in sewage effluent treated soils. J. Environ. Qual. 5:87–90.

----, W. E. Sopper, E. A. Myers, R. R Parizek, and J. B. Nesbitt. 1974. Renovation of secondary effluent for reuse as a water resource. Environ. Protection Tech. Series. EPA 660/2-74-016. p. 495.

Karlen, D. L., M. L. Vitosh, and R. J. Kunze. 1976. Irrigation of corn with simulated municipal sewage effluent. J. Environ. Qual. 5:269–273.

Keeney, D. R. 1973. The nitrogen cycle in sediment water systems. J. Environ. Qual. 2:15–29.

Ketchum, B. H., and R. F. Vaccaro. 1977. The removal of nutrients and trace metals by spray irrigation and in a sand filter bed. p. 413–434. In R. C. Loehr (ed.) Land as a waste management alternative. Ann Arbor Science Publishing, Inc., Ann Arbor, Mich.

Lance, J. C. 1977. Phosphate removal from sewage water by soil columns. J. Environ. Qual. 6:279–284.

----, and F. D. Whisler. 1972. Nitrogen balance in soil columns intermittently flooded with secondary sewage effluent. J. Environ. Qual. 1:180–186.

----, ----, and R. C. Rice. 1976. Maximizing denitrification during soil filtration of sewage water. J. Environ. Qual. 5:102–107.

Larson, W. E. 1979. Utilization of sewage wastes on land. Research Progress Rep., Minnesota Agric. Exp. Stn., St. Paul, Minn.

Latterell, J. J., R. H. Dowdy, C. E. Clapp, W. E. Larson, and D. R. Linden. 1982. Distribution of phosphorus in soils irrigated with municipal wastewater effluent: A five year's study. J. Environ. Qual. 11:124–128.

Leland, D. E., D. C. Wiggert, and T. M. Burton. 1978. Winter spray irrigation of secondary municipal effluent. J. Water Pollut. Control Fed. 51:1850–1858.

Mansell, R. S., H. M. Selim, and J. G. A. Fiskell. 1977. Simulated transformations and transport of phosphorus in soil. Soil Sci. Am. J. 124:102–109.

Munns, D. N., and R. L. Fox. 1976. The slow reaction which continues after phosphate adsorption: Kinetics and equilibrium in some tropical soils. Soil Sci. Soc. Am. J. 40:46–57.

Nutter, W. L., R. C. Shultz, and G. H. Brister. 1979. Renovation of municipal wastewater by spray irrigation on steep forest slopes in the Southern Appalachians. p. 77–85. In w. e. Sopper and S. N. Kerr (ed.) Utilization of municipal wastewater and sludge on forest and disturbed land. The Pennsylvania State Univ. Press, University Park, Pa.

Overman, A. R., and A. Nguy. 1975. Growth response and nutrient uptake by forage crops under effluent irrigation: Corn, sorghum-sudangrass, Kenaf. Commun. Soil Sci. Plant Anal. 6:81–93.

Palazzo, A. J. 1981. Seasonal growth and accumulation of nitrogen, phosphorus, and potassium by orchardgrass irrigated with municipal wastewater. J. Environ. Qual. 10:64–68.

----, and H. L. McKim. 1978. The growth and nutrient uptake of forage grasses when receiving various application rates of wastewater. Vol. II. p. 157–163. In H. L. McKim (coor.) State of knowledge in land treatment of wastewater. U.S. Army Cold Regions Res. and Eng. Lab., Hanover, N.H.

Parizek, R. R. 1973. Site selection criteria for wastewater disposal—soils and hydrogeologic considerations. p. 95–147. In W. E. Sopper and L. T. Kardos (ed.) Recycling treated municipal wastewater and sludge through forest and cropland. The Pennsylvania State Univ. Press, University Park, Pa.

Peters, R. E., and C. R. Lee. 1978. Field studies of advanced treatment of municipal wastewater by overland flow. Vol. II. p. 45–50. In H. L. McKim (coor.) State of knowledge in land treatment of wastewater. U.S. Army Cold Regions Res. and Eng. Lab., Hanover, N.H.

Pound, C. E., and R. W. Crites. 1973. Characteristics of municipal effluents. p. 49–62. In Proc. of the joint conference on recycling municipal sludges and effluents on land. Natl. Assoc. of State Univ. and Land Grant Colleges, Washington, D.C.

Quisenberry, V. L., and R. E. Phillips. 1976. Percolation of surface applied water in the field. Soil Sci. Soc. Am. J. 40:484–489.

Robbins, C. W., and J. H. Smith. 1977. Phosphorus movement in calcareous soils irrigated with wastewater from potato processing plants. J. Environ. Qual. 6:222–225.

Sawhney, B. L., and D. E. Hill. 1975. Phosphate sorption characteristics of soils treated with domestic wastewater. J. Environ. Qual. 4:342–346.

Shaffer, K. A., D. D. Fritton, and D. E. Baker. 1979. Drainage water sampling in a wet, dual-pore soil system. J. Environ. Qual. 8:241–246.

Shuford, J. W., D. D. Fritton, and D. E. Baker. 1977. Nitrate and chloride movement through undisturbed field soil. J. Environ. Qual. 6:255–259.

Sommers, L. E., D. W. Nelson, and L. B. Owens. 1979. Status of inorganic phosphorus in soils treated with municipal wastewater. Soil Sci. Am. J. 127:340–350.

Sopper, W. E., and S. N. Kerr. 1979. Renovation of municipal wastewater in eastern forest ecosystems. p. 61–76. In W. E. Sopper and S. N. Kerr (ed.) Utilization of municipal wastewater and sludge on forest and disturbed land. The Pennsylvania State Univ. Press, University Park, Pa.

Thomas, G. W., and R. E. Phillips. 1979. Consequences of water movement in macropores. J. Environ. Qual. 8:149–152.

Tofflemire, T. J., R. Arnold, and M. Chen. 1979. Phosphate adsorption capacity and cation exchange capacity of 35 common soil series in New York. Vol. II. p. 89–96. *In* H. L. McKim (coor.) State of knowledge in land treatment of wastewater. U.S. Army Cold Regions Res. and Eng. Lab., Hanover, N.H.

Urie, O. H. 1973. Phosphorus and nitrate levels in groundwater as related to irrigation of jack pine with sewage effluent. p. 176–183. *In* W. E. Sopper and L. T. Kardos (ed.) Recycling treated municipal wastewater and sludge through forest and cropland. The Pennsylvania State Univ. Press, University Park, Pa.

Vaisman, I., J. Shalhevet, T. Kipnis, and A. Feigin. 1981. Reducing groundwater pollution from municipal wastewater irrigation of Rhodes grass grown on sand dunes. J. Environ. Qual. 10:434–439.

Van der Ploeg, R. R., and F. Beese. 1977. Model calculations for the extraction of soil water by ceramic cups and plates. Soil Sci. Soc. Am. J. 41:466–470.

Walker, W. G., J. Bouma, D. R. Keeney, and F. R. Magdoff. 1973a. Nitrogen transformations during subsurface disposal of septic tank effluent in sands: I. Soil transformations. J. Environ. Qual. 2:475–480.

----, ----, ----, and P. G. Olcott. 1973b. Nitrogen transformations during subsurface disposal of septic tank effluents in sands: II. Ground water quality. J. Environ. Qual. 2:521–525.

Chapter 17

Management of Soil Systems for Industrial Wastes[1]

JOHN C. COREY[2]

The issues associated with industrial waste management become more complex each year with the proliferation of chemicals and with increased knowledge of the behavior of materials in the environment. In recent years waste management issues have been extensively discussed in the public arena (Litchfield et al., 1976; Epstein and Chaney, 1978; Maugh, 1979). The result has been improved legislation to manage waste. This legislation has defined the options available to industry for various classes of wastes. The most frequently used options are landfill, incineration, biological treatment in domestic sewage treatment plants, deep well disposal, or land application.

Improper waste management can have very significant impacts. To vividly demonstrate this statement, I only need to mention the words "Love Canal." "Superfund" is a legislative approach to correct past mistakes where land was improperly used for waste management. Detailed regulations developed as a consequence of the U.S. Resource Conserva-

---

[1] The information contained in this article was developed during the course of work under Contract DE-AC09-76SR00001 with the U.S. Dep. of Energy.
[2] Research manager, Environmental Sciences Div., Savannah River Lab., E.I. du Pont de Nemours and Co., Aiken, SC 29808.

Copyright © 1983 ASA, SSSA, 677 South Segoe Road, Madison, WI 53711. *Chemical Mobility and Reactivity in Soil Systems.*

tion and Recovery Act of 1976 are an attempt to prevent the improper management of materials in the future.

In spite of the occasions where soil systems have been improperly utilized as a waste management agent, soil systems can and should play an important role in the management of certain industrial wastes. This article discusses the legislative, economic, and technical factors influencing the selection of soil systems for industrial waste management.

WASTE MANAGEMENT LEGISLATION

The objective of waste management legislative action is to protect the environment and provide guidance to all companies within an industry so each is complying with the same disposal requirements. Numerous examples of local, state, and federal legislation could be quoted that restrict indiscriminate waste disposal by industry. A primary example is the U.S. Resource Conservation and Recovery Act of 1976. This Act is aimed at improving the management of hazardous wastes which are defined as "a solid waste, or combination of solid wastes which, because of its quantity, concentration, or physical, chemical, or infectious characteristics may:
- cause, or significantly contribute to an increase in mortality or an increase in serious irreversible, or incapacitating reversible illness, and
- pose a substantial present or potential hazard to human health or the environment when improperly treated, stored, transported, disposed of or otherwise managed.

Hazardous waste for the purpose of this regulation possesses one of the following characteristics: ignitable, corrosive, reactive, or toxic."

The Act specifies acceptable standards for surface impoundments or landfills for hazardous materials. Standards were developed to minimize movement from the disposal site to the groundwater through the use of natural (soil) or artificial barriers (man-made materials) and do not rely on soil properties other than for retarding liquid movement. If an industrial waste is not classed as hazardous, the plant operator may dispose of the waste using less stringent disposal practices. Examples of hazardous wastes that are acceptable for direct application to soil are those that can be made less hazardous or non-hazardous by biological degradation or chemical reactions occurring in or on the soil.

ECONOMICS OF WASTE MANAGEMENT

In evaluating the alternative options for managing waste, industry will generally select the most cost-effective, all other factors being equal. The costs of alternative management options are highly site specific (Barrier et al., 1978). The cost-effectiveness of viable options is beyond the scope of this paper but is a critical step in the selection of the disposal method by an industry.

TECHNICAL OPTIONS TO WASTE MANAGEMENT

Factors influencing the transport and plant availability of organic and inorganic chemicals through soil and their significance have been discussed earlier in this book. The principal factors are soil pH, organic matter content, cation exchange capacity, rainfall patterns, temperature, aeration, time, soil permeability, and material to be applied. The interaction of these items and others has been used by Phillips et al. (1977) for determining the suitability of a site to accept industrial wastes.

Soil plays an important role when an industrial waste is deposited in a landfill or placed on the soil to promote degradation, or break-down, of a waste, such as done in landfarming. Landfarming reduces the potential for groundwater contamination more than the landfill technique, because it reduces the quantity of material available for transport to the groundwater. In both cases the mobility and fate of the waste depends on the properties of the waste in relation to the soil properties enumerated above. Landfilling is essentially placing the industrial wastes in trenches or layers and covering them up.

Landfarming is particularly effective for wastes that are amenable to biodegradation into plant nutrients, humus, carbon dioxide, water, and innocuous salts. The technique involves three basic steps: (i) application of wastes on or into the surface layer of soil, (ii) mixing the wastes and soil periodically by cultivation to provide aeration and conditions conducive for microbial decomposition, and (iii) addition of amendments such as fertilizer to accelerate the decomposition process.

Landfarming has been practiced by industry in areas where land is readily available. Materials that have been applied to land in this manner include coal refuse (Goodwin and Gleason, 1978; Santhanam et al., 1979), petroleum refinery sludges (Grove, 1978; Phung and Ross, 1979; Huddleston and Meyers, 1979; Overcash and Pal, 1979; Raymond et al., 1979; Knowlton and Rucker, 1979; Huddelston, 1979), pharmaceutical wastes (King and Vick, 1978; Swan, 1979), vegetable wastes (Stephenson and Guo, 1977), dairy products (Pico, 1978; Watson et al., 1978; Anonymous, 1980), steel (Dawson, 1980), and chemicals (Rogers and Allen, 1978; Overcash et al., 1978; Barrier et al., 1978).

The petroleum industry has documented its experience with landfarming in the open literature more extensively than most others. Other industries are utilizing landfarming but generally describe results in reports to regulatory agencies and their own company's management. The experience in the petroleum industry over the past 25 years (Grove, 1978; Knowlton and Rucker, 1979; Huddleston, 1979) has been good. Landfarming is preferred by the industry for managing waste sludges and petroleum-containing solutions because of the minimum energy requirement for implementation and operation. The oil industry has considered and obtained data on decomposition rate, vegetative response, odor, and flammability, thereby setting an excellent example for other industries to follow.

The frequency of application of oily wastes varies widely from a single application to a site to multiple applications, as frequent as once per week. Application rates generally range from less than 200 barrels/acre/year to more than 600 barrels/acre/year. The decomposition rate is site specific but can be as high as 50% per year.

Subsurface samples indicate when landfarming is operated correctly. Data indicate trace metal analysis of vegetation growing on oiled areas is generally similar to control locations. Once oily wastes are blended with the soil, they are generally no longer flammable and odor is reduced to minimal levels.

Modifications of the landfarming technique are under development for other types of wastes. Rogers and Allen (1978) discuss the experimental disposal pit for pesticides under investigation by C. V. Hall and his colleagues at Iowa State. The pit prevents the movement of material into the water tables while allowing biodegradation of the material. This approach could be utilized for a variety of mobile biodegradable materials that would otherwise be required to be placed in costly hazardous waste landfills.

Another waste management option more recently developed is fixation of chemicals. Salas (1979) describes the use of chemical fixation and solidification to produce a nontoxic, environmentally safe material. Similar techniques have been used with waste from petrochemical, textile, automotive, steel, and chemical industries.

An example from the pulp and paper industry provides a very positive note to close this presentation. Eberhardt et al. (1978) describe how a waste material became a product sold for its fertilizer value. To achieve this conversion required (i) extensive engineering developments to convert the waste into a commercially acceptable material, (ii) detailed plant nutrient evaluation and testing, and (iii) appropriate biological safety tests.

The situation which prompted some action involved a sulfite pulp mill in Pennsylvania that was landfilling waste-activated sludge from the secondary treatment of plant wastes. This wet, sticky activated sludge was 16 to 18% solids and caused continuous housecleaning problems when trucking this wet material to the landfill. The large amount of water in the sludge aggravated operating conditions at the landfill by creating odor and leachate problems. Because the sludge had a significant nutrient value, a better solution was sought.

The approach utilized by the Proctor and Gamble Paper Products Company included nutrient analyses, commercial market evaluation, a grant-in-aid to the Pennsylvania State University for agronomic testing, and engineering evaluation for product development. The final result is a uniformly sized material with a consistent nutrient content, that is produced at a rate of 13.3 t d^{-1} and sold as a fertilizer. As expected, the heavy metal content of the material is low since the waste stream originates from trees. The net cost to the company for converting a waste to a useful commodity is similar to landfill costs.

CONCLUSIONS

A number of waste management options are available to industry for handling their wastes. Landfarming is a viable and useful technique for disposing of a small fraction of industrial wastes. The waste materials best suited for landfarming are those which are nontoxic initially or readily biodegradable to nontoxic material. Biodegradation is generally most effective if the soil system remains aerobic, the soil is not frozen, and the pH is near neutral.

Hazardous materials should only be landfarmed under very special situations such as concrete degradation pits for pesticides. Instead, hazardous chemicals should be placed in contained repositories consistent with specifications for hazardous wastes outlined in the U.S. Resource Conservation and Recovery Act of 1976, chemically neutralized to inert substances, or destroyed by incineration, photo-oxidation, etc.

LITERATURE CITED

Anonymous. 1980. Sludge management in food processing: in search of a better whey. Sludge 3:30–35.
Barrier, J. W., H. L. Faucett, and L. J. Henson. 1978. Economic assessment of FGD sludge disposal alternatives. Journal of the Environmental Engineering Division, Proc. of the Am. Soc. of Civil Eng. 104:951–965.
Dawson, R. A. 1980. Sludge management in the iron and steel industry: steelmakers eye sludge recycling as more Federal rules approach. Sludge 3:36–43.
Eberhardt, W. A., J. L. Lewis, R. A. Scharp, and C. A. Barton. 1978. Conversion of sulfite pulping waste to an agricultural product. Water Pollut. Control Fed. J. 50:1893–1904.
Epstein, E., and R. L. Chaney. 1978. Land disposal of toxic substances and water-related problems. Water Pollut. Control Fed. J. 50:2037–2042.
Goodwin, R. W., and R. J. Gleason. 1978. Options for treating and disposing of scrubber sludge. Combustion 50:37–41.
Grove, G. W. 1978. Use land farming for oily waste disposal. Hydrocarbon Process. 57:138–140.
Huddleston, R. L. 1979. Solid-waste disposal: landfarming. Chem. Eng. 86(5):119–124.
———, and J. D. Meyers. 1979. Treatment of refinery oily wastes by land farming. *In* Water—1978. AIChE Symposium Series 75:327–339.
King, L. D., and R. L. Vick. 1978. Mineralization of nitrogen in fermentation residue from citric acid production. J. Environ. Qual. 7:315–318.
Knowlton, H. E., and J. E. Rucker. 1979. Landfarming shows promise for refinery waste disposal. Oil Gas J. 77:108–112.
Litchfield, J. H., J. L. Graham, R. M. Soderquist, J. E. Germain, W. L. Stover, R. A. Morrell, and G. W. Grove. 1976. Industrial wastes. Water Pollut. Control 48:1217–1318.
Maugh, T. H. 1979. Toxic waste disposal—a growing problem. Science 204:819–823.
Overcash, M. R., H. C. Klose, D. Rock, R. Marshburn, and D. Pal. 1978. Pretreatment land application of textile plant wastes. *In* Water—1977. AIChE Symposium Series 75:163–174.
———, and D. Pal. 1979. Plant-soil assimilative capacity for oils. *In* Water—1978. AIChE Symposium Series 75:357–361.
Phillips, C. R., J. S. Nathwani, and H. Mooij. 1977. Development of a soil-waste interaction matrix for assessing land disposal of industrial wastes. Water Res. 11:859–868.

Phung, H. T., and D. E. Ross. 1979. Soil incorporation of petroleum wastes. *In* Water—1978. AIChE Symposium Series 75:320–326.

Pico, R. F. 1978. Industrial wastes: dairy wastes. Water Pollut. Control Fed. J. 50:1291–1293.

Raymond, R. L., J. O. Hudson, and V. W Jamison. 1979. Land application of oil. *In* Water—1978. AIChE Symposium Series 75:340–349.

Rogers, C. J., and R. Allen. 1978. Developing technology for detoxification of pesticides and other hazardous materials. Am. Chem. Soc. Symp. Ser. 73:100–111.

Salas, R. K. 1979. Disposal of liquid wastes by chemical fixation/solidification: the Chemfix process. Toxic and Hazardous Waste Disposal 1:321–348.

Santhanam, C. J., R. R. Lunt, S. L. Johnson, C. B. Cooper, P. S. Thayer, and J. W. Jones. 1979. Health and environmental impacts of increased generation of coal ash and FGD sludges. Environ. Health Perspectives 33:131–157.

Stephenson, J. P., and P. H. M. Guo. 1977. State of the art review of processes for treatment and reuse of potato wastes. Environment Canada and Economic and Technical Review Report EPS 3-WP-77-7.

Swan, R. 1979. Pharmaceutical industry sludge: drug makers face waste management headache. Sludge 2:21–25.

Watson, K. S., A. E. Peterson, and W. S. Walker. 1978. Effect of whey application on chemical properties of soils and crops. *In* Water—1977. AIChE Symposium Series 74:176–185.